Lecture Notes in Networks and Systems

Volume 478

The series "Lecture Notes in Networks and Systems" publishes the latest developments in Networks and Systems—quickly, informally and with high quality. Original research reported in proceedings and post-proceedings represents the core of LNNS.

Volumes published in LNNS embrace all aspects and subfields of, as well as new challenges in, Networks and Systems.

The series contains proceedings and edited volumes in systems and networks, spanning the areas of Cyber-Physical Systems, Autonomous Systems, Sensor Networks, Control Systems, Energy Systems, Automotive Systems, Biological Systems, Vehicular Networking and Connected Vehicles, Aerospace Systems, Automation, Manufacturing, Smart Grids, Nonlinear Systems, Power Systems, Robotics, Social Systems, Economic Systems and other. Of particular value to both the contributors and the readership are the short publication timeframe and the world-wide distribution and exposure which enable both a wide and rapid dissemination of research output.

The series covers the theory, applications, and perspectives on the state of the art and future developments relevant to systems and networks, decision making, control, complex processes and related areas, as embedded in the fields of interdisciplinary and applied sciences, engineering, computer science, physics, economics, social, and life sciences, as well as the paradigms and methodologies behind them.

Indexed by SCOPUS, INSPEC, WTI Frankfurt eG, zbMATH, SCImago.

All books published in the series are submitted for consideration in Web of Science.

For proposals from Asia please contact Aninda Bose (aninda.bose@springer.com).

Mousmi Ajay Chaurasia · Chia-Feng Juang
Editors

Emerging IT/ICT and AI Technologies Affecting Society

Editors
Mousmi Ajay Chaurasia
Muffakham Jah College Engineering
and Technology
Hyderabad, India

Chia-Feng Juang
National Chung Hsing University
Taichung, Taiwan

ISSN 2367-3370   ISSN 2367-3389 (electronic)
Lecture Notes in Networks and Systems
ISBN 978-981-19-2942-7   ISBN 978-981-19-2940-3 (eBook)
https://doi.org/10.1007/978-981-19-2940-3

This Springer imprint is published by the registered company Springer Nature Singapore Pte Ltd.
The registered company address is: 152 Beach Road, #21-01/04 Gateway East, Singapore 189721,
Singapore

Preface

This interdisciplinary book aims to focus on the exchange of relevant trends and research results as well as the presentation of practical experiences gained while developing and testing elements of technology-enhanced learning. Information and communications technology (ICT) is a means of searching, storing, archiving, processing and presenting information electronically through a number of media and technologies. The convergence of information systems and global communication infrastructures. ICT is the use of computers in instruction and communication for the purposes of learning and teaching that takes place mainly through information networks with the help of communication technology. It highlights the analytics and optimization issues impacting society and technology, for example on security, sustainability, identity, inclusion, working life, corporate and community welfare and well-being of people. It bridges the gap between pure academic research and walk towards practical publications.

The book includes 21 chapters highlighting the applications of intelligent algorithms, ICT applications in various domains. The first chapter "Sensing Layer—Layered Architecture for Efficient Operation of Intelligent Power System" by Namratha presents intelligent power systems with layered architecture. Its Sub-Layers, Design and Functionality and Protocols of the Sensing Layer (OFP-PSSL) have been presented in detail. In chapter "An Enhanced Model for Fake News Detection in Social Media Using Hybrid Text Representation", Dharmesh et al. explained the model where they input data of news text in a hybrid representation using TF-IDF with N-grams model combined with latent semantic indexing. The main objective of the hybrid text representation is to represent news text by considering the three important factors of text, viz. the important words in the text, their sequence of occurrence and their semantic meaning. In the chapter "A Smart Healthcare Design of Carryable Drip Frame with Fuzzy-PID Controllers", Yuan-Yi et al. proposed an adaptive fuzzy-PID controller and a fuzzy-PID neural network controller. Comparison with existing studies, the experimental results show that the fuzzy-PID neural network controller has the best performance characteristics, which improve the accuracy and performance of the system control. In the chapter "Kernel Learning Estimation: A Model-Free Approach to Tracking Randomly Moving Object", Yuankai Li et al. presented kernel learning methods with application to randomly moving target tracking systems, including kernel-based algorithms for target detection, kernel-based algorithms for generative tracking and for discriminant tracking and multi-kernel learning methods with multiple kernel functions. In the chapter "Transforming Medical Data into Ontologies for Improving Semantic Interoperability", Ayesha et al. focused on the significance of the semantic web in the medical industry and discussed different methods of building or transforming open EHR-based medical data to OWL individuals. Finally, the chapter gives an insight into ontology mapping for semantic interoperability with its benefits and also considerable challenges.

In the chapter "A Novel Fusion Scheme for Face Recognition in Challenging Conditions", Shekhar deals with fusion scheme termed as the PCA+LBP+LPQ. Prior to amalgamation, z-score normalization is carried out on the respective descriptor. The LBP and LPQ features are attained region-wise from corresponding map images. The amalgamated size is on the bigger side; therefore, PCA services are exploited again for compact size. For matching, SVMs are availed, and four datasets deployed are ORL, GT, JAFFE and Faces94. The PCA+LBP+LPQ pulls of superb recognition rates than either of PCA, LBP and LPQ. In the chapter "High-Performance Computing with Artificial Intelligence Benefits for the Civilization Impacted by the COVID-19 Pandemic", Chandrasekhar et al. proposed the techniques of using powerful high- performance computing (HPC) with artificial intelligence (AI) techniques to benefit the society impacted by the COVID-19 pandemic. The COVID-19 helps to aggregate a variety of geographically distributed computational resources, such as supercomputers, computer clusters, data sources, storage systems, scientific instruments and present them as a unified, dependable resource for solving large-scale computations and data-intensive computing applications from HPC to help COVID-19 scientists execute complex computational study to assist with battling the infection. In the chapter "Quality of Life Estimation Using a Convolutional Neural Network Technique", Manjunatha et al. delivered a novel approach for assessing a user's life quality score, which employs a deep learning architecture. A convolutional neural network and a support vector machine (SVM) are coupled to develop this solution for multimodal data. Three tests were conducted to determine the accuracy of the estimation method. In the research on living standard estimation, eight life factors were chosen to make up life quality. Finally, the eye movement was used to determine mental health. It is shown that the estimate is achievable, and the suggested approach employing different modes of activity data proved effective for estimating make up life quality, as well as for extracting high-dimensional information about a human's life quality, such as their pleasure-level towards social activities to improve their comprehensive life quality. In the chapter "E-learning During COVID-19—Challenges and Opportunities of the Education Institutions", Anusha discusses the concept and role of e-learning during the pandemic and the various challenges and opportunities of e-learning encountered by educational institutions. Three broad challenges identified in e-learning are inaccessibility, self-inefficacy and technical incompetency. It does not have any geographic barriers, it is flexible, creative and its critical learning incorporation increased utilization of online resources and reinforced distance learning. In the chapter "Real-Time Human-Machine Interaction Through Voice Augmentation Using Artificial Intelligence", Sumaiya et al. presented the interactions of voice augmentation functions to achieve human to machine interaction (HMI). The quality of the services can be improved with the help of artificial intelligence through deep learning models to illustrate the possibilities of its potential. This work includes audio processing, artificial intelligence, text-to-speech conversion, speech to text, audio conversion, various machine learning algorithms and models, implementation using the Internet of Things and applications in various fields, to evaluate the model.

In the chapter "Gamification in Education and Its Impact on Student Motivation—A Critical Review", Mifzala et al. reviewed interdisciplinary databases for quantitative experimental studies examining educational gamification and providing information on current research lines. Their comprehensive research study updates that gamification can be advantageous at all academic levels, from elementary school to college. The systematic research on gamified learning can increase students' motivation and intellectual accomplishment. The idea of incorporation of gaming elements into the classroom may serve as a motivational tool for students to learn and understand in a better way. In the chapter "Suspicious Human Behaviour Detection Focusing on Campus Sites", Mahmood Ali et al. dealt with suspicious behaviour detection model (SHBDM) effective usage with CNN model pre-trained on the ImageNet dataset, known as Inception V3 (VGG-16), for image feature extraction. This system is implemented in Python on an open-source platform. The use of Inception V3+LSTM resulted in an improved precision accuracy of 88.8% when compared with VGG-16 + LSTM and simple CNN model. This shows an automated CCTV surveillance system can serve as security providers for identifying the behaviour of suspicious humans or intruders based on their actions. In the chapter "The Security in Networks Through Short Normalized Attack Graph Modeling", Gouri proposed a new attack graph model based on shortest path normalization. This method is to provide the security for the dynamic network conditions. As the conventional attack graph models are limited to only static networks, they cannot ensure a secure data communication in the network. The proposed attack graph model tackles static network issue by accomplishment of a dynamic stochastic modelling on the network. To ensure the robustness, this model is employed in two different cases; they are halt condition and varying condition. Simulation is conducted through several attributes, and the performance is measured through computational time and link probability.

In the chapter "A Conception of Blockchain Platform for Milk and Dairy Products Supply Chain in an Indian Context", Dharun et al. presented a blockchain technology in the milk and dairy product supply chain to ensure the safety and quality of the milk produced. This focuses to keep checks and maintains balances to fight adulteration of milk and dairy products in the Indian scenario. In the chapter "Blockchain Technology in Financial Sector and Its Legal Implications", Divyashree et al. aimed to explore how blockchain technology is used in the financial sector and analysed the legal implications. It explains the need of blockchain technology in the financial sector to form a centralized framework to bring in line the latest business models overcoming traditional framework. In the chapter "Artificial Intelligence in Education", Venkata et al. presented the transformation methods of education and put forward the current directions in incorporating artificial intelligence. The authors contemplate on this issue and attempt to address how advancement of AI is essential with advancement in education. The pursuit to examine this and to add productive surface for a fruitful consideration in probing the power of artificial intelligence-based e-learning applications. Education must deploy AI to obtain the basic education goals, i.e. individualized, effective, transformative, output-based, integrative and long-lasting understanding. In the chapter "An Impact of COVID-19 a Well-Being Perspective for a New World", Thanveer presents the grey side of COVID-19 pandemic that will impact the well-being of the society with respect to physical fitness, psychology and other social evaluation measures. It focuses on a new perspective on the usage of digital devices that effects mental health.

In the chapter "Machine Vision Systems for Smart Cities: Applications and Challenges", Shamik et al. reviewed machine vision, smart cities and real-world machine vision applications in smart cities. It uses a combination of low-power sensors, cameras and AI algorithms to observe the city's operation. Machine vision has advanced in terms of recognition and tracking thanks to machine learning. It provides efficient capture, image processing and object recognition for vision applications. Governments benefit greatly from the use of machine vision and other smart applications. These technologies allow city administrators to easily integrate and utilize resources. This chapter highlights the benefits and difficulties of smart cities through a comprehensive literature review.

In the chapter "Emerging Non-invasive Brain–Computer Interface Technologies and Their Clinical Applications", Cory et al. covered current developments in non-invasive brain–computer interfaces (BCI) and their use for a variety of clinical applications. It discusses an overview of EEG hardware and non-invasive BCI systems and covers common electrophysiological recording techniques and signal processing algorithms often employed in BCIs. In addition, it also explains the implementation for specific clinical applications, including attention deficit hyperactivity disorder identification, stroke rehabilitation and sleep enhancement. In the chapter "Artificial Intelligence-Monitored Procedure for Personal Ethical Standard Development Framework in the E-Learning Environment", Rabia et al. proposed early patterns of unethical behaviour in E-learning platforms which the supervisor (parent/teacher) has to lookout for, such as inappropriate guidance on exams, misuse of references on chapter and projects, writing help and support and other improper mentoring, false representation in data collection and reporting. This chapter summarizes how unethical behaviour is found with suitable examples in eLearning and with the help of an artificial intelligence monitored PESD Framework and HD Loop, the situation could be improved. Finally, In the last chapter "The Horizontal Handover Mechanism Using IEEE 802.16 E Standard", Fahmina discusses the design of a heterogeneous network or an enhanced version of a 3G network. The performance metrics used for analysis include throughput, which was calculated using receiver signal strength for different data rates, average jitter and end to end delay. The purpose of the proposal is to discuss the communication difficulties and issues related to 3G plus generation devices, and its solutions for mobile ad hoc networks are proposed in the article.

This book covers state-of-the-art research, application development as well as emerging topics pertaining to ICT and effective strategies for its implementation for engineering and managerial applications. The primary objective of this book is to bring forward thorough, in-depth and well-focused developments of IT/ICT and AI benefitting society.

Hyderabad, India
Taichung, Taiwan

Mousmi Ajay Chaurasia
Chia-Feng Juang

Contents

Editors and Contributors

About the Editors

Dr. Mousmi Ajay Chaurasia is working as Professor and Head in the Information Technology Department at Muffakham Jah College of Engineering and Technology, Hyderabad. She is Senior Member of IEEE and serves in different capacities in IEEE section, region and global level. She is a recipient of the 2020 Significant Volunteer Award from IEEE Hyderabad Section. She has one Indian and two Australian patents to her name. She was the *Editor* for *Contactless Healthcare Facilitation and Commodity Delivery Management During COVID 19 Pandemic*, published by Springer Nature publications. She has worked in South Korea and Saudi Arabia and handled various projects funded by respective governments. She has been part of several international conferences and workshops. Her research interests are in the areas of artificial intelligence, big data and evolutionary computation.

Dr. Chia-Feng Juang (Fellow, IEEE) received his B.S. and Ph.D. degrees in control engineering from the National Chiao-Tung University, Hsinchu, Taiwan, in 1993 and 1997, respectively. Since 2001, he has been with the Department of Electrical Engineering, National Chung Hsing University, Taichung, Taiwan, where he became Full Professor in 2007 and has been Distinguished Professor since 2009. His current research interests include computational intelligence, intelligent control, computer vision and intelligent robots. He was the recipient of the Outstanding Automatic Control Engineering Award from Chinese Automatic Control Society, Taiwan, in 2014; the Outstanding Electrical Engineering Professor Award from Chinese Institute of Electrical Engineering, Taiwan, in 2019; and the Outstanding Research Award from Ministry of Science and Technology, Taiwan, in 2021. He was elevated to CACS Fellow in 2016 and IEEE Fellow in 2019. He is IEEE Computational Intelligence Society Distinguished Lecture in 2020–2022. He was Associate Editor for the *IEEE Transactions on Fuzzy Systems* and is Associate Editor for the *IEEE Transactions on Cybernetics, Asian Journal of Control* and *Journal of Information Science and Engineering* and Area Editor for the *International Journal of Fuzzy Systems*.

Contributors

Abhay Rabia Christ University, Bangalore, India

Abi Abirami Christ University, Bangalore, India

Aditya Shastry K. Nitte Meenakshi Institute of Technology, Bangaluru, India

Akash U. S. Department of Electronics and Communication Engineering, Dayananda Sagar Academy of Technology and Management, Bangalore, India

Ali Mohammed Mahmood MuffakhamJah College of Engineering and Technology, Hyderabad, Telangana, India

Ameen Ayesha Department of IT, Deccan College of Engineering and Technology, Hyderabad, India

Ansar Mifzala Research Scholar, Department of Commerce, CHRIST (Deemed to be University), Bangalore, India

Anusha B. Department of Commerce (PG), Krupanidhi Degree College, Bengaluru, India

Banu Ayesha Department of CSE, Vaagdevi College of Engineering, Warangal, India

Chandrashekhar B. N. Department of Information Science & Engineering, Nitte Meenakshi Institute of Technology, Yelahanka, Bangalore, Karnataka, India

Chang Yang National Yang Ming Chiao Tung University, Hsinchu, Taiwan

Chang Yuan-Yi Department of Electrical Engineering, National Chung Hsing University, Taichung, Taiwan

Dharanendra Gowda G. M. Department of Electronics and Communication Engineering, Dayananda Sagar Academy of Technology and Management, Bangalore, India

Divyashree K. S. School of Law, Christ University, Bangalore, Karnataka, India

Gangadharan G. R. National Institute of Technology, Tiruchirappalli, India

George Ginu Asst. Professor, Department of Commerce, CHRIST (Deemed to be University), Bangalore, India;
School of Commerce, Finance and Accountancy, CHRIST (Deemed to be University), Bengaluru, India

George Julie Dairy Consultant, Kochi, Kerala, India

He Congying National Yang Ming Chiao Tung University, Hsinchu, Taiwan

Jahan Thanveer Vaagdevi College of Engineering, Warangal, Telangana, India

Jain Anurag Systemics Cluster, School of Computer Science, University of Petroleum & Energy Studies, Dehradun, India

Joy Anson Kangirathingal School of Commerce, Finance and Accountancy, CHRIST (Deemed to be University, Bengaluru, India

Joy Justin Christ University, Bangalore, India

Kala Aravind Sharma Department of Electronics and Communication Engineering, Dayananda Sagar Academy of Technology and Management, Bangalore, India

Karanwal Shekhar Department of CSE, Graphic Era University (Deemed), Dehradun, UK, India

Karthika M. Christ University, Bangalore, India

Ko Li-Wei National Yang Ming Chiao Tung University, Hsinchu, Taiwan

Kureethara Joseph Varghese Christ University, Bangalore, India

Li Yuankai University of Electronic Science and Technology of China, Chengdu, China

Lin Ro-Wei National Yang Ming Chiao Tung University, Hsinchu, Taiwan

Lou Jiaxin University of Electronic Science and Technology of China, Chengdu, China

Manjunatha B. A. Nitte Meenakshi Institute of Technology, Bangaluru, India

Mishra Achyutananda School of Law, Christ University, Bangalore, Karnataka, India

Moturu Venkata Rajasekhar Assistant (Academics and Research), Indian Institute of Management Visakhapatnam, Visakhapatnam, Andhra Pradesh, India

Namratha Manohar J. Muffakham Jah College of Engineering and Technology, Hyderabad, Telangana, India

Nethi Srinivas Dinakar Assistant (Academics and Research), Indian Institute of Management Visakhapatnam, Visakhapatnam, Andhra Pradesh, India

Noorain Sara MuffakhamJah College of Engineering and Technology, Hyderabad, Telangana, India

Othayoth Poornima Kapadan Christ University, Bangalore, India

Patil Gouri R. Muffakham Jah College of Engineering and Technology, Hyderabad, India

Phang Chun-Ren National Yang Ming Chiao Tung University, Hsinchu, Taiwan

Puliyanmakkal Jiran Kurian Christ University, Bangalore, India

Qaseem Mohammad S. Nawab Shah Alam Khan College of Engineering and Technology, Hyderabad, Telangana, India

Sanjay H. A. Department of Information Science & Engineering, M.S. Ramaiah Institute of Technology, Bengaluru, Karnataka, India

Settipalli Lavanya National Institute of Technology, Tiruchirappalli, India

Singh Dharmesh National Institute of Technology, Tiruchirappalli, India

Sreekanth B. V. Department of Electronics and Communication Engineering, Dayananda Sagar Academy of Technology and Management, Bangalore, India

Stevenson Cory National Yang Ming Chiao Tung University, Hsinchu, Taiwan

Su Cheng-Hua National Yang Ming Chiao Tung University, Hsinchu, Taiwan

Sumaiya M. N. Department of Electronics and Communication Engineering, Dayananda Sagar Academy of Technology and Management, Bangalore, India

Tan Xiaosu University of Electronic Science and Technology of China, Chengdu, China

Taranum Fahmina Muffakham Jah College of Engineering and Technology, Osmania University, Telangana, India

Tiwari Shamik Systemics Cluster, School of Computer Science, University of Petroleum & Energy Studies, Dehradun, India

ur Rahman Ateeq Shadan College of Engineering and Technology, Hyderabad, Telangana, India

Vincent Dharun Christ University, Kengeri Campus, Bangalore, India

Wang Yuan University of Electronic Science and Technology of China, Chengdu, China

Wen Chih-Yu Department of Electrical Engineering, National Chung Hsing University, Taichung, Taiwan

Wu Ming-Feng Division of Chest Medicine, Department of Internal Medicine, Taichung Veterans General Hospital, Taichung, Taiwan;
Department of Medical Laboratory Science and Biotechnology, Central Taiwan University of Science and Technology, Taichung, Taiwan

Sensing Layer—Layered Architecture for Efficient Operation of Intelligent Power System

J. Namratha Manohar

Abstract Power systems today are smart and intelligent in that they are responding fast to the deviations from normal operations and dynamically adjusting the inputs to bring back the power system to a stable state. The power systems today have been integrated with several sources of energy, smart devices and meters, and above all are large data-intensive with the advent of modern computing devices. The power system is thereby not only a pure electrical engineering field but encompasses information technology. The best way to handle the increasing complexity is to develop a layered architecture of intelligent power system, which enables development, installation and efficient operation of power system. The chapter presents the various layers and sub-layers of the layered architecture. The purpose of the various layers has been presented. The sub-layers, objectives, design and functionality and protocols of the sensing layer (OFP-PSSL) have been detailed. The advantages and applications of the layered architecture have also been presented.

Keywords Layered architecture · Sensing layer · Sub-layers · Objectives · Design and functionality · Protocols

1 Introduction

The advancement in the global scenario of technology, industry and market of electric power system has called for a decentralized structure, for efficient operation and management. The future power systems will be intelligent power systems [1], consisting of components which are equipped with high-speed parameter measurement sensors, computational and communication facilities in addition to performing the basic power system function for which they are designed for. Future power systems will be hybrid systems, with centralized generation, distributed generation and renewable generation. To manage this complexity of electrical system expansion technology-wise and geographically, the layered architecture is the best model.

J. Namratha Manohar (✉)
Muffakham Jah College of Engineering and Technology, Hyderabad, Telangana, India
e-mail: j.namratha@mjcollege.ac.in

© The Author(s), under exclusive license to Springer Nature Singapore Pte Ltd. 2023
M. A. Chaurasia and C.-F. Juang (eds.), *Emerging IT/ICT and AI Technologies Affecting Society*, Lecture Notes in Networks and Systems 478,
https://doi.org/10.1007/978-981-19-2940-3_1

A layered architecture (LA) [2] approach for the development and maintenance of power system is essential as the modern trend is open-source architecture (OSA). OSA is defined as an architecture in which equipment of different manufacturers can be functionally integrated with ease and achieve better performance. The LA supports the development of plug-and-play equipment and components. Plug-and-play components are those that can be integrated with the existing system, without the need to make major modifications, in that they are backward compatible.

Layered architecture provides best solution to meet the increasing challenges faced by the powers system. The primary challenge of abiding by the Paris Agreement [3] has to be foremost objective of all advancements toward meeting other challenges, as increase in customer demand, reduction of power losses, curtailing theft, integrate hybrid energy sources and enhancing efficiency of the power system. Another important latest concern is the impact of bidirectional flows in modern power systems [4]. The introduction of distribution systems and renewable energy systems by consumers, who are able to generate more energy than they need, thereby able to feed their excess energy to the grid, has posed challenges of managing the bidirectional flows and their impact on system reliability and stability.

The intelligent power system has many components, which have hardware and software embedded in them; therefore, layered architecture enables development, installation, maintenance and extending the functionality of the components; as it would segregate the sub-functions of the component. That is, say, we have two layers—hardware layer and software layer, and if the software is updated, it would not be necessary to replace the hardware, thus saving time and reducing complexity of updating and cost as well.

The chapter presents the various layers and the sub-layers of the layered architecture for intelligent power system. The early controls in power systems were implemented to control the power generation, in accordance with the change in load pattern. This control was and is still essential, until storage technologies advance further; as electricity is still a commodity which needs to be consumed immediately on generation. For generation—load balance control, as sensitive equipment to measure the power systems parameters were not available, and several techniques as load–frequency control [5], area control, economic dispatch were adopted. The advent of mechanical devices as relays has enabled to introduce the protection of power systems from major failures. However, these devices could only detect the occurrence of certain faults as over-current and dis-connect the healthy part of the PS from unhealthy part, but for complete restoration of the power system, manual intervention is required. Advancement of technology from mechanical to electro-mechanical-to electronics and now computer based has resulted in evolving of sophisticated power systems monitoring and control devices (MCD). The advancements in computer technology by way of developing artificial intelligence techniques have made it possible to make power system also intelligent, as several smart devices are now available. The present-day power systems have sensors with hardware and software, which need to be integrated with the computing device and communication protocols. There is therefore need to develop the objectives, functionality and protocols of the power system layered architecture (PSLA).

The intelligent power system has several devices which are smart meters. In the context of this chapter, the terms intelligent power system, power system and smart grid will be used alternatively for appropriate description of the context in the sections.

The chapter is organized as follows: Sect. 1: introduction, Sect. 2 intelligent power system layered architecture—introduction, Sect. 3: sensing layer of layered architecture of power system, Sect. 4: protocols of PSLA sensing layer, Sect. 5: benefits of intelligent power systems, Sect. 6: real-time application of intelligent power system and Sect. 7: conclusion and further research.

2 Intelligent Power System Layered Architecture—Introduction

An intelligent power system is a system that responds to events in the shortest time, transforming the largely passive electric network to active adaptive network [6] and dramatically improving controllability in order to increase the capacity of power lines, improve quality of electric energy, sustainability, ensuring efficient and reliable functioning of the electric power system [7, 8].

To mold the electric network into an intelligent network, the working conditions [6] of electric power equipment of generation, transmission and distribution stages, relay protection system, control systems need to be modified to embed not only upgraded hardware but also software. This has been possible with the advent of microprocessors, microcontrollers, FACTS devices and various embedded systems and software. The various functional elements that have become complex have been made to consist of several sub-components. The sub-components are nodes and compliment the need for a layered architecture [2]. Figure 1 depicts the layered architecture of an intelligent power system [1].

From Fig. 1, we see that the layers in the intelligent power system architecture are organized as in Table 1:

The various layers and sub-layers have been explained in the subsequent sections.

2.1 Power System Layer

The power system layer relates to all components of the power generation, transmission and distribution stages. It deals with the physical design and principle of operation of the components.

Fig. 1 Layers of the
intelligent power system
architecture

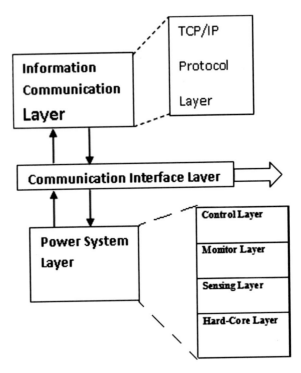

Table 1 Layers of intelligent
power system architecture

Major layer	Sub-layer
1. Power system layer	Hard core layer Sensing layer Monitor layer Control layer
2. Communication layer	Communication interface layer Information transmission layer

2.1.1 Hard Core Layer

Design of the physical characteristics of the equipment for effective and efficient functioning is the main objective of this layer. The design parameters should not only take into consideration the mechanical and electrical characteristics [1] for effective power system operations but also data capturing, computation and communication.

2.1.2 Sensing Layer

Sensing layer deals with the logical aspects of the sensors [5] that measure the physical quantities to be monitored and controlled such as voltage, current, frequency,

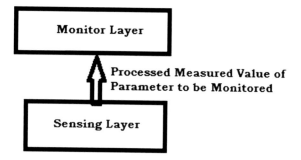

Fig. 2 Interaction between sensing layer and monitor layer

active and reactive power, power factor and temperature. The physical quantities are measured directly, or a parameter proportional to the quantity is captured. These parameters indicate the status of transmission and enable to determine the parameters to be monitored, to improve power system quality, stability and efficiency. Real-time data is captured and transmitted to the monitoring layer.

2.1.3 Monitor Layer

The monitor layer receives the value of the parameter measured at the sensing layer, as shown in Figure 2. The measured value is compared with the standard value, and the deviation is forwarded to the control layer for corrective action. The monitor layer encompasses the function of data acquisition [1] at remote terminals. The placement of the sensors in the physical power system is a very important decisive factor to be able to effectively monitor and control the operation of the power system.

2.1.4 Control Layer

The control layer initiates the control action [1] that triggers events as closing/opening a circuit breaker. The layer can also be defined as the "action layer". Control signals are sent through communication channels such as fiber cable, coax to activate the control equipment such as circuit breakers isolators and sectionalizers.

2.2 Communication Layer

Intelligent power systems are electrical power grids that apply information, advanced networking and real-time monitoring and control technologies to lower costs, save energy and improve security, interoperability and reliability. The communication layer in intelligent power system is responsible to communicate the data and information for the technological operation and maintenance of the power system. The

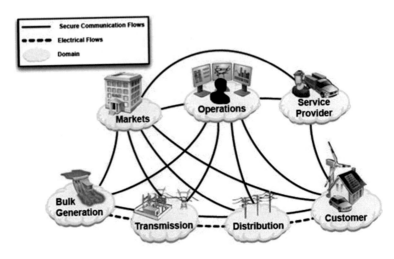

Fig. 3 Communication channels in power system

communication channels have to be established between the various players of the power system. The players here are the energy providers, consumers and the market. Figure 3 [9] shows the various communication channels.

3 Sensing Layer of Layered Architecture of Power System

3.1 Power System Sensing Layer (PSSL)—Introduction

Any system to adjust its input–output for stabilization requires that the important parameters be monitored. Monitoring requires measurement of the parameters which is done by sensors. In power systems, sensors measure and transmit data to automatically adjust electricity flows according to supply and demand. Energy managers can use this information to adjust the grid and respond to problems in real time.

Sensors measure the base parameters—voltage and current of the power system, and the derived parameters as power, energy and power factor can be computed. There are devices which measure the derived parameters also. Determining the quality of the power system is very important, and with the present-day technology, it is easy, convenient and possible. Sensing, processing the signal and networking capabilities need to be bundled along within the sensor [9]. The capabilities depend on the degree of analytics and data preprocessing that we want to perform at the sensor itself. The hardware and the software have to meet the various standard specifications and protocols, for smooth integrated operation with the associated components of the power system.

Sensors in power system are used at various stages as power generation, transmission, distribution and locations as substation and customers end. The various sensors in power system are: current transformers (CTs), voltage transformers (VTs), phasor measurement units (PMUs), merging units (MUs), temperature sensors, humidity sensors, accelerometers, rain gauges, Internet Protocol (IP) network cameras, pyranometers and pyrheliometers (solar irradiance), weather stations, sonic anemometers, partial discharge sensors, gas sensors, ultrasound and ultra-high frequency sensors, torque sensors, discharge rate sensors, load-leveling sensors, occupancy sensors and power quality monitors.

The present-day power systems are intelligent and smart; therefore, the sensors technology has also been upgraded to be in synchronization with the power system and the information communication technology. Protection and control systems need to react [10] to faults and unusual transient behavior and ensure recovery after such events. Real-time network simulation and performance analysis will be needed to provide decision support for system operators and the inputs to energy and distribution management systems.

Smart sensors (SSs) can provide real-time data [11] and status of the grids for real-time monitoring, protection and control of grid operations. Figure 4 [12] depicts the intelligent power grid framework with integrated smart devices.

Some important parameters of the power system measured by the sensor and communicated to the monitor layer are shown in Fig. 5 [13]. The monitor layer processes the measured value, and command is sent to the control layer for appropriate corrective action. The action parameter shown in Fig. 5 is the events that take place based on the decision of processing the corresponding parameter.

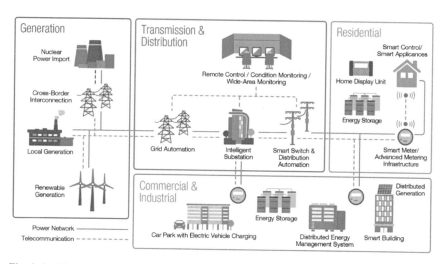

Fig. 4 Intelligent power grid framework with integrated smart devices

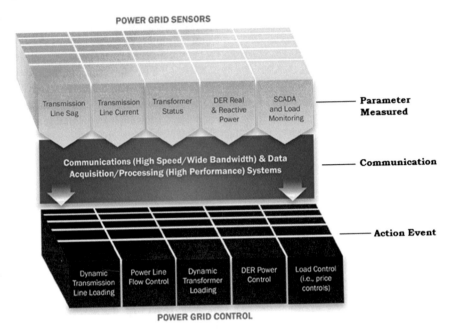

POWER GRID SENSORS

Transmission Line Sag | Transmission Line Current | Transformer Status | DER Real & Reactive Power | SCADA and Load Monitoring — **Parameter Measured**

Communications (High Speed/Wide Bandwidth) & Data Acquisition/Processing (High Performance) Systems — **Communication**

— **Action Event**

Dynamic Transmission Line Loading | Power Line Flow Control | Dynamic Transformer Loading | DER Power Control | Load Control (i.e., price controls)

POWER GRID CONTROL

Fig. 5 Parameter measured and action taken

3.2 Objectives of PSLA—Sensing Layer

The objective of the sensing layer must be to successfully achieve the objectives of the intelligent power system.

In order to meet so, the sensing layer must be able to measure all the critical parameters of the power system as voltage, current and frequency and send appropriate signals to the control devices as the circuit breaker and other power electronic self-regulating devices. The generally desirable properties of sensors are:

- high sensitivity
- high selectivity
- wide sensing range
- high signal-to-noise ratio
- low power consumption
- robust
- reliable
- optimum cost.

3.3 Sensor Placement—Decisive Factor

Location of the sensor at a strategic point in the power system is a very important decisive factor. The primary objective of sensor placement was fault detection, but with the advancement in power system technology, the list of objectives is expanded. The objectives could be one or more of the following:

- Fault identification.
- Measuring of parameters as current and voltage, to compute derived parameters as power and power factor.
- Analyze the status of power system performance.

With power system simulation software available now, several algorithms can be implemented to determine the optimum positioning of sensors.

3.4 Power System Sensing Layer (PSSL) —Sub-layers

The two sub-layers of the sensing layer are:

1. Sensing physical device (SFD), 2. Sensing computational module (SCM).

The intelligent power system layered architecture with the sub-layers of the sensing layer is depicted in Fig. 6.

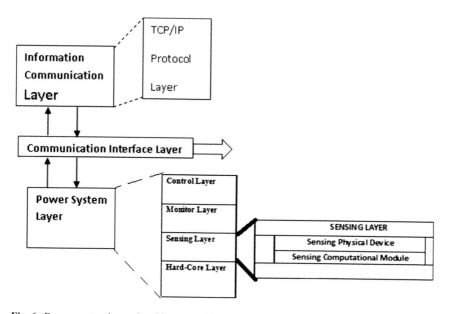

Fig. 6 Power system layered architecture with sub-layers of sensing layer

3.5   Sensing Physical Device (SFD)—Design and Functionality

Figure 7 depicts the interaction between the two sub-layers of the PSLA sensing layer. The sensing physical device (SPD) measures the physical parameter to be monitored and controlled and the signals transferred to the sensing computational module (SCM). The SCM performs the computations and sends the action signal to the control device to take appropriate action as tripping of the circuit breaker, adjusting the transformer settings or to any appropriate power electronic device.

Design of a SPD is the design of the hardware aspects as physical dimension—size, material, shape, location for embedding in the power system depending on the parameter it needs to measure. The hardware design is based on the parameter it needs to measure and the principle adopted to measure. Some of the electrical power sensor technologies to measure electrical characteristics are:

- **Hall Effect**—These sensors measure the presence and magnitude of magnetic field. The output voltage is directly proportional to the strength of the field. Aging of insulation can be determined by sensing methods as degree of polymerization, insulation resistance, power frequency dissipation factor and polarization index measurement.
- **Inductive sensors**—These sensors use a coil of wire that goes around the power carrying wire to measure voltage current phase and wattage.
- **Direct measurement sensors**—These sensors are in line with the power-carrying wires and convert it to a value proportional to the signal strength for measurement or display.
- **Voltage response measurements**—These sensors that use the electrical characteristics of a circuit are determined from the amplitude and phase of a test current flowing through a circuit. To measure power quality, electrical power sensors measure the phase difference between voltage and current, as well as the resulting total harmonic distortion (THD).

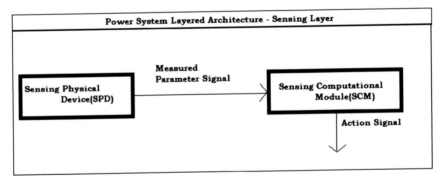

Fig. 7 Interaction between the sub-layers of the PSLA—sensing layer

3.6 Sensing Computational Module (SCM)

Sensing computational module design involves considering the software aspects as development of algorithm, writing of the code for processing the measured parameter, taking decision of action and sending signals to the control device as said tripping of circuit breaker setting of the transformer or any power electronic device.

Data is collected from sensors distributed across the grid through existing communications channels [13]. The data received from the various sensors is processed by the computational module. The result of the computation is the action signal that is sent to the monitoring layer and thereby the control layer to initiate appropriate corrective action.

Data collection and aggregation requires automation, different collection and storage mechanisms and different formats and protocols. Several advances have been made in sensor network technologies.

As essential is the communication between the sensor and power system, so also communication between sensors is important for self-organization of the power system to a stable state. Wireless sensor networks with distributed computing and smart sensors would give optimum performance. Advancements in digital signal processors had made it possible to have self-organizing sensor network architecture. Intelligence in this architecture is possible with the advent of complex algorithms for performance assessment and optimization, inspired by biological systems.

4 Protocols of PSLA Sensing Layer

The protocols for PSLA sensing layer can be in two categories

i.   Protocols for sensor measurement and representation of measured quantity.
ii. Protocols for communication of control signal to the action taking devices in the power system.

The Technical Committee on Sensor Technology (TC-9) of the IEEE Instrumentation and Measurement Society defined smart transducers (sensors and actuators) that provide functions beyond those necessary for generating an accurate representation of a sensed quantity in IEEE 1451 that is a family of Smart Transducer Interface Standards for Sensors and Actuators [10]. This functionality typically simplifies the integration of the transducers into applications in a networked environment [8]. The general model of the smart sensors was presented by Eugene Y. Song in their chapter on smart sensors [14].

5 Benefits of Intelligent Power Systems

The benefits of intelligent power systems can be categorized into the following four groups:

1. **TECHNICAL:**

    - Facilitates easy monitoring and control of the power system second by second.
    - Reliable high-quality power can be delivered over a stable grid [15]
    - Power system performance metrics can be improved.
    - Provides accurate and prompt information.
    - Increases system reliability and security of supply.
    - Achieve reduced power losses at all stages.
    - Increase power system efficiency.
    - Manage and curtail power theft.
    - Integrate different sources of energy with the grid.
    - Obtain vast accurate data for power system state analysis.

2. FINANCIAL

    - Reduce transmission and distribution costs.
    - Reduce fuel costs.
    - Optimize energy unit rate.

3. ADMINISTRATIVE

    - Improved visibility of utility operations to senior management.
    - Improved access to historical data for strategic planning purposes.

4. SOCIETY

    - Ability to meet or exceed mounting customer performance requirements.
    - Improvement in customer satisfaction.
    - Provide consumers with greater information and options on how they use their power.
    - Significantly reduce the environmental impact of the whole electricity supply system by reduced emissions.

6 Real-time Application of Intelligent Power System

The intelligent power system's effective and efficient functionality is made possible with artificial intelligence techniques. Some of the applications [16] are presented as follows:

- Fault prediction and clearance is of the major applications of artificial intelligence (AI) [17, 18] in power system.

- Toshiba ESS is conducting research on the use of Internet of Things (IoT) and artificial intelligence to improve the efficiency and reliability of geothermal power plants [19].

The applications at various stages of power system are given in Table 2

Smart devices such as Amazon Alexa, Google Home and Google Nest enable customers to interact with their smart devices and monitor their energy consumption [20]. The digital transformation of home energy management and consumer appliances allows automatic meters to use AI to optimize energy consumption and storage. For example, it can trigger appliances to be turned off when power is expensive or electricity to be stored via car and other batteries when power is cheap or solar rooftop energy is abundant.

Table 2 Power system stage and the application

Power system stage	Application
Power system operation	• Unit commitment • Economic dispatch • Hydro-thermal coordination • Maintenance scheduling • Congestion management • Load/power flow • State estimation • Location and sizing of facts devices • Control of facts devices • Restoration and management • Fault diagnosis and reliability • Network security, etc.
Power system planning	• Generation expansion planning • Transmission expansion planning • Reactive power planning • Power system reliability, etc.
Power system control	• Voltage control • Load–frequency control • Stability control • Power flow control • Dynamic security assessment, etc.
Electricity markets	• Bidding strategies • Market analysis and clearing, etc.
Distribution system application	• Operation and planning of distribution system • Demand-side management and demand response • Network reconfiguration • Operation and control of smart grid, etc.

7 Conclusions and Further Research

The layered architecture of an intelligent power system has been presented. The purpose of each layer has been briefly presented. The design and functionality of the sub-layers of the sensing layer were also presented. There is scope for extensive further detailed study of the various layers, their functions and services.

Websites

- https://ieeexplore.ieee.org/document/7986956
- https://info.ornl.gov/sites/publications/Files/Pub57485.pdf
- https://www.ncbi.nlm.nih.gov/pmc/articles/PMC6459191/
- https://electrical-engineeringportal.com/communications-power-system-protection
- https://onlinelibrary.wiley.com/doi/epdf/10.1002/wcm.2258

References

1. Namratha Manohar J (2015) Modeling of a layered architecture for efficient operation of intelligent power system. Int J Eng Res Technol (IJERT). ISSN: 2278-0181, IJERTV4IS060141, vol 4, issue 06, June 2015. www.ijert.org
2. Bharadwaj V, Ramana Reddy YV, Chandramouli A, Reddy S. A layered architecture to model interdisciplinary complexity in the deregulated power industry
3. United Nations, Paris Agreement n Climate Change. https://unfccc.int/sites/default/files/english_paris_agreement.pdf
4. Lopes JAP et al. Wires energy and environment. The future of power systems: challenges, trends and upcoming paradigms. Wiley https://wires.onlinelibrary.wiley.com/doi/epdf/10.1002/wene.368
5. Cohn N.. The evolution of real time control applications to power systems. 14'7 Noble Road, Enkintown, PA 19046, USA, Copyright CC) Real Time Control Guadalajara. Mexico 198~, Session [Plenary Chapters]
6. Ufa RA et al (2020) Intelligent electric power systems with active adaptive electric networks: challenges for simulation tools. Appl Sci 10:2951. https://doi.org/10.3390/app10082951. www.mdpi.com/journal/applsci
7. Liu ZH, Wang Y, Chen JM, Guo YR, Wang XG, Li ZQ, Du DX, Li X (2014) Modeling and simulation research of large-scale AC/DC hybrid power grid based on ADPSS, Kowloon, Hong Kong
8. Negnevitsky M, Voropai N, Kurbatsky V, Tomin N, Panasetsky D (2013) Development of and intelligent system for preventing large-scale emergencies in power systems, Vancouver, Canada
9. Ali M, Bizon N (2017) Communications for electric power system, Chapter 14, Book reactive power control in AC power systems, Edition: 1, Publisher: Springer International Publishing. https://www.researchgate.net/publication/315890769_Communications_for_Electric_Power_System
10. Sethi P, Sarangi SR (2017) Hindawi Internet of Things: architectures, protocols, and applications. J Electrical Comput Eng Article ID 9324035. https://doi.org/10.1155/2017/9324035

11. Bade K (2020) Smart Grid—an intelligent power system: review and future challenges. Int Res J Eng Technol (IRJET) e-ISSN: 2395-0056, vol 07, issue 04. www.irjet.net p-ISSN: 2395-0072 © 2020, IRJET
12. Overview of Smart Grid technology and its operation and application (for existing power system). www.elprocus.com/overview-smart-grid-technology-operation-application-existing-power-system/
13. Chapter 3: enabling modernization of the electric power system technology assessments cyber and physical security designs, architectures, and concepts electric energy storage flexible and distributed energy resources. Quadrennial Technology Review 2015, US Department of Energy
14. Standard for a Smart Transducer interface for sensors and actuators-common functions, communication protocols, and Transducer Electronic Data Sheet (TEDS) Formats, IEEE Standard, 1451.0–2007, Oct 2007
15. Song EY, Fitz Patrick GJ, Lee KB (2017) Smart sensors and standard-based interoperability in smart grids. IEEE Sensors J 17(23). https://doi.org/10.1109/JSEN.2017.2729893. https://doi.org/10.1109/JSEN.2017.2729893
16. Taft J (2006) The intelligent power grid, IBM, © 2006 IBM
17. Makala B, Bakovic T (2020) Artificial intelligence in the power sector , Note 81 Apr 2020. www.ifc.org/thoughtleadership. Fresh Ideas on Private Sector Development bu IFC World Bank Group
18. 5 ways the energy industry is using artificial intelligence. Research Briefs, CBInsights, 8 Mar 2018. https://www.cbinsights.com/research/artificial-intelligence-energy-industry/
19. Richter A (2018) Toshiba energy systems & solutions corporation has launched a research program on Internet-of-Things and artificial intelligence technology to improve the efficiency of geothermal power plants. Think GeoEnergy, 16 Aug 2018. https://www.thinkgeoenergy.com/improving-efficiency-of-geothermal-plans-with-artificial-intelligence-and-iot-technology/
20. Saberi O, Menes R (2020) Artificial intelligence and the future for smart homes. EM Compass Note 78, IFC, Feb 2020

An Enhanced Model for Fake News Detection in Social Media Using Hybrid Text Representation

Dharmesh Singh, Lavanya Settipalli, and G. R. Gangadharan

Abstract In today's digital age, the majority of people obtain their news via the Internet and social media. However, it is difficult to identify which sources are reliable and which are disseminating false information. So far, many models have been derived to detect fake news or misinformation from social media news using machine learning (ML) or deep learning (DL) techniques. This chapter aims to enhance the detection accuracy by representing the text of social media news in a hybrid format using Natural Language Processing (NLP) techniques. We model the input data of news text in a hybrid representation using TF-IDF with N-grams model combined with Latent Semantic Indexing. The main objective of the hybrid text representation is to represent news text by considering the three important factors of text, viz. the important words in the text, their sequence of occurrence, and their semantic meaning. We applied different ML and DL techniques for news classification and compared the performance of fake news detection models with and without hybrid text representation. The obtained results evidence that there is a significant improvement in detection accuracies when the news text is represented in a hybrid format as proposed in our approach.

Keywords Fake news detection · Latent Semantic Indexing · Vectorization · Text representation

1 Introduction

The vast amount of information has led to a rapid rise in fake news and misinformation. Spreading fake news is an intentional deed to lessen the reputation of any organization or agency or a person and/or to get financially or politically benefited [1]. The Internet has become a hot seat for the spread of fake news and misinformation and the reliability of the news source is also questionable. Spreading fake news shares dishonest or outright fabricated news repeatedly to make it a sensation via

D. Singh · L. Settipalli · G. R. Gangadharan (✉)
National Institute of Technology, Tiruchirappalli, India
e-mail: geeyaar@gmail.com

increasing readership and Internet click revenue. To mislead and deceive visitors, Web sites with fake material usually utilize a combination of Web site spoofing and authentic news styling approaches. These Web sites strongly look and operate like real news sites. The detection of fake news is a difficult task as it spreads quickly through social and online media [2].

Humans have a difficult time in detecting fake news, and the only way to spot false news is to have a thorough understanding of the subject. Even with knowledge, determining whether the material in the article is true or false would be difficult. The open nature of the Internet and social media, as well as recent advances in computer science, make it easier to create and propagate fake news. Several studies have been investigating the magnitude or spread of misinformation online using popular source-based approaches. Many fake news detection models that developed using machine learning (ML) or deep learning (DL) techniques [3] encounter the problem of reduced performance due to the complexity in representing news text.

In this chapter, we propose a hybrid approach for representing news text to enhance the performance of fake news detection in social media. This novel hybrid text representation approach converts the news text into vector format using Term Frequency-Inverse Document Frequency (TF-IDF) for N-grams of text. The text representation is combined with Latent Semantic Indexing (LSI) to analyze the statistics of co-occurrence between fake and real news due to its high quality of statistical analysis [4]. Once the vectors are formed, various ML and DL classifiers are applied for classifying the news as fake and real.

The rest of the chapter has the following orientation. Section 2 reviews the existing models that were developed to solve the problem of fake news detection. The proposed methodology for fake news detection using a hybrid text representation with machine learning and deep learning classifiers is described in Sect. 3. Section 4 depicts the datasets used and the pre-processing steps. Section 5 describes the analysis of results, and the concluding remarks are given in Sect. 6.

2   Related Work

Many researchers proposed machine learning or deep learning models for detecting fake news in social media among which some are based on contextual learning, semantic indexing, or vectorization concepts. Ahmed et al. [5] proposed a fake news detection model through the N-gram analysis and different machine learning techniques. Using N-gram analysis, the news text is reduced by representing the text in vector format. Then six different classification approaches including Stochastic Gradient Descent (SGD), Support Vector Machines (SVM), Linear Support Vector Machines (LSVM), K-Nearest Neighbor (KNN), and Decision Trees (DT) are applied. They have compared the investigative results and concluded that the TF-IDF with LSVM performed better in detecting fake news with an accuracy of 92%.

Singhania et al. [6] employed an automated fraud detection model using a novel deep learning approach. They developed a Hierarchical Attention Network with three levels (3HAN) for text representation in order to enhance the detection accuracy of fake news. 3HAN approach contained three levels whereas the first level analyzes the words, the second level analyzes the sentences, and the third level analyzes the headlines. Based on the hybrid analysis in a hierarchical bottom-up manner, they represented news text as a news vector. They have conducted empirical analysis on a large fake news dataset that is taken from the real-world scenarios and observed that the fake news detection model achieved an accuracy of 96.77% with an effect of 3HAN.

Aphiwongsophon and Chongstitvatana [7] compared the performance of three machine learning techniques including Naïve Bayes, SVM, and Neural Network in detecting fake news. Their investigative results left the concluding remarks as the SVM and Neural Networks can achieve better accuracy compared to Naïve Bayes. However, SVM and Neural Networks are complex algorithms for detecting fake news from a huge amount of social media text compared to Naïve Bayes. Aswini et al. [8] developed a fake news detection model based on a stance detection approach. Unlike the other existing models of fake news detection, the stance approach finds the relationship between two pieces of text, a news article and news headline pairs. They classified the news text among four classes, viz. "agree," "disagree," "discuss," or "unrelated" using Neural Networks and achieved an accuracy of 92.46%.

Girgis et al. [9] built a prediction model to decide whether the news is fake or not using an ML classifier based on the analysis of news content. This work compared the prediction performance of three approaches vanilla, GRU, and LSTMs of RNN, and concluded that GRU obtained the best results. Hence, they developed a hybrid model which combines CNN with GRU for enhanced performance in detecting fake news. They conducted experiments of the proposed hybrid model on the LAIR dataset.

Monther and Ali [10] ranked the attributes from news text based on Info Gain Eval and Correlation Attribute Eval algorithms and chose the most relevant attributes that contribute to improving the accuracy and reducing the training time. The selected attributes are used by the model to train different machine learning algorithms such as BayesNet (BN), Naïve Bayes (NB), Random Forests (RF), and Logistic Regression (LR). The experiment analysis of the proposed model concluded that LR performed with an accuracy of 99.4% with the proposed attribute selection approach. Ajao et al. [11] developed a hybrid fake news detection model using CNN and LSTM without considering the previous knowledge about the news text. In their later work [12], they analyzed the characteristics of fake news by relating it with the respective sentiments.

They developed an automated fake news detection model with the assumption that the news and sentiments underlying posting a text online are related to each other. They compared the state-of-the-art models with and without sentiment analysis and concluded that there is a significant improvement in the performance of detecting fake news with sentiment analysis. Mahir et al. [13] analyzed the Twitter post using different machine learning approaches such as NB, SVM, LR, and RNN. Then, they compared the classification performance of various machine learning approaches used, and based on the experimental results, they concluded that SVM and NB classifiers outperformed the other classification models.

Hiramath and Deshpande [14] compared the performance of LR, NB, SVM, RF, and DNN classification techniques in classifying the fake news of social media. Poddar et al. [15] compared two vectorizers such as Count and TF-IDF to find the appropriate one which improves the accuracy of fake news detection. They compared the classification performance by applying various ML techniques such as NB, SVM, LR, and DT on the vectorized news text. The experimental results highlighted that SVM with TF-IDF produced the most accurate prediction. Abdullah et al. [16] performed bimodal (domain name, author name) analysis to identify fake news based on their source and previous history. They classified 12 different categories of news articles using the hybrid approach of CNN and LSTM.

Agarwal and Dixit [17] proposed a fake detection model by conducting feature extraction and assigning credible scores to the news text information. The analysis of features was done by building an ensemble network for depictions of the new report, author details, and the corresponding titles simultaneously. Then, various ML techniques including NB, KNN, SVM, LSTM, and CNN are used to classify the news and they concluded from the results that LSTM produced better accuracy. Ibrishimova and Li [18] considered fake news information as a disinformation tool and defined fake news as relative bias and factual accuracies. Based on the fluid definition, the proposed model developed a fake news detection model using ML techniques. Uppal et al. [19] developed a structural analysis-based model using discourse segmentation and deep learning techniques.

Baarir and Djeffal [20] used TF-IDF and N-grams to extract features from news text and applied Support Vector Machine (SVM) to classify the fake news. Jamal et al. [21] proposed a hybrid deep learning model by combining CNN and RNN. They used long-term dependencies from RNN to analyze and identify fake news in social media using CNN. They evaluated the model by experimenting on two fake news datasets (ISO and FAKES) and compared the results with non-hybrid baseline models. Samadi et al. [22] proposed a deep contextualized representation for fake news data using different embedding models such as GPT2, BERT, Funnel-Transformer, and RoBERTa. Then, the embedded fake news data is classified using three classifiers: CNN, Single-Layer Perceptron (SLP), and Multi-Layer Perceptron (MLP). They conduct experiments on three fake news datasets such as COVID-19, LIAR, and ISOT, to evaluate the performance of the model. Song et al. [23] proposed a temporal propagation-based model for detecting fake news to combine the benefits of structural, content semantic, and temporal analysis. The proposed model analyzed the temporal evolution patterns in the real-world fake news as graph evolutions. These graph evolutions are constructed based on the settings of continuous-time dynamic diffusion networks.

3   Proposed Methodology

There are many reputable Web sites broadcasting authentic news items, such as Twitter, and also some Web sites, such as Snopes and PolitiFact for fact checking published news. Following the retrieval of news headlines and text via web scraping on news URLs, the data cleaning and preparation step begins, which entails converting news text into a numerical vector form that can be utilized as an input for training the models. We preprocessed the extracted news text from Web sites by removing stop words, punctuations, special characters, and other URL links. The steps that followed in our proposed hybrid text representation methodology are shown in Fig. 1.

Our proposed hybrid text representation methodology included two major steps:

- Representing the news text through its vectorization using TF-IDF for bigrams and trigrams.
- Selection of the most relevant features that enhance the accuracy of detection using Latent Semantic Indexing LSI.

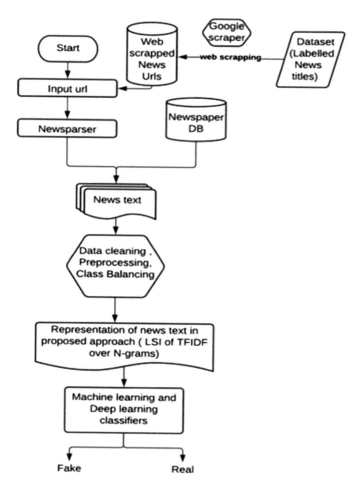

Fig. 1 Flow diagram of the proposed model

The TF-IDF for N-grams along with Latent Semantic Indexing is the uniqueness of our methodology. We vectorize the news text using TF-IDF by considering the social media news text in bigrams or trigrams. Then, the statistics from TF-IDF are analyzed using Latent Semantic Indexing to obtain the co-occurrence between the fake and real ones. Based on the co-occurrence values, the significant attributes are selected as features to train the fake detection model with the reduced dimensionality.

3.1   TF-IDF with N-grams Model for Vectorization

The primary focus of the approach is to represent the news text for analyzing fake news by considering the three important characteristics of the text including the frequency of important words present in the text, their order of appearance, and the number of hidden topics in the text. The text frequency and the order of words' appearance can be analyzed using the concepts of NLP including TF-IDF and N-grams model. The news text is passed to TD-IDF Vectorizer with N-grams analyzer for representing the frequency of occurrence of N-grams of news text in vector matrix form. The reason behind selecting the TF-IDF with the N-grams model is that TF-IDF considers the importance of the word in a text very well; however, it does not consider the order of appearance of those words. This issue gets resolved by using the N-grams model (with n_2) where the TF-IDF score is calculated for every N-word group in the document. Hence, in contrast to singular words (unigrams), the TF-IDF with N-grams is used in our approach for considering the words in sequence.

3.1.1   TF-IDF

TF-IDF (term frequency-inverse document frequency) is an NLP technique that represents the importance of a word in a textual document. TF-IDF can be applicable in NLP tasks such as text mining, user modeling, and information retrieval. TF-IDF estimates the importance of a word in a document by measuring the term weights. The TF-IDF value measures the term weights based on the frequency of a word in the document and the number of documents that contain the word within a corpus.

Term Frequency (TF) can be measured as the ratio of word frequency in a document to the total number of words in the document. Inverse Document Frequency (IDF) can be measured as the ratio of the logarithm value of the number of documents in the corpus to the number of documents that contain the specific term. It measures how important a term is. In such a way, TF-IDF measures how important a term is in the document and helps to adjust for the fact that some words appear more frequently in general.

We can calculate the TF-IDF score of a term i in the document as given in the following equation

$$j = \text{TF}(i, j) * \text{IDF}(i)$$

where

$$\text{TF}(i, j) = \frac{\text{frequency of term } i \text{ in the document } j}{\text{Total words in document } j}$$

and

$$\text{IDF}(i) = \log_2 \left(\frac{\text{total documents}}{\text{documents with term } i} \right)$$

3.1.2 *N*-grams

According to the linguistic context, *N*-grams is a sequence of N continuous items in any given sample of speech or text. The sequence of items can be syllables, words, or, phonemes based on the application. In the context of fake news detection, *N*-grams is the sequence of *N* textual words from the news article. These *N*-grams in the article are considered to identify the appearance of the words in order to detect fake news.

3.2 *Dimensionality Reduction by Feature Selection Using LSI*

Another challenge in fake news detection is processing the huge amount of text. The vector matrix obtained from TF-ID is also very sparse (or high dimension) and noisy (or includes lots of low frequency words). Latent Semantic Indexing helps in reducing the dimensionality of the matrix obtained from TF-IDF vectorization. So the truncated Singular Value Decomposition (SVD), which uses Latent Semantic Indexing in the background implementation, is adopted to reduce dimension. The truncated SVD takes the number of components as the main parameter which describes the number of dimensions to which a matrix is to be reduced.

Along with dimensionality reduction, LSI also analyzes the hidden topics from the text and contributes better in the concise and more precise vector representation of the news text. The resulting vector is then analyzed using various ML and DL techniques that used in the existing models including Naïve Bayes (NB) [7, 10, 13–15], Decision Tree (DT) [20], Logistic regression (LR) [10, 13–15], Random Forests (RF) [10, 14], Support Vector Machine (SVM) [7, 13–15], Convolutional Neural Network with LSTM (CNN-LSTM) [8], and Convolutional Neural Network with GRU (CNN-GRU) [8]. Using the hybrid approach of text representation, it is expected to get better results in our work of fake news and misinformation detection.

3.2.1 Singular Value Decomposition (SVD)

In the context of linear algebra, the Singular Value Decomposition (SVD) is decomposing a real or complex matrix by factorizing. SVD factorizes any complex $m \times n$ matrix M into a generalized form of $U\Sigma V^*$ which is an Eigen decomposition of M into a square normal matrix with an orthonormal Eigen basis. In the generalized form, U and V are complex unitary matrices of sizes $m \times m$ and $n \times n$, respectively, and Σ is a diagonal matrix of size $m \times n$ with non-negative real numbers on the diagonal $\sigma_i = \Sigma_{ij}$ which can also be called as singular values of M. The inference is that for all real values of M, the matrices U and V will also be real orthogonal matrices. In such cases, the SVD can also be denoted as $U\Sigma V^T$.

3.2.2 Latent Semantic Analysis

Latent Semantic Analysis (LSA) is an unsupervised topic modeling approach that infers the hidden topics represented in the given text or document. LSA can be applied for text documents for noise or dimensionality reduction as well. The hidden topic in the text document is not the actual topic of the document that exactly classifies the document into any group such as business, news, or sports. The hidden topic represents the better way in which the text in the document can be depicted. The LSA model uses the Bag of Words (BoW) approach to analyze for the hidden topics in the textual document. Using BoW, LSA drafts a term-document matrix for the text based on the frequency of the terms appearing in the document. In the term-document matrix, the row values indicate the frequency of terms and the column values indicate the documents. Then, to analyze the latent semantic relation among the text in the documents, LSA performs the decomposition of the term-document matrix using Singular Value Decomposition (SVD).

4 Dataset Description and Pre-processing

The environment used for the experiments is a Google colaboratory Jupyter interface with two single core processors of Intel(R) Xeon(R) CPU @ 2.20 GHz running on 64-bit Ubuntu 18.04.5 LTS operating System. The programming language used is Python 3.6. The dataset used for our experimentation is extracted using a web scrapping approach from a Fake news competition on Kaggle [24]. The dataset extracted for our experimentation is published in the English language and contained 23,481 fake news and 21,417 real news. However, due to the limitations on computational capacity, only a subset of this dataset is used for model training. The dataset is shuffled and the maximum possible number of rows according to the capacity limit are extracted which are then class-balanced for consideration in our work.

The original dataset contains the news URLs classified as fake and real according to various fact-checking organizations such as PolitiFact and Snopes. Then, the corresponding news text has been fetched using web scraping and newspaper library of python for further processing. After collecting the URLs of all fake and real news, we fetch news titles or topics of the article containing that news. This is done using Google scraper on the corresponding news URLs. Then, the news text is extracted through a python library (Newspaper3k) which provides an easy way to get the text body using the news URLs by parsing the text in a readable form.

The Newspaper3k package in the Python library uses lxml for parsing the text which is built on top of the requests. Requests package is again another library in python which has Beautiful Soup as a dependency module for parsing using lxml. Newspaper3k is also able to scrape any kind of data that includes author details, images, videos, URLs, publish dates, etc. The Newspaper3k can also produce a summary of the documents to lessen the burden of reading the entire documents. The text that is parsed using Newspaper3k is further preprocessed to remove stop words, special characters, punctuations, external links, etc., before conducting vector transformation using our proposed hybrid text representation approach.

5 Results and Discussion

The performance of the proposed model is validated by considering the news article data using tenfold cross-validation. Then, we have applied different machine learning and deep learning classifiers, namely NB, DT, LR, RF, SVM, CNN with LSTM (CNN-LSTM), and CNN with GRU (CNN-GRU) on each fold of data. The models are evaluated using accuracy score which was measured as the average of tenfold cross-validation accuracies. The accuracy of detecting fake news can be calculated as the number of fake news detected by the model correctly out of the total number of fake news. In our experimentation, non-fake news is the positive case and fake news is the negative case. In terms of True Positives (TP), True Negatives (TN), False Positives (FP), and False Negatives (FN), detecting non-fake news as non-fake news is True Positive and as fake news is False Positive. Detecting fake news as fake news is True Negative and fake news as non-fake news is False Negative. Some examples of false positive and false negatives extracted from the dataset are given in Table 1. The accuracy in detecting fake news can be detected based on the values of TP, TN, FP, and FN.

The accuracy of the proposed model is measured for both bigrams ($N = 2$) and trigrams ($N = 3$) and with all possible values of $n\_components$ of SVD (from 100 to 500), and the accuracies are measured as depicted in Table 2. It can be observed from the results that bigrams performed better than trigrams at $n\_components$ equal to 500. It is also observed that the CNN-GRU model with the proposed hybrid text representation outperformed the other models.

We also compared our approach for bigrams at $n\_components = 500$ with TF-IDF (for bigrams) in terms of precision, recall, f score, and accuracy. The comparison

Table 1 Some examples of false positive and false negative news text from the dataset

News	Actual	Predicted	FP/FN
Washington—The White House on Tuesday said it strongly opposes a House bill to fight the Zika virus, saying its funding is "woefully inadequate" to support the public health response that is needed. The Obama administration has urged Congress to pass legislation that would direct $1.9 billion to fighting the Zika virus that is linked to birth defects including microcephaly	1	0	FP
Sydney—More than 20,000 people rallied in Sydney on Sunday urging the legalization of same-sex marriage, days ahead of a contentious postal survey on the issue that has divided the country. Organizers said the gathering was Australia s largest gay rights demonstration, as a diverse range of people clad in rainbow colors converged on the heart of the city to insist on equal rights	1	0	FP
Black lives matter but only when white cops are responsible for their deaths. When blacks kill other blacks not so much. Kudos to the Fred L. Davis Insurance Company for telling the truth about who's really doing the killing in black communities. That's the statement on a billboard in Memphis, Tennessee. However, there's more to it. The full message reads: Black lives matter. So let's quit killing each other. It's a statement directed against black-on-black crime. Normally, liberals take umbrage at the idea that somebody might evaluate problems caused within the black community by other blacks because it doesn't suit the narrative. However, the author of this message has a background that will make it difficult for people to attack him like that	0	1	FN
Ron Paul, who's a retired doctor, not an economist, has a prediction that in the very near future, the stock market will lose half its value. Guess who would be to blame? Why it's the black guy, of course. A 50% pullback is conceivable, Paul said on Futures Now recently. I don't believe it's ten years off. I don t even believe it's a year off. According to his calculations, it would cut the S&P 500 Index in half, to 1212, and the blue-chip Dow Jones Industrial Average would collapse to 10,837	0	1	FN
Buenos Aires—An Argentine submarine with 44 crew on board was missing in the South Atlantic two days after its last communication, prompting the navy to step up its search efforts late on Friday in difficult, stormy conditions. The ARA San Juan was in the southern Argentine sea 432 km (268 miles) from the Patagonian coast when it sent its last signal on Wednesday, naval spokesman Enrique Balbi said. The emergency operation was formally upgraded to a search-and-rescue procedure on Friday evening after no visual or radar contact was made with the submarine, Balbi said. Detection has been difficult despite the quantity of boats and aircraft involved in the search, Balbi said, noting that heavy winds and high waves were complicating efforts. Obviously, the number of hours that have passed—two days in which there has been no communication—is of note	1	0	FP

Table 2 Performance results of various ML and DL models for bigrams and trigrams at different values of n_components (number of SVD components)

Model	N_components (SVD)									
	100		200		300		400		500	
	$N = 2$	$N = 3$	$N = 2$	$N = 3$	$N = 2$	$N = 3$	$N = 2$	$N = 3$	$N = 2$	$N = 3$
NB	91.32	79.36	90.56	80.03	91.23	82.47	92.24	82.91	91.71	84.51
RF	93.54	81.95	92.99	84.14	93.85	85.69	93.96	85.29	94.02	89.93
LR	91.43	80.67	92.60	83.87	92.93	84.33	93.13	84.60	93.53	86.27
SVM	93.20	81.49	91.54	81.26	92.25	82.28	93.42	84.54	93.80	89.50
DT	89.53	76.23	90.20	79.15	90.53	80.09	91.20	80.29	89.87	81.34
CNN-LSTM	92.60	83.21	93.10	84.37	94.33	86.13	94.60	86.01	94.80	90.20
CNN-GRU	93.80	83.86	94.20	84.91	95.00	86.84	95.40	86.47	95.80	91.33

results are presented in Table 3. The comparison results illustrate the fact that the detection accuracy of the model has a significant improvement when the news text is represented using our proposed hybrid text representation approach.

Table 3 Comparison analysis of various ML and DL models with different text representation approaches

Model	Text representation approach	precision	Recall	F score	Accuracy
NB	TF-IDF	82.13	80.01	81.05	80.54
RF		83.62	81.36	82.47	83.97
LR		83.12	81.94	82.52	83.12
SVM		82.14	81.14	81.63	83.77
DT		80.23	78.34	79.27	79.28
CNN-LSTM		84.15	80.47	82.26	85.42
CNN-GRU		85.36	84.27	84.81	85.93
NB	Proposed	91.61	91.23	91.41	91.71
RF		94.89	92.79	93.82	94.02
LR		92.45	90.55	91.49	93.53
SVM		93.47	91.01	92.22	93.80
DT		91.62	87.19	89.35	89.87
CNN-LSTM		95.71	92.73	94.19	94.80
CNN-GRU		95.93	94.81	95.36	95.80

6 Conclusion

The task of news classification as fake or real manually requires good domain knowledge and experience to determine the content's validity. We tackled the subject of categorizing news and misinformation as fake or true in this research by combining existing NLP techniques to develop a hybrid text representation model and then training on different machine learning and deep learning classifiers. The main objective of this work is to present a new hybrid approach of text representation which could make the fake detection model better differentiate the fake news from social media news. The hybrid text representation approach transformed the news text into numerical features using TF-IDF vectorizer with N-grams model and then selected relevant features through Latent Semantic Indexing. Our hybrid text representation approach showed a great impact in increasing the performance of fake news detection models.

References

1. Horne BD, Adali S (2017) This just in: fake news packs a lot in title, uses simpler, repetitive content in text body, more similar to satire than real news. https://arxiv.org/pdf/1703.09398.pdf
2. Celliers M, Hattingh M (2020) A systematic review on fake news themes reported in literature. In: Hattingh M, Matthee M, Smuts H, Pappas I, Dwivedi YK, Mäntymäki M (eds) Responsible design, implementation and use of information and communication technology. Lecture notes in computer science, vol 12067. Springer, Cham. https://doi.org/10.1007/978-3-030-45002-119
3. Khan T, Michalas A, Akhunzada A (2021) Fake news outbreak 2021: can we stop the viral spread? J Netw Comput Appl 190:1–17
4. Wen Z, Taketoshi Y, Xijin T (2011) A comparative study of TFIDF, LSI and multi-words for text classification. Expert Syst Appl 38(3):2758–2765
5. Ahmed H, Traore I, Saad S (2017) Detection of online fake news using N-gram analysis and machine learning techniques. In: Traore I, Woungang I, Awad A (eds) International conference on intelligent, secure, and dependable systems in distributed and cloud environments (ISDDC 2017). Lecture Notes in Computer Science, vol 10618. Springer, pp 127–138
6. Singhania S, Fernandez N, Rao S (2017) 3HAN: a deep neural network for fake news detection. In: Liu D, Xie S, Li Y, Zhao D, El-Alfy ES (eds) International conference on neural information processing (ICONIP 2017). Lecture notes in computer science, vol 10635. Springer, pp 572–581
7. Aphiwongsophon S, Chongstitvatana P (2018) Detecting fake news with machine learning method. In: Proceedings of the 15th international conference on electrical engineering/electronics, computer, telecommunications and information technology (ECTI-CON 2018), pp 528–531
8. Aswini T, Priyanka T, Simrat A, Nibrat L (2018) Fake news detection: a deep learning approach. SMU Data Sci Rev 1(3):1–20
9. Girgis S, Amer E, Gadallah M (2018) Deep learning algorithms for detecting fake news in online text. In: 13th international conference on computer engineering and systems (ICCES 2018), pp 93–97
10. Monther A, Ali A (2018) Detecting fake news in social media networks. In: 9th international conference on emerging ubiquitous systems and pervasive networks (EUSPN 2018), vol 141, pp 215–222

11. Ajao O, Bhowmik D, Zargari S (2018) Fake news identification on twitter with hybrid CNN and RNN models. In: Proceedings of the 9th international conference on social media and society, pp 226–230
12. Ajao O, Bhowmik D, Zargari S (2019) Sentiment aware fake news detection on online social networks. In: Proceedings of the IEEE international conference on acoustics, speech and signal processing (ICASSP 2019), pp 2507–2511
13. Mahir EM, Akhter S, Huq MR (2019) Detecting fake news using machine learning and deep learning algorithms. In 7th international conference on smart computing & communications (ICSCC), IEEE, pp 1–5
14. Hiramath CK, Deshpande GC (2019) Fake news detection using deep learning techniques. In: 1st international conference on advances in information technology (ICAIT 2019), pp 411–415
15. Poddar K, Amali D, Umadevi KS (2019) Comparison of various machine learning models for accurate detection of fake news. In: Innovations in power and advanced computing technologies (i-PACT 2019), pp 1–5
16. Abdullah YA, Awan MJ et al (2020) Fake news classification bimodal using convolutional neural network and long short-term memory. Int J Emerg Technol 11(5):1–4
17. Agarwal A, Dixit A (2020) Fake news detection: an ensemble learning approach. In: Proceedings of the 4th international conference on intelligent computing and control systems (ICICCS 2020), pp 1178–1183
18. Ibrishimova MD, Li KF (2020) A machine learning approach to fake news detection using knowledge verification and natural language processing. In: Barolli L, Nishino H, Miwa H (eds) International conference on advances in intelligent networking and collaborative systems (INCoS 2019). Advances in intelligent systems and computing, vol 1035. Springer, pp 223–234
19. Uppal A, Sachdeva V, Sharma S (2020) Fake news detection using discourse segment structure analysis. In: 10th international conference on cloud computing, data science & engineering (confluence 2020), pp 751–756
20. Baarir NF, Djeffal A (2021) Fake news detection using machine learning. In: 2nd international workshop on human-centric smart environments for health and well-being (IHSH 2021), pp 125–130
21. Jamal AN, Osama SK, Iraklis V (2021) Fake news detection: a hybrid CNN-RNN based deep learning approach. Int J Inf Manage Data Insights 1(1):1–13
22. Samadi M, Mousavian M, Momtazi S (2021) Deep contextualized text representation and learning for fake news detection. Inf Process Manage 58(6):1–13
23. Song C, Shu K, Wu B (2021) Temporally evolving graph neural network for fake news detection. Inf Process Manage 58(6):1–18
24. Kaggle (2018) Fake news: build a system to identify unreliable news articles. Available via Google. https://www.kaggle.com/c/fake-news/data

A Smart Healthcare Design of Carryable Drip Frame with Fuzzy-PID Controllers

Yuan-Yi Chang, Ming-Feng Wu, and Chih-Yu Wen

Abstract Nowadays, besides bedside intravenous (IV) therapy drip stands, wheeled IV drip stands can be applied for injection. Although the wheeled IV drip stand is more convenient for patient movement, it is not easy to go up and down stairs or through steep ground because of the weight and the length of the IV stand. In this study, the design of carryable IV frame aims to provide patients with safety, convenience in wearing, getting out of the bed, going to eat, and going up and down stairs. Due to bending or movement, the proposed control mechanism of angle compensation can ensure that the potential energy is sufficient to overcome the intravenous pressure for avoiding blood reflux. To dynamically adjust the system parameters, this work develops an adaptive fuzzy-PID controller and a fuzzy-PID neural network controller. Comparing to existing studies, the experimental results show that the fuzzy-PID neural network controller has the best performance characteristics, which improves the accuracy and performance of the system control.

1 Introduction

Currently, two types of frames are commonly used: One is arranged at the front end of a bed body, and the other is attached to a wheelchair. However, the body of a wheel-type drip frame is too long and has a considerable weight, which can impair mobility considerably. This work references the system architecture in our previous works [1, 2] (Fig. 1a, b) and develops a control strategy of the piggyback

Y.-Y. Chang · C.-Y. Wen (✉)
Department of Electrical Engineering, National Chung Hsing University, Taichung, Taiwan
e-mail: cwen@dragon.nchu.edu.tw

M.-F. Wu
Division of Chest Medicine, Department of Internal Medicine, Taichung Veterans General
Hospital, Taichung 407, Taiwan

Department of Medical Laboratory Science and Biotechnology, Central Taiwan University of
Science and Technology, Taichung 406, Taiwan

© The Author(s), under exclusive license to Springer Nature Singapore Pte Ltd. 2023
M. A. Chaurasia and C.-F. Juang (eds.), *Emerging IT/ICT and AI Technologies Affecting
Society*, Lecture Notes in Networks and Systems 478,
https://doi.org/10.1007/978-981-19-2940-3_3

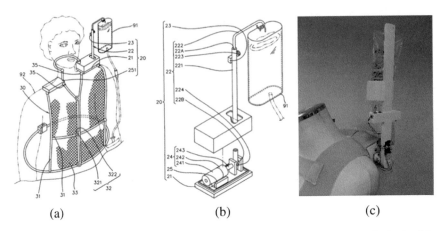

(a)   (b)   (c)

Fig. 1 a, b Conceptual system structure on the shoulder; **c** Proposed system on the back of the shoulder

intravenous drip frame based on a motion sensor, a motor with the proportional–integral–derivative (PID) control, the fuzzy neural network, and angle information, which uses the effect of attitude determination and dynamic response performance of a piggyback intravenous drip frame. Considering convenience and comfort of the system, a new vest structure is developed, which changed the device location from shoulder [2] to back (Fig. 1c), it describes a carryable and automatically balanced intravenous drip frame which allows the patient to move freely. Subjects wearing the new drip frame rise from bed, go to the toilet, eat, or walk up and down stairs, thus increasing the quality of life. The proposed system contains a carrying unit and a supporting unit, where the supporting unit is a vest worn on the body and has a lightweight skeletal joint structure with a carrying unit on its shoulder for hanging the dripping frame. The carrier unit is provided with angle monitoring and compensation controlled using a fuzzy logic controller. The proposed system avoids the problem that the patients' blood may flow back into the drip by suitable regulation of the pressure, and the system may be tilted forward or backward.

As we know, conventional proportional–integral–derivative (PID) controllers are well known and have been most widely used in various fields of the industry control. PID control provides an efficient method for engineers to solve many problems with tuning P, I, and D term. Although conventional PID controller has strong ability, they generally not work well for nonlinear systems and time-delayed linear systems. Fuzzy logic control (FLC) is a tool for tuning the parameters of a system, and its expression of control theory is easy to understand. Fuzzy logic controller has more robust than conventional controller for many linear and nonlinear systems. Therefore, it has many advantages in combining FLC and conventional PID control.

In our previous work [2], a reliable balanced system can be applied to facilitate patients' movements and ensure patient safety with compensating the inclination angle of the drip frame such that the reduction of blood returning and the balance control of the piggyback intravenous drip frame can be achieved. However, the controller design is not self-adaptive. In order to make the control adaptive, on the basis of FLC, PID control and neural network, this work proposes two system structures, which are adaptive fuzzy-PID controller [3–7] and fuzzy-PID neural network controller (F-PIDNN) [8], to dynamically adjust system parameters. F-PIDNN has been proven to be very reliable in HVAC systems [9]. Therefore, this work applies a similar concept to design a F-PIDNN controller for motor speed control. The proposed F-PIDNN integrates the PID control and fuzzy rule with the neural network. The PID neural network [10, 11] consists of proportional (P), integral (I), and derivative (D) neurons and its weights are adjusted by the back-propagation algorithms. The design principles of the PID controllers are described in the following sections.

2 Proportional–Integral–Derivative Controller

A proportional–integral–derivative (PID) control theory [12], which is widely used in the industry. It is mainly used in feedback systems. It has become the main control technology in the industry due to its simple structure, convenient adjustment, good stability, and high reliability. PID control is composed of three units: Proportional, integral, and derivative, which represent the present error, the past accumulated error, and the future error, respectively. (2) depicts the continuous-time PID controller output $u(t)$. The block diagram of PID control system is shown in Fig. 2.

$$u(t) = K_p e(t) + K_i \int_0^t e(\tau)d\tau + K_d \frac{d}{dt}e(t) \tag{1}$$

where $e(t)$ is the output signal of controller, $e(t)$ is error signal, $y(t)$ is the output of sensor, K_p is the proportional parameter, K_i is the integral parameter (K_p/T_i), and K_d is the derivative parameter ($K_p * T_d$). In this work, as the PID controller is implemented in a digital processor, the backward difference method [13] is applied to discretize the controller, which gives the discrete-time PID controller output.

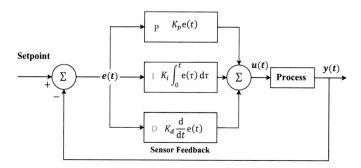

Fig. 2 Block diagram of PID control system

$$u(n) = u(n-1) + K_p[e(n) - e(n-1)] + K_i T_s e(n)$$

$$+ \frac{K_d}{T_s}[e(n) - 2e(n-1) - e(n-2)], \tag{2}$$

where T_s is the sampling time of the analog-to-digital converter.

3 The Adaptive Fuzzy-PID Controller

In industrial automatic control, it may be hard to set the parameter of PID control. Due to the time-varying control object, the system parameters should be adaptively tuned in most situations. For solving the problem, this section describes the proposed adaptive fuzzy-PID algorithm, incorporating the adaptive control theory with a fuzzy-PID controller, as shown in Fig. 3.

Wherein the diagram, r is the setpoint, u is the controlled quantity, and y is the output. First, setting the initial values of the conventional PID controller's parameters (K_{p0}, K_{i0}, K_{d0}). Then, set the fuzzy control rules based on the control experience, and consider the system error (e) and the error derivative (ce) as the inputs. Next, using an inference mechanism to produce an output from a collection of if–then rules, and finally obtain the output values of three control parameters (ΔK_p, ΔK_i, ΔK_d) by defuzzification. The three values can fix conventional PID controller online as described in (3), which yields

$$K_p = K_{p0} + \Delta K_p$$
$$K_i = K_{i0} + \Delta K_i \tag{3}$$
$$K_d = K_{d0} + \Delta K_d$$

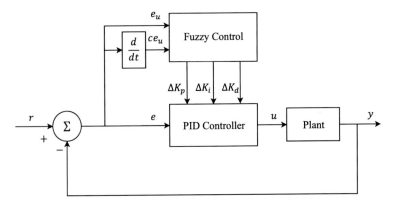

Fig. 3 System structure of the proposed adaptive fuzzy controller

The following subsections describe the fuzzy control logic for adaptively adjusting three conventional PID control parameters with two fuzzy inputs.

3.1 Fuzzy Input Variables

In this section, the angle measurements of crisp input transforming to linguistic variables are error (e) and change in error (ce) in Eqs. (4) and (5), respectively. The transform methods of fuzzifier employ the triangular and trapezoid membership functions, as shown in Fig. 4.

$$e(t) = r(t) - x(t) \tag{4}$$

$$ce(t) = e(t) - e(t-1) \tag{5}$$

$$e_u(t) = e(t)/e_{\max} \tag{6}$$

$$ce_u(t) = ce(t)/ce_{\max} \tag{7}$$

where r is setpoint and x is a measurement value by the sensor. Note that the fuzzy sets of e_u and ce_u are established in domain of discourse -1 to 1, which normalizes the two inputs.

There are five defined fuzzy sets, which are negative big (NB), negative small (NS), zero (ZO), positive small (PS), and positive big (PS). The fuzzy sets of triangular and trapezoid membership functions are shown in Table 1.

Fig. 4 Membership functions of inputs (e_u and ce_u)

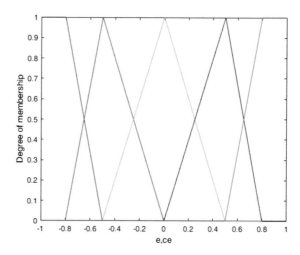

Table 1 Fuzzy sets of inputs (e_u and ce_u)

Linguistic variables	Range	Type
NB	$\{-1, -1, -0.8, -0.5\}$	Trapezoid
NS	$\{-0.8, -0.5, 0\}$	Triangular
ZO	$\{-0.5, 0, 0.5\}$	Triangular
PS	$\{0, 0.5, 0.8\}$	Triangular
PB	$\{0.5, 0.8, 1, 1\}$	Trapezoid

Table 2 Fuzzy rule for ΔK_p

ce_u	e_u				
	NB	NS	ZO	PS	PB
NB	PB	PB	PS	PS	ZO
NS	PB	PS	PS	ZO	NS
ZO	PS	PS	ZO	NS	NS
PS	PS	ZO	NS	NS	NB
PB	ZO	NS	NS	NB	NB

3.2 The Design Rule of K_p

In PID controller, the higher K_p is, the smaller steady-state error is. However, when K_p value is too high, it will generate a large overshoot. Table 2 describes the fuzzy control rules of output (ΔK_p).

3.3 The Design Rule of K_i

The integral control is mainly used to eliminate the steady-state error of the system. Since the integral process may produce integral saturation at the beginning, which may cause a large overshoot, the integral value should be a small value or zero in the early stage. The fuzzy control rules of output (ΔK_i) are established in Table 3.

3.4 The Design Rule of K_d

The differential part mainly changes the dynamic performance of the system. When the K_d value is too large, the brake is advanced and the adjustment time will be longer. On the other hand, when the K_d value is too small, the brake will fall behind, which leads to an overshoot. The fuzzy control rules of output (ΔK_d) are established and summarized as below:

Table 3 Fuzzy rule for ΔK_i

ce_u	e_u				
	NB	NS	ZO	PS	PB
NB	NB	NB	NS	NS	ZO
NS	NB	NS	NS	ZO	PS
ZO	NS	NS	ZO	PS	PS
PS	NS	ZO	PS	PS	PB
PB	ZO	PS	PS	PB	PB

Table 4 Fuzzy rule for ΔK_d

ce_u	e_u				
	NB	NS	ZO	PS	PB
NB	PS	ZO	ZO	ZO	PB
NS	NB	NS	NS	ZO	PS
ZO	NB	NS	NS	ZO	PS
PS	NB	NS	NS	ZO	PS
PB	PS	ZO	ZO	ZO	PB

Referring to Tables 2, 3, and 4, the syntax for defining rules in rule base acts like:

$$\text{If } e_u \text{ is NB and } ce_u \text{ is NB, then } K_p \text{ is PB.} \tag{8}$$

$$\text{If } e_u \text{ is NB and } ce_u \text{ is NB , then } K_i \text{ is NB.} \tag{9}$$

$$\text{If } e_u \text{ is NB and } ce_u \text{ is NB, then } K_d \text{ is PS.} \tag{10}$$

3.5 Fuzzy Output Variables

Defuzzification method uses height defuzzification (or called local mean-of-maximum), presented in Eq. (11). There are linguistic variables converted to an actual value, according to the linguistic rules. The fuzzy sets of outputs (K_p, K_i, K_d) are revealed in Table 5 and Fig. 5.

$$\text{LMOM}z_h = \frac{\sum_{i=1}^{r} \overline{z_i} \Phi_i}{\sum_{i=1}^{r} \Phi_i} \tag{11}$$

Table 5 Fuzzy sets of outputs (ΔK_p, ΔK_i, ΔK_d)

NB	$\{-1, -1, -0.6, -0.3\}$	Trapezoid
NS	$\{-0.6, -0.3, 0\}$	Triangular
ZO	$\{-0.3, 0, 0.3\}$	Triangular
PS	$\{0, 0.3, 0.6\}$	Triangular
PB	$\{0.3, 0.6, 1\}$	Trapezoid
NB	$\{0.3, 0.6, 1, 1\}$	Trapezoid

Fig. 5 Membership functions of outputs (ΔK_p, ΔK_i, ΔK_d)

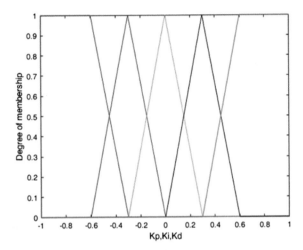

where r is the number of rules. Note that Φ_i is the firing strength, which is the result of the ith fuzzy inference, and $\overline{z_i}$ is the point with the maximum membership in the ith output set.

4   The Fuzzy-PID Neural Network Controller

To reduce the steady-state error and suppress the short-term performance decline, a fuzzy-PID controller is developed to improve system stability. In the system structure, the parameters (K_p, K_i, K_d) are first obtained, according to the adaptive fuzzy control calculation. Then, two input variables (i.e., e and fuzzy-PID output) are fed into neural network. In the end, a control output (u) is built up to control a DC motor. Figure 6 depicts the structure of fuzzy-PID neural network (F-PIDNN) controller.

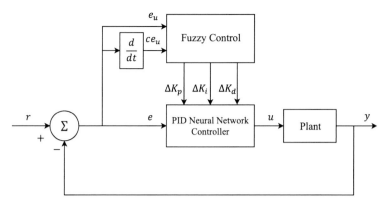

Fig. 6 System structure of fuzzy-PID neural network controller

4.1 PID Neural Network

This section represents the PID neural network, which is 2-3-1 structure. It is composed of two neurons in the input layer, three in the hidden layer, and one in the output layer, respectively. In the input layer, the two neurons are the error and the system output from an adaptive fuzzy-PID controller. In the hidden layer, the three neurons are, respectively, represented as P-neuron, I-neuron, and D-neuron, where P-neuron performs the proportional function, I-neuron performs the integral function, and D-neuron performs the derivative function. In the output layer, the neuron carries the system control signal.

Figure 7 shows the PID neural network structure. The system consists of two parts: One is the forward propagation, which computes the neuron signal; the other is the training stage, which employs back-propagation (BP) algorithm to train connecting weights for optimizing the system output. Note that reference inputs 1 and 2 ($r1, r2$)

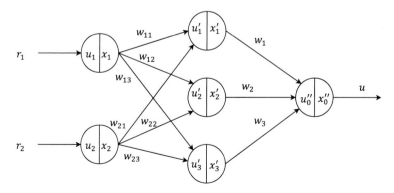

Fig. 7 PID neural network structure

are error and output of fuzzy controller, respectively. The following subsections detail the system parameters.

4.2 Input Layer

In the input layer, the neuronal input–output function is given by

$$u_1(t) = e(t)/e_{\max} \tag{12}$$

$$u_2(t) = \text{Fout}/\text{Fout}_{\max} \tag{13}$$

$$x_i(k) = \begin{cases} 1 & u_i(k) > 1 \\ u_i(k) & -1 \le u_i(k) \le 1 \; i = 1, 2, \\ -1 & u_i(k) < -1 \end{cases} \tag{14}$$

Therefore, we have, where $u_i(k)$ and $x_i(k)$ are, respectively, the convergent input signals and the output signals at the sampling time k, and F_{out} is the adaptive fuzzy-PID output.

4.3 Hidden Layer

In the hidden layer, the neuronal input–output functions are the key in the PID neural network. Referring to Fig. 7, the input of hidden layer neuron is

$$u'_j(k) = \sum_{i=1}^{2} w_{ij} x_i(k), \; j = 1, 2, 3, \tag{15}$$

where $x_i(k)$ is the output of input layer and w_{ij} is the connection weight between the input layer and the hidden layer. Accordingly, the neuronal input–output functions of P-neuron, I-neuron, and D-neuron, respectively, yield.

A. P-neuron:

$$x'_1(k) = \begin{cases} 1 & u'_1(k) > 1 \\ u'_1(k) & -1 \le u'_1(k) \le 1 \\ -1 & u'_1(k) < -1 \end{cases} \tag{16}$$

B. I-neuron:

$$x_2'(k) = \begin{cases} 1 & x_2'(k) > 1 \\ x_2'(k-1) + u_2'(k) - 1 \le x_2'(k) \le 1 \\ -1 & x_2'(k) < -1 \end{cases} \tag{17}$$

C. D-neuron:

$$x_3'(k) = \begin{cases} 1 & x_3'(k) < 1 \\ u_3'(k) - u_3'(k-1) - 1 \le x_3'(k) \le 1 \\ -1 & x_3'(k) > -1 \end{cases} \tag{18}$$

4.4 Output Layer

As depicted in Fig. 8, the input of the output layer neuron is

$$u_0''(k) = \sum_{j=1}^{3} w_j x_j'(k) \tag{19}$$

where $x_j'(k)$ is the output of hidden layer and w_j is the weight from the nodes of the hidden layer to the nodes of the output layer. With $u_0''(k)$ and the final input–output function, the output of the output layer is given by

$$x_0''(k) = \begin{cases} 1 & u_0''(k) > 1 \\ u_0''(k) & -1 \le u_0''(k) \le 1 \\ -1 & u_0''(k) < -1 \end{cases} \tag{20}$$

Therefore, the actual output is

$$u = x_0'' \tag{21}$$

4.5 Back-Propagation (BP) Algorithm

The back-propagation algorithm is one of important training methods for neural network. It employs the gradient descent algorithm to optimize the weights (w_{ij}, w_j) due to reverse calculation. The target of a PID neural network is to minimize the mean squared error. Consider the error function

$$J = \frac{1}{2}[y_d(k) - y(k)]^2 \tag{22}$$

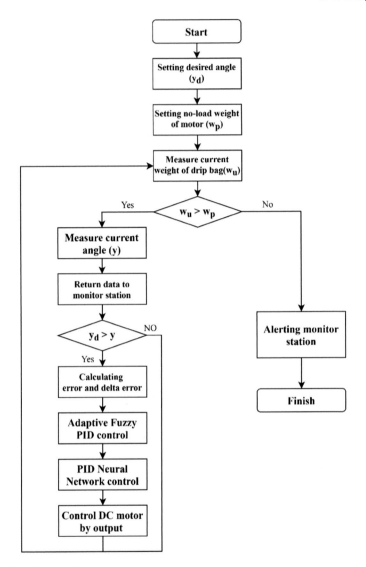

Fig. 8 Overall system flowchart

where k is the training step, $y_d(k)$ is the desired output, and $y(k)$ is the actual output.

After training k steps, the new weight values between the hidden layer and the output layer are

$$w_j(k+1) = w_j(k) - \eta \frac{\partial J}{\partial \mathrm{w}_j} + \alpha \Delta w_j(k-1) \tag{23}$$

where $j = 1, 2,$ and $3; \eta$ is the learning rate, ranging from 0 to 1; α is the momentum factor, ranging from 0 to 1.

With the partial derivative with respect to w_j, we have

$$\frac{\partial J}{\partial w_j} = \frac{\partial J}{\partial y} \cdot \frac{\partial y}{\partial x_0''} \cdot \frac{\partial x_0''}{\partial u_0''} \cdot \frac{\partial u_0''}{\partial w_j}. \tag{24}$$

Hence, according to Eqs. (19), (20), (22), every part of partial derivative is

$$\frac{\partial J}{\partial y} = -[y_d(k) - y(k)]$$

$$\frac{\partial y}{\partial x_0''} = \text{sgn}\left(\frac{y(k+1) - y(k)}{x_0''(k) - x_0''(k-1)}\right)$$

where the sign function is used to approximate the result of $\partial y / \partial x_0''$.

$$\frac{\partial x_0''}{\partial u_0''} = 1s$$

$$\frac{\partial u_0''}{\partial w_j} = x_j'(k)$$

Thus,

$$\frac{\partial J}{\partial w_j} = -\delta_j(k) \cdot x_j'(k) \tag{25}$$

where $\delta_j(k)$ is

$$[y_d(k) - y(k)] \cdot \text{sgn}\left(\frac{y(k+1) - y(k)}{x_0''(k) - x_0''(k-1)}\right) \tag{26}$$

After training the weight parameter, the new weight values between the input layer and the hidden layer are

$$w_{ij}(k+1) = w_{ij}(k) - \eta \frac{\partial J}{\partial w_{ij}} + \alpha \Delta w_{ij}(k-1) \tag{27}$$

where $i = 1, 2, j = 1, 2, 3; \eta$ is the learning rate $(0 \le \eta \le 1)$. $1.\alpha$ is the momentum factor, $(0 \le \eta \le 1)$.

By applying the partial derivative definition, we obtain

$$\frac{\partial J}{\partial w_{ij}} = \frac{\partial J}{\partial y} \cdot \frac{\partial y}{\partial x_0''} \cdot \frac{\partial x_0''}{\partial u_0''} \cdot \frac{\partial u_0''}{\partial x_j'} \cdot \frac{\partial x_j'}{\partial u_j'} \cdot \frac{\partial u_j'}{\partial w_{ij}}. \tag{28}$$

Hence, according to Eqs. (15–18), (25), every part of partial derivative is

$$\frac{\partial u_0''}{\partial x_j'} = w_j$$

$$\frac{\partial x_j'}{\partial u_j'} = \text{sgn}\left(\frac{x_j'(k+1) - x_j'(k)}{u_j'(k) - u_j'(k-1)}\right),$$

where the sign function is used to approximate the result of $\partial x_j'/\partial u_j'$.

$$\frac{\partial u_j'}{\partial w_{ij}} = x_i(k)$$

Accordingly,

$$\frac{\partial J}{\partial w_{ij}} = -\delta_{ij}(k) \cdot x_i(k), \qquad (29)$$

where $\delta_{ij}(k)$ is

$$\delta_j(k) \cdot w_j \cdot \text{sgn}\left(\frac{x_j'(k+1) - x_j'(k)}{u_j'(k) - u_j'(k-1)}\right). \qquad (30)$$

The algorithm will stop learning as the mean square error is less than the setting error value at sample point k and then return the updated weight values. The F-PIDNN procedures are summarized in Algorithm 1.

Algorithm 1 F-PIDNN Procedures.

---

1. Set traditional PID control parameters.

2. Get inputs from sensors.

3. Initialize the previous_error parameter and set the integral value be 0.

4. Initialize the weights.

5. F-PIDNN control:

   start:

      ■   Correct the output of Fuzzy-PID controller (F_out)

      ■   Input and normalize e and F_out.

      ■   Forward propagation from the input layer, to the hidden layer, and then to the output layer.

      ■   Return the actual output.

      ■   Map the output.

      ■   Limit the output within the range set by the upper and lower limits.

      ■   Control the motor by the output signal.

      ■   IF the error of objective function doesn't satisfy the set error value, start Back-propagation for training new weights.

   previous_error = e
     if (input < setpoint)
        goto **start**
     else
        stop control
     end
   goto Step **2**.

---

5 System Implementation

Figure 8 shows the flowchart of a control system. The system design concept is to avoid blood reflux while hanging a drip. Note that the whole system will stop as the drip bag is empty. Figure 9 is the vest with the balance system on the left and right shoulders. The clothes are made of external polyester and internal nylon by sandwich structure, which provides breathable and strong knots. The connections on vest use a Velcro, which provides more convenient to dress vest and set the system device on back.

Fig. 9 Design of the vest
and the determination of
frame height

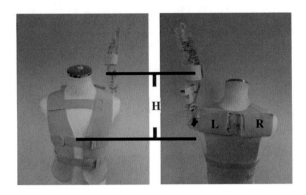

5.1 Hardware

The device consists of two parts: The control unit and the computing unit, including
Arduino Nano, MPU-6050, weight scales, L293D, and DC gear motor. For determining the frame height, the potential energy needs to be calculated. Generally,
venous pressure is 35 (cm H_2O). Suppose the drip infusion density is 1.1 (gw/cm^3)
and the drip frame has a minimized height H, the potential energy (P). If the setting
position of drip frame is higher, its burden on user is stronger. It also could be
influenced by the environment (e.g., door). Based on the above analysis,

$$P = \rho h = 1.1(\text{gw/cm}^3) \cdot H(\text{cm}) > 35(\text{cm H}_2\text{o}) = 35(\text{gw/cm}^3).$$

Thus, as shown in Fig. 9, we have $H(\text{cm}) > 31.8$ cm.

In Fig. 10, MPU-6050 [14] and weight scales (HX711 module with load cell) [15]
measure drip frame angle and drip weight, respectively. Arduino Nano is a board to
active motor and run algorithm. L293D [16] is a driver board for motor. The DC gear
motor is used to change tilt angle. Finally, Arduino will deliver these data to a smart
phone with Bluetooth.

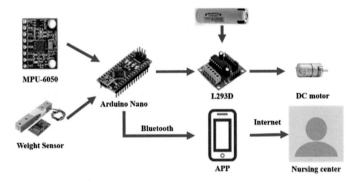

Fig. 10 System hardware

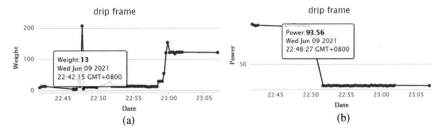

Fig. 11 Chart of weight; the chart of power

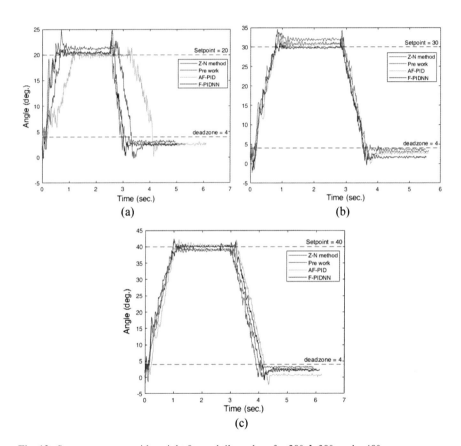

Fig. 12 System response with weight 0 g and tilt angles of **a** 20°, **b** 30°, and **c** 40°

5.2 Software

In the App, three states (Warning, Less, and Enough) are applied to show the current weight and describe how full the drip chamber is. Moreover, two states (Low and Enough) are presented to show the current battery power. After connecting Bluetooth

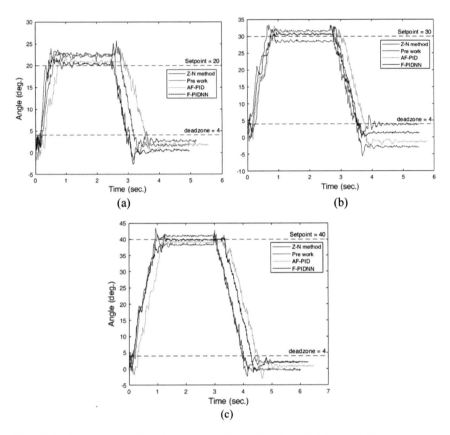

Fig. 13 System response with weight 200 g and tilt angles of **a** 20°, **b** 30°, and **c** 40°

and pushing the "start" button, the information of weight and battery power will be sent by the microcomputer through Bluetooth. After pushing the "stop" button, the operation of data collection will be temporarily stopped.

Figure 11a and b, respectively, illustrates the changes of weight and battery power. When the App gets the data from the microcomputer, the users can push the button "upload" to forward the data to a database through the Internet. Without loss of generality, ThingSpeak [17], an open-source application platform for the Internet of Things, is used to characterize and visualize the uploaded data.

6 Performance Evaluation

This section describes performance evaluation of the proposed system and performs a comparison between four control algorithms, including Ziegler–Nichols (Z–N) method PID [18], fuzzy-PID in [2], the proposed adaptive fuzzy-PID, and fuzzy-PID neural network (F-PIDNN). The previous work is a fuzzy-PID controller with two inputs, which are weight and angle obtained by HX711 and MPU-6050 sensors. For getting exact angle, the adaptive complementary filter [19] is used. The F-PIDNN is constructed by a conventional fuzzy controller and a PID neural network controller. The connection weights in neural network are training by back-propagation algorithm.

The work employs three characteristics to assess the performance of controllers, which are max overshoot (M_p), steady-state error and rise time, respectively. The definitions are described as follows:

1. **Percent max overshoot** ($\%M_p$): That is a value used to describe a signal when exceeding its setpoint, which is

$$\%M_p = \frac{c_{max} - c_{ss}}{c_{ss}} * 100 \tag{31}$$

   where c_{max} is max value of system response and c_{ss} is the steady-state value of system response.
2. **Steady-state error**: the difference between the final value of the response and the setpoint.
3. **Rise time**: Time for system response to rise from 10 to 90% of its final value.
4. **Settling time**: The time required for balancing the system with a tilt angle to reach the steady state.

For evaluating the system performance, the system is fixed on the table. Two experiments are conducted with the four controllers. Experiment 1 examines the system performance with the drip bag weight of 0 g and angles of 20, 30, and 40 degrees, respectively. Experiment 2 explores the system performance with the drip bag weight of 200 g and the same angle settings in Experiment 1. The motor has two rotation modes in the experiment. Mode 1 refers to the motor rotating front for drip frame angle closing setpoint, which is called "front rotation." The performance of system only calculates in the front rotation mode. Mode 2 defines the motor rotating back after the angle arrives the setpoint, waits a period of time, and then balance the system, which is called "reverse rotation." There is a dead zone 4° for stopping the drip frame in reverse rotation mode. Table 6 shows the initial values of system parameters.

6.1 Experiment 1

Given tilt angles of 20°, 30°, and 40°, Fig. 12 and Tables 7, 8, and 9 describe angle adjustment with a zero weight, which indicates that among the four methods,

Table 6 Settings of initial values

Setpoint (°)	20, 30, 40	
Weigh (g)	0, 200	
System delay time (s)	0.001	
Waiting time (s)	2	
Z–N method	$K_u = 5$	
Adaptive F-PID	$K_{p0} = 1.8, K_{i0} = 7.2, K_{d0} = 0.1125$	$e_{max} = $ Setpoint $ce_{max} = 2.0$
F-PIDNN	$w_{1j} = 1, w_{2j} = -1, j = 1, 2, 3. \eta = 0.25, \alpha = 0.05$	

Table 7 Controllers performance with weight 0 g—Overshoot (%)

Controller type	20 degrees (%)	30 degrees	40 degrees
Z–N method [18]	17.1	11.3	11.8
F-PID [2]	10.2	6.1	7
Proposed adaptive F-PID	6.5	5.6	5.1
Proposed F-PIDNN	3.7	3.7	2

Table 8 Controllers performance with weight 0 g—Rise time (RT) and Settling time (ST) (second)

Angle	20 degrees		30 degrees		40 degrees	
method	RT	ST	RT	ST	RT	ST
Z–N method [18]	0.63	1.17	0.72	1.39	0.87	1.51
F-PID [2]	0.37	0.91	0.78	1.29	0.84	1.16
Proposed adaptive F-PID	1.06	1.15	0.74	1.44	0.99	1.33
Proposed F-PIDNN	0.37	0.59	0.67	1.39	0.9	1.37

Table 9 Controllers performance with weight 0 g—Steady-state error (°)

Controller type	20 degrees	30 degrees	40 degrees
Z–N method [18]	1.34°	1.09°	2.02°
F-PID [2]	0.31°	1.9°	0.82°
Proposed adaptive F-PID	0.20°	0.78°	0.55°
Proposed F-PIDNN	0.41°	0.01°	0.25°

the proposed F-PIDNN method has a good convergent speed, a compatible performance in rise time, and a superior performance in terms of settling time, overshoot suppression, and steady-state error.

Table 10 Controllers performance with weight 200 g—Overshoot (%)

Controller type	20 degrees (%)	30 degrees (%)	40 degrees (%)
Z–N method [18]	5.9	12.2	13.6
F-PID [2]	9.7	7.2	6.6
Proposed adaptive F-PID	9.6	6.9	5.4
Proposed F-PIDNN	5.3	2.6	3.9

6.2 Experiment 2

In the second set of experiments with a 200 g weight, the system performance was evaluated by varying the tilt angle from 20 to 40. Given the tilt of angles, Fig. 13 and Tables 10, 11, and 12 show that the Z–N method has the largest values of settling time, percentage undershoot, and steady-state error among the four test methods. Similarly, the experimental results show that the proposed F-PIDNN method performs well in various weight conditions in terms of settling time, maximum overshoot, and steady-state error.

Table 11 Controllers performance with weight 200 g—Rise time (RT) and Settling time (ST)

Angle method	20 degrees		30 degrees		40 degrees	
	RT	ST	RT	ST	RT	ST
Z–N method [18]	0.58	1.14	0.7	1.45	0.86	1.6
F-PID [2]	0.47	0.79	0.56	1.38	0.87	1.36
Proposed adaptive F-PID	0.72	1.34	0.77	1.43	1.09	1.84
Proposed F-PIDNN	0.37	0.72	0.67	1.04	0.91	1.65

Table 12 Controllers performance with weight 200 g—Steady-state error (°)

Controller type	20 degrees	30 degrees	40 degrees
Z–N method [18]	2.36°	1.58°	2.89°
F-PID [2]	1.85°	1.26°	0.86°
Proposed adaptive F-PID	0.84°	0.98°	0.71°
Proposed F-PIDNN	1.09°	0.47°	0.02°

7 Conclusion

This work proposes two new controllers, an adaptive F-PID controller and a F-PIDNN controller. The adaptive F-PID controller demonstrates the smoother control than the conventional fuzzy control. To intelligently determine the system parameters, a F-PIDNN controller is developed to dynamically train the parameters. The experimental results demonstrate that the proposed F-PIDNN controller has better self-learning ability and robustness, which improve control accuracy and response speed. Accordingly, a reliable balanced system can be applied to facilitate patients' movements and ensure patient safety with compensating the inclination angle of the drip frame such that the reduction of blood returning and the balance control of an intravenous drip frame can be achieved. Its derivative values can cover applications and services that are extremely in need of smart medical devices, such as information and communications systems and healthcare industries.

In this work, the definition of settling time is the required time for balancing the system within the dead zone of the tilt angle (less than 4 degrees). Since the settling time is related to several factors, (e.g., the motor characteristics and the range of the blind zone), in the future, we will carefully explore these related factors to further optimize the system performance.

References

1. Wu M-F, Wen C-Y (2015) Device and method for a piggyback intravenous drip frame with balance control, ROC Patent (Invention No. I480075)
2. Wu M-F, Chen C-S, Chen I-S, Kuo T-H, Wen C-Y, Sethares WA (2020) Design of carryable intravenous drip frame with automatic balancing. Sens Spec Issue Med Appl Sens Syst Dev 20(793):1–26
3. Sharma K, Palwalia DK (2017) A modified PID control with adaptive fuzzy controller applied to DC motor. In: 2017 International conference on information, communication, instrumentation and control (ICICIC), pp 1–6
4. Juang CF, Lu CF (2005) Power system load frequency control by genetic fuzzy gain scheduling controller. J Chin Inst Eng 28(6):1013–1018
5. Juang CF, Lu CF (2006) Load frequency control by hybrid evolutionary fuzzy PI controller. In: IEE Proceeding generation, transmission and distribution, vol 153, no 2, pp 196–204 (SCI, EI)
6. Lu CF, Hsu CH, Juang CF (2013) Coordinated control of flexible AC transmission system devices using an evolutionary fuzzy lead-lag controller with advanced continuous ant colony optimization. IEEE Trans Power Syst 28(1):385–392
7. Juang CF, Chou CY, Lin CT (2021) Navigation of a fuzzy-controlled wheeled robot through the combination of expert knowledge and data-driven multiobjective evolutionary learning, Early Access. IEEE Trans Cybernet. https://doi.org/10.1109/TCYB.2020.3041269
8. Yongquan Y, Ying H, Bi Z (2003) A PID neural network controller. In: Proceedings of the international joint conference on neural networks, vol 3, pp 1933–1938
9. Dehghani A, Khodadadi H (2017) Designing a neuro-fuzzy PID controller based on smith predictor for heating system. In: 2017 17th international conference on control, automation and systems (ICCAS), pp 15–20

10. Zhou M, Zhang Q (2015) Hysteresis model of magnetically controlled shape memory alloy based on a PID neural network. 2015 IEEE international magnetics conference (INTERMAG), pp 1–1
11. Maraba V, Kuzucuoglu A (2011) PID neural network based speed control of asynchronous motor using programmable logic controller. Adv Electr Comput Eng (AECE), 23–28
12. Ang KH, Chong G, Li Y (2005) PID control system analysis, design, and technology. IEEE Trans Control Syst Technol 13(4):559–576
13. Mukhopadhyay S et al (1992) New class of discrete-time models for continuous-time systems. Int J Cont 55:1161–1187
14. MPU-6050 Datasheet https://invensense.tdk.com/wp-content/uploads/2015/02/MPU-6000-Datasheet1.pdf
15. HX711 Datasheet https://cdn.sparkfun.com/datasheets/Sensors/ForceFlex/hx711_english.pdf
16. L293D Datasheet https://www.arduino.cc/documents/datasheets/H-bridge_motor_driver.PDF
17. ThingSpeak https://thingspeak.com/
18. Ziegler JG, Nichols NB (1942) Optimum settings for automatic controllers. Trans ASME 64:759–768
19. Pititeeraphab Y, Jusing T, Chotikunnan P, Thongpance N, Lekdee W, Teerasoradech A (2016) The effect of average filter for complementary filter and Kalman filter based on measurement angle. In: 2016 9th biomedical engineering international conference (BMEiCON), pp 1–4

Kernel Learning Estimation: A Model-Free Approach to Tracking Randomly Moving Object

Yuankai Li, Yuan Wang, Xiaosu Tan, and Jiaxin Lou

Abstract Kernel learning estimation (KLE) is a kernel-based method, where the original spatial data is mapped into a high-dimensional Hilbert space by a nonlinear mapping, hiding the nonlinear mapping in a linear learning framework. The kernel function of the method can be used to replace the complex inner product operation in the high-dimensional space and avoid the Curse of Dimensionality caused by high-dimensional calculation effectively. The kernel-based method has advantages on learnability, computational complexity, precise linearization and generalization performances, providing a promising way to solve the problem of nonlinear target tracking. In traditional tracking methods, nonlinear tracking models are usually built as a priori to predict the current state of target motion, emphasizing on tracking accuracy and real-time performance. However, kernel-based method provides a general way of linearization processing, which can be independent of specific models to achieve highly efficient data-driven computation. Introducing the kernel learning mechanism into target tracking problem is expected to improve the environmental adaptability. In this paper, a review on kernel learning method with application to randomly moving target tracking is presented, including kernel-based algorithms for target detection, kernel-based algorithms for generative tracking and for discriminant tracking, and multi-kernel learning methods with multiple kernel functions. Further research is prospected in optimization of kernel function, long-term robust tracking, feature extraction, target occlusion and other potential aspects on moving target tracking using kernel learning theory.

Y. Li (✉) · Y. Wang · X. Tan · J. Lou
University of Electronic Science and Technology of China, Chengdu, China
e-mail: yuankai.li@uestc.edu.cn

Y. Wang
e-mail: 463932587@qq.com

X. Tan
e-mail: 2495646072@qq.com

J. Lou
e-mail: 2801197250@qq.com

© The Author(s), under exclusive license to Springer Nature Singapore Pte Ltd. 2023   55
M. A. Chaurasia and C.-F. Juang (eds.), *Emerging IT/ICT and AI Technologies Affecting Society*, Lecture Notes in Networks and Systems 478,
https://doi.org/10.1007/978-981-19-2940-3_4

Keywords Kernel learning method · State estimation · Pattern recognition · Target tracking · Target detection

1 Introduction

Kernel learning estimation (KLE) method, which is built based on kernel function and statistical learning theory, is a newly developing branch of pattern recognition and applied to the techniques of support vector machines (SVMs) originally [1, 2]. The kernel-based method maps the data from original space into a high-dimensional feature space by using nonlinear mapping function, transforming nonlinear problem into linear analysis in sample space. It is an effective method to deal with nonlinear pattern recognition problems. Also, the kernel-based method can provide a way of efficient calculation. By using the kernel function, it can hide the nonlinear mapping in a linear learning machine for synchronous calculation and replace complex inner product operation in high-dimensional space that provides a new way to solve the problem of high-computational complexity in high-dimensional space, avoiding the Curse of Dimensionality.

Driven by emerging technologies, techniques of target tracking are developing rapidly, related to many disciplines such as pattern recognition, dynamic systems and artificial intelligence. It has already achieved significant applications in intelligent decision-making [3], unmanned driving [4–6], biomedicine [7], human–computer interaction, reconnaissance and other application areas, and still has great development prospects. The problem of moving target tracking essentially is a typical nonlinear pattern recognition problem. The tracking model is established through the prior knowledge of moving target, followed by predicting the motion state or motion pattern of the target at the current time or frame. However, quality of the tracking model directly impacts on the tracking performances. Deviation of the tracking model, such as mismatching of the target motion state or unreasonable setting of the searching area, may lead to remarkable tracking errors or target losses.

The moving target tracking algorithm based on kernel learning method, by introducing nonlinear mapping structure with a learning mechanism, is independent of the specific model and can guarantee tracking quality under the nonlinear change of the target motion state, giving the tracking algorithm better adaptability to the environment.

This paper presents a review on the moving target tracking algorithms based on kernel learning methods, including discussions on the target detection algorithms, generative and discriminant target tracking algorithms using kernel methods with their research progress. For design and optimization of the kernel functions, multi-kernel learning method with typical structures of multiple kernel functions is also discussed. In the end, a prospect on research and applications of the kernel learning target tracking method is provided, with development trends in target-scale change, kernel optimization, accuracy of feature extraction and other critical problems.

The rest of the paper is organized as follows. Section 2 presents the fundamental principle of nonlinear kernel mapping. Kernel learning estimation with application to moving target tracking is discussed in detail in Sect. 3, where target detection, generative and discriminant tracking algorithms are shown, respectively. In Sect. 4, multiple kernel learning estimation is presented. Section 5 gives a discussion and prospect to the future research on kernel learning method, and Sect. 6 concludes the paper.

2 Nonlinear Kernel Mapping

The key idea of the kernel-based method is to map the data from the original space to a reproducing kernel Hilbert space (RKHS) through a fixed nonlinear mapping operator, performing data processing in a high-dimensional feature space.

Define the nonlinear mapping operator as kernel function κ. For all x and x', the kernel function satisfies

$$\kappa\left(x, x'\right) = \varphi(x), \varphi\left(x'\right) \tag{1}$$

where $\varphi(\cdot)$ is a feature vector transformed from original space to feature space.

When the kernel function is continuous, positive definite and symmetric, it is called Mercer kernel [8]. If the kernel function meets the following two conditions:

(1)   For any vector of $x \in X$, $\kappa(x, z)$, its function belongs to the vector space F;
(2) $\kappa(x, z)$ is reproducible.

It is called a reproducible kernel function. A certain reproducible kernel function defines a reproducible Hilbert space. For a reproducible kernel function, the kernel $\kappa(x, \cdot)$ composes the function

$$g(\cdot) = \sum_{i=1}^{l} a_i \kappa(c_i, \cdot) \tag{2}$$

and for all $i, c_i \in X$, there is

$$g, \kappa(x, \cdot) = \sum_{i}^{l} a_i \kappa(c_i, x) = h(x) \tag{3}$$

Through definition of Mercer kernel, the reproducible kernel can be expressed as

$$\kappa\left(x, x'\right) = \sum_{i=1}^{\infty} \varsigma_i \varphi_i(x) \varphi_i\left(x'\right) \tag{4}$$

Fig. 1 Schematic diagram
of the nonlinear mapping
$\varphi(\cdot)$

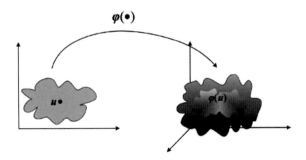

where ς_i and φ_i are nonnegative feature value and vector, respectively. The mapping φ can be expressed as

$$\begin{cases} \varphi : X \to F \\ \varphi(x) = \left[\sqrt{\varsigma_1}\varphi(x_1), \sqrt{\varsigma_2}\varphi(x_2), \ldots\right] \end{cases} \tag{5}$$

For data u in the original input space, $\varphi(u)$ can be obtained by mapping $\varphi(\cdot)$ into the feature space. Figure 1 shows the mapping relationship of the data from the original space to the Hilbert space.

In Hilbert space, $\varphi(\cdot)$ is the mapping from the original space to the feature space, and $\varphi(x)$ is the mapped feature vector. For $x, x' \in X$, the kernel function satisfies

$$\varphi(x)^T \varphi(x') = \kappa(x, x') \tag{6}$$

That is, when the mapping function $\varphi(\cdot)$ is difficult to express explicitly, the calculation of the kernel function is used to replace the complex calculation to the inner product of the nonlinear mapping in the RKHS, thereby simplifying the calculation process and improving the calculation efficiency.

3 Moving Target Tracking Using Kernel Learning Estimation

Target tracking is a challenging branch in the field of intelligent systems. It aims to realize the recognition and decision-making of target motion modes. It has broad research and application prospects and has been applied in intelligent monitoring, unmanned driving, biological medicine, human–computer interaction, military and other fields. Kernel method is deeply applied in the target tracking technique due to its linearization, data driven and other advantages, and has achieved fruitful results.

Target tracking includes three main parts: appearance model, motion model and search strategy, involving image processing, data processing and machine learning. Considering that target detection is a prerequisite of target tracking, we describe the process of moving target tracking as the following general steps: detecting directive information of target motion, predicting target movement trend, and performing higher-level behavior analysis and decision-making for the randomly moving target.

3.1 Target Detection

The quality of target detection determines the performance of target tracking. Presently, target detection algorithms based on kernel methods mainly include three types of algorithms: kernel principal component analysis (KPCA), kernel Fisher linear discriminant (KFLD) and kernel independent component analysis (KICA).

Inspired by the kernel mapping theory, Scholkopf combined it with the principal component analysis (PCA) and proposed the KPCA [9] algorithm in 1998, allowing the kernel method break away from support vector machine to cooperate with other algorithms for the first time. That is of great significance for the kernel method been successfully applied to the field of target feature extraction and target detection.

Fisher linear discriminant (FLD), also known as linear discriminant analysis (LDA), was proposed by Mika [10] in 1999. The principle of it is to find a most suitable projection axis, so that the projection distance between all kinds of samples on this axis is as far as possible, and the projections of samples in each category are as compact as possible, intending to achieve the best classification effect. Also, Mika extended the FLD with kernel method and proposed the KFLD algorithm to solve the nonlinear problems. A feature space is established to judge the linear Fisher criterion for KFLD, which is helpful to obtain more accurate result for target feature extraction under nonlinear conditions.

In 2002, Bach, based on independent component analysis (ICA), proposed the KICA [11] algorithm, which further enriched kernel detection methods. KICA deals with functions in the kernel Hilbert space based on the entire nonlinear function space and uses kernel mapping technique to search the space with high efficiency. The use of function space enables it to adapt to a variety of data samples, thus making the algorithm more robust to time-variant sample distribution.

3.2 Generative Tracking

Generative tracking is a type of tracking method in which online learning is used to model the target feature and through the feature model to search and match and find the target position. A representative generative tracking method is mean shift (MS) [12].

3.2.1   Representative Algorithm: Mean Shift

During the initial stage of tracking, the MS algorithm needs to determine the search window independently to select the target, therefore, it is a semi-automatic tracking method. The algorithm calculates the histogram distribution of the initial frame search window and calculates the histogram distribution of the Nth frame window in the same way, making the search window to move along the direction of the maximum histogram density, and finally gets the position of the target. The standard steps are as follows:

(1)   Target model of the initial frame

Divide the feature space into multiple eigenvalues according to pixel color values. At the initial frame, the probability of the u-th eigenvalue in the search window of the initial frame is

$$\tilde{q}_u = C \sum_{i=1}^{n} k\left(\frac{x_0 - x_i}{h}^2\right) \delta[b(x_i) - u] \tag{7}$$

where x_0 and x_i are the central pixel coordinates of the $N\,th$ frame search window and the coordinates of the i-th pixel, respectively; $k(\cdot)$ is the kernel function, and h is the bandwidth of the kernel function.

(2)   Target model of the N-th frame

Calculate the probability of the u-th eigenvalue in search window of the N-th frame with

$$\tilde{p}_u(y) = C_h \sum_{i=1}^{n_k} k\left(\frac{y_0 - x_i}{h}^2\right) \delta[b(x_i) - u] \tag{8}$$

where y_0 is the coordinate of the pixel in the center of the search window.

(3)   Similarity function

Similarity function represents degree of similarity between the initial frame model and the N-th frame model. Define the function as

$$\tilde{\rho}(y) \equiv \rho(\tilde{\rho}(y), \tilde{q}) = \sum_{u=1}^{m} \sqrt{\tilde{p}_u(y)\tilde{q}_u} \tag{9}$$

(4)   Mean shift vector

By maximizing the similarity function, the mean shift vector can be obtained as

$$m_{h,G}(y) = y_1 - y_0 \frac{\sum_{i=1}^{n_k} x_i w_i g\left(\frac{y_0 - x_i}{h}^2\right)}{\sum_{i=1}^{n_k} w_i g\left(\frac{y_0 - x_i}{h}^2\right)} - y_0 \tag{10}$$

in which y_0 is the central coordinate of the search window and y_1 is that of the found new search window.

3.2.2 Related Studies

The mean shift algorithm was first proposed by Fukunaga [12] in 1975. It was not widely concerned until 1995 when Cheng introduced the kernel function and weight coefficient [13] to the original MS and extended the MS theory. Comaniciu applied the MS algorithm to video image processing and realized smoothing, segmentation and tracking of visual objects [14–17].

Usually, the bandwidth of the kernel function in MS tracking algorithm is fixed, leading to large tracking errors when the target size varies. For this reason, based on scale space theory [18], Coliins proposed a scale variable MS tracking algorithm [19]. Sometimes, the color feature map in the MS algorithm cannot accurately describe the distribution of the target color in the image, leading to vacancy of the feature space. To solve this problem, the spatial color histogram is introduced into the target tracking model [20], combining the target color with the image color in the current frame to improve the accuracy of target recognition. Also, to improve the anti-occlusion ability of MS algorithm, filtering algorithms are integrated into the MS [21]. To track multi-target objects and human joints, confidence fusion propagation algorithm is introduced into the MS algorithm [22] for realizing tracking of single target, multi-target and human joints.

With good performance on processing time and robustness, MS algorithm has achieved rich results in the field of target tracking for several decades. However, in the generative tracking methods, environmental information is usually ignored so that the entire tracking process is quite easy to be disturbed. It yields a key problem of how to extract the target from background effectively that motivates the studies of discriminant tracking.

3.3 Discriminant Tracking

Discriminant tracking uses the correction theory of machine learning to model the target feature and background feature, respectively, taking the peak in the confidence map as target position and training corresponding filter to distinguish the target from background. It has two representative classes of kernel-based algorithms: circulant structure of tracking-by-detection with kernels (CSK) and kernelized correlation filters (KCF).

3.3.1 Representative Algorithm I: CSK

Correlation filter has a fast operation rate in frequency domain, and the tracking speed can reach hundreds of frames per second. It is an effective way to realize real-time moving target tracking. In 2010, Bolme [23] introduced correlation filtering into minimum output sum of square error (MOSSE) algorithm, expanding target tracking from time domain to frequency domain. Based on the MOSSE, Henriques [24] proposed the CSK algorithm. It uses correlation theory of cyclic matrix and shift the training samples and candidate samples cyclically to increase the number of two samples, meanwhile, use the training samples to train the classifier and then use the trained classifier to detect target in the candidate region which is constructed by the candidate samples, and finally, take the maximum response of the correlation calculation as the target location. Kernel mapping function is adopted for the CSK to fasten the calculation speed.

To find the response position, CSK algorithm needs to transform from frequency domain to time domain, which will lead to marginal effect similar with the leakage phenomenon of aperiodic signals in systems. Moreover, application range of CSK algorithm is narrow, and it is only suitable for gray scale feature space. It lacks the establishment of target appearance model and is easily disturbed by complex task background and environmental noises. In order to establish the target appearance model more accurately, Danelljan et al. use the color name (CN) [25] based on the CSK algorithm to enhance the description of the appearance model, mapping three-dimensional color space into a high-dimensional CN space. In the way, the algorithm can more effectively establish the target appearance model and has a better tracking effect on the problem of target partial occlusion, light change, complex background, etc. However, due to the complexity of computing in a high-dimensional space, the calculation rate of CSK decreases after introducing three-dimensional color space.

When the target scale is changed or occluded, the CSK algorithm will degenerate obviously. On the basis of the algorithm, other target tracking algorithms based on correlation filter and their improved algorithms are proposed, which have achieved better tracking effect and tracking robustness than CSK algorithm.

3.3.2 Representative Algorithm II: KCF

To further improve tracking accuracy and tracking precision, Henriques proposed the kernelized correlation filters [26] in 2014. The KCF algorithm mainly includes four typical steps that are ridge regression [27], diagonalization of cyclic matrix, kernel correlation filtering and fast target detection.

Ridge Regression

For linear regression model, the objective function is

$$f(x) = W^T X \tag{11}$$

Find the weight W_i in sample X to minimize the prediction of target Y_i, that is

$$\min_{w} \sum_i (f(x_i) - y_i)^2 + \lambda w^2 \tag{12}$$

where λ is the parameter that controls over-fitting. Take the derivative of (12) and set it to zero, and then, the optimal solution is obtained as

$$w = \left(X^H X + \lambda I \right)^{-1} X^H y \tag{13}$$

where X is the sample matrix, and y is the label.

Cyclic Matrix Calculation

Note $X_{n \times 1} = \{x_1, x_2, \ldots x_n\}^T$ as the positive sample and generate the training sample (negative sample) by cyclic shifting to the positive sample. Use the negative sample to train the classifier, and one-dimensional operator P in cyclic shifting is

$$P = \begin{bmatrix} 0 & 0 & 0 & \cdots & 1 \\ 1 & 0 & 0 & \cdots & 0 \\ 0 & 1 & 0 & \cdots & 0 \\ \vdots & \vdots & \ddots & \ddots & \vdots \\ 0 & 0 & \cdots & 1 & 0 \end{bmatrix} \tag{14}$$

By multiplying the vector x constantly with the operator P in cyclic shifting, for one-dimensional image, a cyclic matrix $C(x)$ with combinations of n vectors is obtained. For two-dimensional image, training sample of the two-dimensional image is obtained by cyclic shifting in the region of interest, and finally, the cyclic matrix of the two-dimensional matrix is obtained.

Kernel Correlation Filtering

To improve the classifier performance, map the original spatial data into a high-dimensional space with weight w that is a linear combination of input samples as

$$w = \sum_i \alpha_i \varphi(x_i) \tag{15}$$

By introducing $\varphi^T(x)\varphi(x') = k(x, x')$ for kernelization, the optimal solution of (15) can be transformed into

$$\alpha = (K + \lambda I)^{-1} y \tag{16}$$

where K is also a cyclic matrix. By using the properties of the cyclic matrix, and (16) can be rewritten as

$$\hat{\alpha} = \frac{\hat{y}}{\hat{k}^{xx} + \lambda} \tag{17}$$

where k^{xx} is the first row of K and $\hat{y}$ is the discrete Fourier transform of y. In general, the classifier can be solved by calculating $\hat{\alpha}$.

Fast Target Detection

Suppose that the target state of the previous frame is x, and the input sample of the current frame is z, and the kernel matrix K is introduced to obtain the response of the test sample, that is

$$\hat{f}(z) = \hat{k}^{xz} \odot \hat{\alpha} \tag{18}$$

Performing inverse Fourier transform to (18) obtains the sample response. Take the maximum response value as the target position, and update the classifier parameters and the target model by

$$\begin{cases} x^i = (1 - \theta)x^{i-1} + \theta z^i \\ \alpha^i = (1 - \theta)\alpha^{i-1} + \theta \sigma^i \end{cases} \tag{19}$$

where x^i is the target position model and α^i is the classifier parameters predicted for the i frame, respectively; z^i and σ^i are the detected target position model and classifier parameters, respectively; θ represents the learning rate.

3.3.3   Related Studies

Traditional target tracking algorithms, such as mean shift algorithm, support vector machine [28], Kalman filtering [29], particle algorithm based on Bayesian filtering [30, 31], tracking learning detection (TLD) [32] based on target characteristics, and compressive tracking (CT) [33] algorithm based on compression perception, appear deficiencies in tracking performance and in computation efficiency. KCF algorithm, absorbing advantages of CSK, shows high tracking precision and calculation speed with improved classifier training and sample detection.

To solve the problems of scale change, target occlusion and target loss in the process of target tracking, a tracking method with multi-scale correlation filters is proposed in [34], where two ridge regression models having strong plasticity and strong stability are adopted. The model with good plasticity tracks the target position to construct image pyramid, putting the position as the center. The model with good stability predicts the target-scale change, realizing multi-scale detection and tracking. In [35], a depth scaling kernelized correlation filtering (DSKCF) algorithm is developed, which extends the RGB tracking algorithm in KCF to form an RGB-D tracking algorithm by integrating depth features. The algorithm gives an improved solution for the problems of target-scale change and target occlusion. To deal with the marginal effect caused by cyclic shift in KCF algorithm, Danelljan introduced the spatial setting factor to constrain the weight of filters, alleviating the marginal effect at the expense of calculation complexity of the algorithm [36].

For many complex scenes, it is usually difficult for a single kernel to satisfy the requirements of various tracking performances, and multi-kernel fusion for the KCF algorithm provides an extension scheme.

4 Multiple Kernel Learning

Kernel function design is an important part of the kernel-based target tracking algorithm, which determines the tracking accuracy. To improve the tracking adaptivity to complex environments, fusion of multiple kernel functions is studied recently. Many works [37–39] have shown that multi-kernel model has better flexibility and robustness than model with single kernel. Presently, multi-kernel learning methods can be categorized into three main types, which are synthetic kernel, multi-scale kernel and infinite kernel.

Synthetic kernel method is the fundamental method of multi-kernel learning. One can use many ways to construct the synthetic kernel, such as multi-kernel linear combinatorial synthesis method, multi-kernel extended synthesis method, non-stationary multi-kernel learning, local multi-kernel learning and non-sparse multi-kernel learning. Due to simplicity of the synthetic kernel method, it has been applied in many practical problems, such as target eigenvalue extraction and processing [40, 41], classification [42–45], image segmentation [46] and system identification [47].

Synthetic kernel method generally uses linear combination of kernel functions to generate an integrated kernel function, which is questionable for desired tracking accuracy when facing unbalanced distribution of samples. Multi-scale kernel method introduces the concept of scale space and merges multi-scale kernel functions into a more flexible new kernel function. The multi-scale kernel method needs to train bandwidth for each kernel function. The kernel function with smaller bandwidth is used to track samples with drastic changes, and the function with larger bandwidth is to track samples with gentle changes. Using SVM regression or RKHS to extend the multi-scale kernel method is beneficial to improve the tracking accuracy. The multi-scale kernels are classified in [48], where a typical multi-scale kernel method is provided.

The aforementioned two kernel methods use limited number of kernel functions for linear combination or fusion. However, for large-scale data processing, the multi-kernel processing methods with limited number of kernels become hard to achieve desired tracking performance with optimal decision. Therefore, a trend of expanding from finite to infinite kernel functions for multi-kernel learning emerges recently, but there are few reports about its application in the field of target tracking.

5   Discussion and Prospect

Presently, target tracking method based on kernel learning estimation is developing fast. Compared with other types of tracking methods, kernel-based target tracking is free of nonlinear error and promising to be model-free and data-driven, however for future use, it still needs further research to satisfy requirements under practical conditions. The potential work is mainly reflected in the following aspects.

(1)   Reduce the dependency to tracking model.

Traditional Bayesian tracking algorithms strongly depend on a prior math model and model parameters. Introducing kernel method into classical tracking algorithms can expand the algorithm to nonlinear applications precisely. With kernel function replacing inner product operation in sample space and target dynamics modeled by trained measurements, high-dimensional computing complexity and dependency on tracking model can be reduced greatly.

(2)   Improve the accuracy of feature extraction.

The quality of feature extraction is the key factor affecting the efficiency of visual target tracking. Convolution features have advantages over artificial features, but the design of training network still needs further research. Multi-feature fusion is helpful to describe the target much more precisely, but the increasing computation complexity needs to be controlled to balance tracking accuracy and response speed.

(3)   Improve the balance of base-kernel design.

Single kernel function is difficult to obtain satisfied performances in complicated tracking tasks. Multi-kernel learning method can optimize kernel function through combination of multiple base kernels and adjust weight coefficients and parameters of the base kernels to deal with complex application scenarios. However, multi-kernel learning method increases computational complexity. Learning efficiency is also necessary to be considered in algorithm design, and the balance of multi-kernel learning framework also needs for further exploration.

(4)   Improve the adaptability to environmental dynamics.

Changes and occlusion of the target will decrease the tracking accuracy or lead to target missing. It is significant for tracking algorithm to maintain high precision of tracking when target shape or scale is changing, flipping or occluded. Trajectory prediction and target blocking are effective to improve environmental adaptability.

(5) Improve the persistence of stable tracking.

For long-term tracking of random-moving target, uncertainties have much more chance to appear, leading to weak robustness and decreasing tracking accuracy. The algorithms for short-term tracking are generally lack of framework from long-term tasks to construct continuity of stable tracking, which is meaningful to be addressed in further research.

6 Conclusion

Kernel learning estimation method gives a feasible bridge from linear to nonlinear estimation problems and forms a unified framework of solutions in pattern analysis fields. By applying nonlinear kernel mapping to the tracking problem and replacing complicated inner product operation by designed kernel functions, kernel-based target tracking method can achieve good performance in capabilities of nonlinear processing, data-driven and generalization. The method has significant research prospects and potential applications in decision intelligence, pattern recognition, autonomous unmanned vehicles and other information processing systems.

References

1. Juang CF, Chiu SH, Chang SW (2007) A self-organizing TS-type fuzzy network with support vector learning and its application to classification problems. IEEE Trans Fuzzy Syst 15(5):998–1008
2. Chen GC, Juang CF (2013) Object detection using color entropies and a fuzzy classifier. IEEE Comput Intell Mag 8(1):33–45
3. Smeulders AWM, Chu DM, Cucchiara R et al (2014) Visual tracking: an experimental survey. IEEE Trans Pattern Anal Mach Intell 36(7):1442–1468
4. Wang D (2017) Design of intelligent video surveillance system based on motion detection. North University of China, Taiyuan
5. Gündüz G, Acarman AT (2018) A lightweight online multiple object vehicle tracking method. In: Proceedings of IEEE intelligent vehicles symposium, Changshu, China, pp 427–432
6. Lu Y, Dai H, Hu Y et al (2020) Research on collaborative target tracking algorithm of UAV based on machine vision. Electronic Devices 43(5):1096–1099
7. Gao T (2012) Medical image analysis based on multi-target tracking. Xidian University
8. Aronszajn D (1950) Theory of reproducing kernel. Trans Am Math Soc 68:337–404
9. Scholkopf B, Smola AJ, Muller KR (1998) Nonlinear component analysis as a kernel eigenvalue problem. Neural Comput 10(5):1299–1319
10. Mika S, Ratsch G, Weston J et al (1999) Fisher discriminant analysis with Kenels. In: Proceedings of IEEE signal proceeding society workshop, pp 41–48

11. Bach FR, Jordan MI (2003) Kernel independent component analysis. In: Proceedings of IEEE international conference on acoustics, speech, and signal processing, HK, China
12. Fukunaga K, Hostetler L (1975) The estimation of gradient of a density function with applications in pattern recognition. IEEE Trans Inf Theory 21(1):32–40
13. Cheng Y (1995) Mean shift, mode seeking, and clustering. IEEE Trans Pattern Anal Mach Intelli 17(8):790–799
14. Comaniciu D, Ramesh V, Meer P (2000) Real-time tracking of non-rigid objects using mean shift. In: Proceedings of the IEEE conference on computer vision and pattern recognition, pp 142–149
15. Comaniciu D, Meer P (1999) Mean shift analysis and application. In: Proceedings of the IEEE international conference on computer vision, pp 1197–1203
16. Comaniciu D, Meer P (2002) Mean shift: a robust application toward feature space analysis. IEEE Trans Pattern Anal Mach Intell 24(5):603–619
17. Comaniciu D, Meer P (1997) Robust analysis of feature spaces: color image segmentation. In: Proceedings of the IEEE conference on computer vision and pattern recognition, pp 750–755
18. Lindeberg T (1998) Feature detection with automatic scale selection. Int J Comput Vision 30(2):194–203
19. Collins RT (2003) Mean shift blob tracking through scale space. In: Proceedings of the IEEE conference on computer vision and pattern recognition, pp 234–240
20. Birchfield ST, Rangarajan S (2005) Spatiograms versus histograms for region-based tracking. In: Proceedings of the IEEE conference on computer vision and pattern recognition, pp 1158–1163
21. Welch G, Bishop G (2006) An introduction to the Kalman filter,UNC-Chapel Hill, NC
22. Park M, Liu Y, Collins R (2008) Efficient mean shift belief propagation for vision tracking. In: Proceedings of the IEEE conference computer vision and pattern recognition
23. Bolme DS, Beveridge JR, Draper BA et al (2010) Visual object tracking using adaptive correlation filters. In: Proceedings of IEEE computer society conference on computer vision and pattern recognition, pp 2544–2550
24. Henriques JF, Caseiro R, Martins P et al (2012) Exploiting the circulant structure of tracking-by-detection with kernels. In: Proceedings of European conference on computer vision, pp 702–715
25. Danelljan M, Shahbaz KF, Felsberg M et al (2014) Adaptive color attributes for real-time visual tracking. In: Proceedings of IEEE conference on computer vision and pattern recognition, pp 1090–1097
26. Henriques JF, Caseiro R, Martins P et al (2015) High-speed tracking with kernelized correlation filters. IEEE Trans Pattern Anal Mach Intell 37(3):583–596
27. He X (2005) Multivariate linear model and ridge regression analysis. Huazhong University of Science and Technology, Wuhan
28. Liang CW, Juang CF (2015) Moving object classification using local shape and HOG features in wavelet-transformed space with hierarchical SVM classifiers. Appl Soft Comput 28:483–497
29. Liang CW, Juang CF (2015) Moving object classification using a combination of static appearance features and spatial and temporal entropies of optical flows. IEEE Trans Intell Transp Syst 16(6):3453–3464
30. Arulampalam MS, Maskell S, Gordon NA (2002) Tutorial on particle filters for online nonlinear/non-Gaussian Bayesian tracking. IEEE Trans Signal Process 50(2):174–188
31. Juang CF, Chang CW, Hung TH (2021) Hand palm tracking in monocular images by fuzzy rule-based fusion of explainable fuzzy features with robot imitation application. IEEE Trans Fuzzy Syst 29(12):3594–3606
32. Kalal Z, Mikolajczyk K, Matas J (2011) Tracking-learning-detection Kernel. IEEE Trans Pattern Anal Mach Intell 34(7):409–422
33. Zhang K, Zhang L, Yang MH (2012) Real-time compressive tracking. In: Proceedings of European conference on computer vision, pp 864–877
34. Xia X, Zhang X, Li J (2017) Kernel correlation filter target tracking method combined with scale prediction. Electronic Design Eng 25(2):130–136

35. Hannuna S, Camplani M, Hall J et al (2019) DSKCF: a real-time tracker for RGB-D data. J Real-Time Image Proc 16(5):1439–1458

36. Danelljan M, Hager G, Khan FS et al (2015) Learning spatially regularized correlation filters for visual tracking. In: Proceedings of IEEE international conference on computer vision, pp 4310–4318

37. Gonen M, Alpaydin E (2011) Multiple Kernel learning algorithms. J Mach Learn Res 12:2211–2268

38. Lanckriet G, Cristianini N, Bartlett P et al (2004) Learning the kernel matrix with semidefinite programming. J Mach Learn Res 5(1):27–72

39. Lee WJ, Verzakov S, Duin RP (2007) Kernel combination versus classifier combination. In: Proceedings of the international workshop on multiple classifier systems. Czech Republic, Prague, pp 22−31

40. Mak B, Kwok JT, Ho S (2004) A study of various composite kernels for kernel eigenvoice speaker adaptation. In: Proceedings of the IEEE international conference on acoustics, speech, and signal processing. Montreal, Canada, pp 325−328

41. Fu SY, Yang GS, Hou ZG, Liang Z, Tan M (2008) Multiple kernel learning from sets of partially matching image features. In: Proceedings of the international conference on pattern recognition

42. Zheng S, Liu J, Tian JW (2005) An efficient star acquisition method based on SVM with mixtures of kernels. Pattern Recogn Lett 26(2):147–165

43. Fung G, Dundar M, Bi J, Rao B (2004) A fast iterative algorithm for fisher discriminant using heterogeneous kernels. In: Proceedings of the international conference on machine learning. Banff, Canada, pp 40−47

44. Damoulas T, Girolami MA (2009) Combining feature spaces for classification. Pattern Recogn 42(11):2671–2683

45. Damoulas T, Girolami MA (2009) Pattern recognition with a Bayesian kernel combination machine. Pattern Recogn Lett 30(1):46–54

46. Gustavo V, Luis C, Jordi M et al (2006) Composite kernels for hyperspectral image classification. IEEE Trans Geosci Remote Sens Lett 3(1):93–97

47. Gustavo V, Manel R, Rojo-Alvarez J et al (2007) Nonlinear system identification with composite relevance vector machines. IEEE Signal Process Lett 14(4):279–282

48. Kingsbury N, Tay DBH, Palaniswami M (2005) Multi-scale kernel methods for classification. In: Proceedings of the IEEE workshop on machine learning for signal processing. Washington, DC, USA, pp 43−48

Transforming Medical Data into Ontologies for Improving Semantic Interoperability

Ayesha Banu and Ayesha Ameen

Abstract Healthcare systems today have mostly become patient-centric and digitally expressed in form of electronic health record (EHR). The medical data collected using clinical codes from multiple sources are saved in structured form or free text. This data heterogeneity has extensively increased due to the exponential growth of healthcare data which makes data extraction complex, creates interoperability issues, and hinders healthcare development. Web Ontology Language (OWL) combined with Semantic Web technologies adds simplicity to searching, integrating, reusing, sharing information, and addressing interoperability issues. It is very much essential to transform medical data into ontologies to achieve improved semantic interoperability. This chapter focuses on a review of the significance of the semantic web in the medical industry and discusses different methods of building or transforming open EHR-based medical data to OWL individuals. Finally, the chapter gives an insight into ontology mapping for semantic interoperability with its benefits and also considerable challenges.

Keywords Medical data · Ontologies · Web ontology language · Semantic interoperability

1 Introduction

From its advent, Internet has revolutionized the computer and communications world like never before and added loads of data at an incredible momentum. Moreover, this data is generated from multiple sources and in a wide variety of formats making it heterogeneous and complex. The features of this data, when exploited appropriately, can bring out valuable insights, helping in improved decision making [1]. The focus

A. Banu
Department of CSE, Vaagdevi College of Engineering, Warangal 506005, India

A. Ameen (✉)
Department of IT, Deccan College of Engineering and Technology, Hyderabad, India
e-mail: Hyd.ameenayesha@gmail.com

© The Author(s), under exclusive license to Springer Nature Singapore Pte Ltd. 2023　　71
M. A. Chaurasia and C.-F. Juang (eds.), *Emerging IT/ICT and AI Technologies Affecting Society*, Lecture Notes in Networks and Systems 478,
https://doi.org/10.1007/978-981-19-2940-3_5

is especially originating in health care where the EHRs are generating large volumes of medical data which require to be analyzed in real time to make several decisions regarding patients such as symptom analysis, disease diagnosis, treatment suggestion, medication prescription, etc. In spite of large amounts of data in health care, its heterogeneity arises difficulties in utilizing the data and interoperability. This gives rise to many challenges pertaining to the structure and data format, naming irregularities, and data organization. A new framework needs to be developed for the reconciliation of these data sources [2].

The major challenge for the healthcare industry is semantic interoperability. Any health information exchange (HIE) system must have the ability to interpret the information from the receiving system holding the same sense as projected by the transmitting system and support seamless information sharing. The Semantic Web comprises all the necessary technologies required to achieve semantic interoperability [3]. Semantic Web technologies are proving to be the most significant for solving many challenges faced by the healthcare community. This also motivates researchers, scientists to find semantic-based solutions that can better handle healthcare data [4].

Most of the medical data like patient's name, date of birth, age, blood pressure, or any test result are available in a structured format. Data like clinical recordings, emails, scan images, or physician notes about a patient come in an unstructured format. This data pertaining to health care is commonly represented in XML format. The XML lacks data representation support since the main focus is on the structure level. In a highly heterogeneous environment of health care, it is difficult for medical applications and search engines to understand and integrated at are presented in these formats. To solve these interoperability issues, an infrastructure is provided by Web Ontology Language: OWL and Semantic Web technologies. Hence, transforming medical data into OWL Ontologies is very much essential for Improving Semantic Interoperability [5]. The following sections of this paper are organized as: Section 2 explains semantic modeling and ontology concepts. Section 3 focuses on the various semantic interoperability issues. The method for building any medical ontology with some examples is included in Sects. 4 and 5 and introduces the process of ontology construction. Medical data can be transformed from its existing formats to OWL ontology which is explained in Sect. 6. Section 7 focuses on the benefits and challenges of semantic interoperability and the conclusions are given in Sect. 8.

2 Semantic Modeling and OWL Ontology

The Semantic Web provides a standard for data annotation and integration. Adding semantics to the web content makes the data easily understandable and ac-accessible through the Internet. The languages used to model and represent the data include Extensible Markup Language (XML), Resource Description Framework (RDF), Resource Description Framework Scheme (RDFS), SPARQL, and Web Ontology Language (OWL) [6].

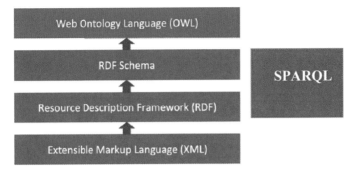

Fig. 1 Simplified semantic web stack

XML is text-based and describes documents that are readable by both humans and machines using a certain set of rules and gives flexibility to users to describe the data using their own tags. However, this flexibility leads to confusion and understanding issues for applications. This happens because every object is described using different vocabularies or many objects can be expressed using a single vocabulary. Due to this ambiguity, it becomes difficult for the computing machines to differentiate between the semantics of the data. As XML's primary focus is on the syntax of the data the re-exists no way to express the semantics and this becomes a major challenge for semantic interoperability. To address this problem, there is a high requirement of transforming medical data from its current format into OWL for gaining semantic support [5]. The simplified Semantic Web Stack [7] shown in Fig. 1 illustrates the way in which OWL is stacked upon XML, RDF, and RDFS. SPARQL is a semantic query language specially designed to work with data in RDF format for data retrieval and manipulations [8].

The RDF is a W3C standard designed as a foundation for processing metadata. It is a standard model for data interchange on the Web. It supports interoperability between applications on the web and allows them to exchange machine-processable information across the web. RDF is the first layer in the Semantic Web layer stack where information becomes machine-understandable. RDF is used to write statements about web resources. A statement comprises subject, predicate, and object. Subjects describe the things in the statements. The relationship between subject and object is specified by the predicate. Both subject and object can be a resource. Resources can be any web page, collections, or parts of web pages, or any real-world objects that are not directly part of WWW.

RDF describes the resources but does not define the semantics related to resources. Resource Development Framework Schema (RDFS) is used to specify the semantics related to resources. Application-specific classes and properties in the form of ontologies are specified by RDFS. The properties and classes are organized in the generalization and specialization hierarchies. Properties domain and range are specified in ontologies along with classes and subclass definitions. Although RDFS provides the basis for defining ontologies, it is incapable of supporting complex relationships that must be represented using an ontology.

Ontology vocabulary is the next layer in Semantic Web-layered architecture as it supports more complex relationship that exists among the classes in ontology. More complex relationships like symmetry, transitivity, and cardinality constraints can be represented using ontology vocabulary. Ontologies are represented in the form of Web Ontology Language (OWL) in this layer. OWL ontologies are more expressive and robust. OWL provides a much larger vocabulary when compared to RDFS and specifies when you cannot and can use a certain vocabulary. OWL pursues to model data in a machine-understandable and unambiguous way to provide data interoperability and more automation. OWL also permits the user to define constraints related to their applications. OWL has the means to check the logical consistency of the constructed ontologies to ensure that the ontology does not violate any consistency constraints and user-defined constraints present in the domain.

SPARQL is a recursive acronym for SPARQL Protocol and RDF query language and is pronounced as "sparkle." SPARQL is a query language and protocol. Query language contains the definition of syntax to frame queries protocol defines the communication between SPARQL client and SPARQL endpoint or processor. An endpoint is a service that processes and accepts SPARQL queries and returns the result in different formats based on the input queries.

Most of the current developments address their limitations in semantic interoperability by adopting ontologies in their approaches. Ontologies represent the domain knowledge formally and support better interoperability because the data is allowed to be linked at the semantic level. When ontologies are used to support semantic interoperability, it provides a fixed set of concepts and their meanings and relations are constant and agreed by all users. If the meaning of any word used in any interoperable context changes, then the pointer that specifies the meaning of the word is changed to preserve semantic interoperability [10].

3 Semantic Interoperability Issues

Semantic interoperability is defined by the European Commission recommendation as a process that ensures that the meaning of data that is exchanged by any application or system is easy to understand even if the application is not specifically developed for that purpose [9]. When the applications are developed by different resources and when exchange of data is performed, many issues are raised in understanding the semantics of data. It is very much essential to identify the issues, trace the reason behind them and solve them for better interoperability. Some of the major issues are classified as shown in Table 1.

Most of the current developments address their limitations in semantic interoperability by adopting ontologies in their approaches. Ontologies represent the domain knowledge formally and support better interoperability because the data is allowed to be linked at the semantic level. When ontologies are used to support semantic

Table 1 Classification of semantic interoperability issues

Semantic interoperability issues	
1. Meaning	In linguistics, a group of words that have the same spelling and are also pronounced in the same way but differ in their meanings is called **homonyms**. These are major issues of concern for interoperability
	There also exist multiple words that refer to the nearly same concept or meaning in linguistics called **synonyms** Both these homonyms and synonyms give rise to ambiguity
2. Granularity	The level of detail and completeness with which data is collected and stored in the databases can also be important issue for interoperability
3. Temporal	Whenever the clinical definitions keep changing over time the meaning of The data also has every possibility to change
	Whenever new concepts are introduced as a part of technological Advancements there will be rise in terminological changes
4. Structural	Interoperability issues may arise when the same information categorization is done differently by different designs
	The wrong position of data in the vocabulary structures may also lead to Interoperability issues

interoperability, it provides a fixed set of concepts and their meanings and relations are constant and agreed by all users. If the meaning of any word used in an interoperable context change, then the pointer that specifies the meaning of the word is changed to preserve semantic interoperability [10].

4 Solving Semantic Interoperability Issues Using Ontologies

Ontologies can be used to provide semantic interoperability by providing a static set of concepts whose meaning and the relationship among the concepts do not change and are agreed upon by the definition of ontology. An ontology is a formal, explicit specification of a shared conceptualization. Ontologies characterize conceptualization because they are an abstract model of a domain consisting of a set of concepts, relations, instances, and axioms. Conceptualization is a simplified, abstract view of the world that is represented for some purpose. In the definition of ontology explicit specification indicates that all the elements of conceptualization are unambiguously described. Formal language must be used to refer to the elements of conceptualization. The term formal in ontology definition specifies that the language used for representing conceptualization must be formal; i.e., it must be machine-readable and interpretable. Shared specifies that the knowledge content is approved by the members of a community and it also specifies that the knowledge is not owned by a single person, but accepted by a group.

Semantic interoperability is not able to integrate large data set in research because of the absence of similar unambiguous meaning of data at the distributor and at the receiver of the data. In many systems, the semantic meaning of the data differs which necessitates the need for understanding the syntactic and structural representation of the data here the ontologies come in for help for all the reasons mentioned in the previous paragraph. Ontologies provide a flexible approach for sharing semantics and integrating data across heterogeneous platforms. Ontologies-based data representation enables a better understanding of data and allows inferences to be carried on the data in complex situations [9].

In the context of healthcare systems, ontologies are used for maximizing the following:

- Various granularities (level of details and completeness) in data.
- The inferred meaning from the coded data.
- The capability to handle temporal fluctuations in clinical practice, definitions.
- Structural studies carried out by health professionals, policies of governance, and privacy.

5 How to Construct an Ontology?

An ontology represents precious resources that are slowly, but continuously gaining significance and use throughout a set of disciplines [11]. Ontologies creation and management do not come for free; they must be careful analysis; design decisions must be taken by the group of people who are creating the ontologies. An area that explores the principle, methods, and tools for initiating, developing, and maintaining ontologies is "ontology engineering." The ultimate purpose of ontology engineering is "to provide a basis of building models of all things in which computer science is interested."

5.1 Ontology Engineering

Ontology engineering is also called ontology building is a field that studies the methods and methodologies, languages, and tools used for building ontologies. Methods and methodologies include a representation, formal naming, and definition of concepts, properties, and relations between the concepts, data, and entities. Ontology engineering includes a knowledge construction of the domain using formal ontology representation.

Ontology engineering aims to make explicit knowledge contained in software applications and organizational procedures for a particular domain [12]. Ontology engineering helps to overcome the semantic obstacles related to business terms and software classes. There are a few challenges that are faced in ontology engineering, the first challenge is to guarantee that building ontology is consistent with the domain knowledge and the concepts present in the domain. Next, is to ensure that adequate specificity and concept coverage for the domain of interest, thereby lessening the problem of content completeness. Last, is to make sure that the build ontology supports its use case.

Ontology engineering is a field that ranges from the methodologies for ontology development, to the methods dedicated to supporting model's modification and evolution through their life cycle [13]. Methodologies split the ontology building process into a different number of stages and propose a set of activities for each stage. The significance of a particular activity within a methodology primarily depends on the complexity of ontology to be built, the characteristic of the ontology-based application, the accessibility of information sources, and the skill of the ontology engineers.

Various methodologies for ontology building have been proposed in past. Till now, there is no standard methodologies exist for ontology construction. There exist several methodologies and best practices. At first, a methodology that specifies the general activities of ontology building is discussed for the general understanding of what comprises a methodology, then a detailed study of a few existing methodologies is presented.

Stages and Key Activities in Ontology Engineering Methodologies

The set of basic activities covered by the majority of ontology engineering methodologies can be briefed according to Simperl et al.'s survey [14] as follows:

1. Feasibility study: It studies whether the ontology or the adaptation of ontology-based application in the given situation is the best way to solve the given problem at hand. It performs a detailed analysis of the problem, opportunities, potential solutions, and economic feasibility.
2. Domain analysis: It performs a detailed investigation on the domain in which the ontology is to be developed. It analyzes all motivational scenarios, raises a set of competency questions, studies all existing solutions to similar problems considers them for reuse to solve the current problem.
3. Conceptualization: It finds out all the important concepts and their relationships in the domain. It considers the opportunity of extension and integration of existing solutions.
4. Implementation: It implements the formal model into representation language.
5. Maintenance: It adopts the ontology as per the novel requirements. This task takes place after the construction of the ontology.
6. Use: After the development of the ontology, this task predicts the use of ontology in several applications and its probable alignment with other models.

These activities can be classified into ontology management, ontology development, and support and ontology use stages as in Fig. 2. Feasibility study, cost–benefit analysis, and preliminary identification of the type of ontology to be developed are performed in the ontology management phase, whereas domain analysis, conceptualization, implementation are carried out in the ontology development and support stage along with the development and documentation of the ontology and its engineering process. After the ontology is developed, ontology uses stage groups of the

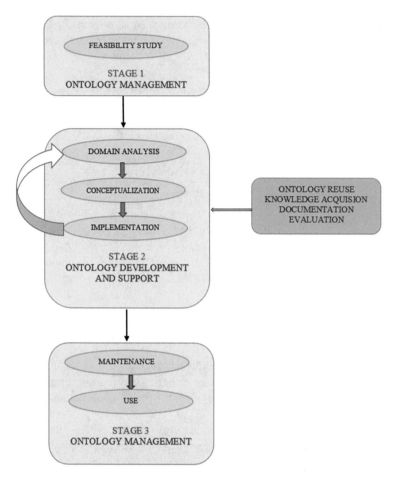

Fig. 2 Ontology engineering activities grouped into stages

activities that are needed to maintain and update the ontology. Maintenance, use of the ontology is carried out in the ontology use stage. Throughout the tasks of domain analysis, conceptualization, and implementation ontology engineering methodologies put focus on the significance of knowledge acquisition, reuse of existing ontologies, evaluation of the ontology, and documentation of the development process.

Ontology engineering methodologies also define the role of individuals in the ontology building process. There are three basic types of people involved in ontology building. The first is the domain expert who provides the knowledge about the domain to be modeled. Second, is the ontology engineer who has expertise in the field of representing knowledge in the form of ontology and has skills over various ontology languages and tools. Last, is the user who is going to use the ontology constructed for a particular purpose.

6 Building Medical Ontology

There does not exist any one correct protocol for building medical ontology or any other domain ontology. There are always multiple ways. We can build ontology manually with some domain knowledge [15] abiding by certain fundamental rules of ontology design and following the seven-step iterative process given by Deborah L McGuiness [16]. Constructing ontology manually is not only difficult but is also time-consuming. This challenge can be addressed by using a semiautomatic way of encoding ontologies where many parts of ontology can be taken from existing resources [17]. Table 2 shows the list of different ontologies constructed for the medical domain. For example, the MeSH Ontology is a Medical Subject Headings ontology developed by the National Library of Medicine which includes the subject headings appearing in MEDLINE/PubMed, the NLM Catalog, and other NLM databases. This ontology is used for indexing, cataloging, and searching of biomedical and health-related information. There are several such ontologies built-in medical domain as shown in Table 2.

Most of these ontologies use standard libraries consisting of thousands of medical terms, their properties, and relationships. Some of such libraries include SNOMEDCT, RxNorm, MeSH, MEDDRA, and NCIT. Unified Medical Language System (UMLS) includes several clinical terms and is integrated with more than 100 different vocabularies (thesauri). It is considered to be an important reference for medical terms.

7 Transforming Medical Data from Existing Formats to Ontologies

Most of the medical data or healthcare data is in XML format where the major focus is on the structure of the data. Such XML data can be transformed into owl ontologies for enhancing semantics and supporting better interoperability.

1. The different methods for transforming XML to OWL are illustrated by Hacherouf et al. [18] as shown in Fig. 3. The figure shows two levels, namely instance level and schema level. The instance level mainly focuses on the individuals of the ontology, and the schema level on the other hand gives the higher-end picture. The different types of XML data include XSD, DTD, and XML files that are transformed into OWL ontologies. The different kinds of arrows indicate the different processes of using correspondence rules for transforming XML data into an OWL ontology.

The figure shows two levels, namely instance level, and schema level. The instance level mainly focuses on the individuals of the ontology and schema level on the other hand gives the higher-end picture. The different types of XML data include XSD, DTD, and XML files that are transformed into OWL ontologies. The different kinds of arrows indicate the different processes of using correspondence rules for transforming XML data into OWL ontology.

Table 2 Medical domain ontologies

Ontology name	Domain concepts	Source available
DO: Disease ontology	Descriptions terms related to human disease, corresponding disease concepts from medical vocabulary and phenotype characteristics	https://github.com/DiseaseOntology/HumanDiseaseOntology/blob/main/src/ontology/HumanDO.owl
SYMP: Symptom ontology	A symptom is an apparent change reported by a patient regarding any disturbance in his health indicating any disease	https://bioportal.bioontology.org/ontologies/SYMP
MeSH Ontology	The Medical Subject Headings (Mesh) appear in MEDLINE/PubMed, the NLM Catalog, and other NLM databases	http://bike.snu.ac.kr/sitees/default/files/meshonto.owl
Human ontology	Concepts are taken from the Anatomy track of OAEI Ontology Alignment Evaluation Initiative	http://oaei.ontologymatching.org/2013/anatomy/index.html
GO: Gene ontology	Includes the biological domain concepts per-training to Biological Process, Cellular Components, and Molecular Function	http://purl.obolibrary.org/obo/go.owl. http://purl.obolibrary.org/obo/go/extensions/go-plus.owl
OBI: Ontology for biomedical investigations	This ontology is an integration of life-science And clinical investigations concepts	https://github.com/obiontology/obi/blob/master/obi.owl
OGMS: Ontology for general medical science	It includes Concepts related to Disease and Diagnosis of Carcinomas and other Pathological Entities	https://bioportal.bioontology.org/ontologies/OGMS
VO: Vaccine ontology	Concepts related to vaccines and also support Automated reasoning	http://purl.obolibrary.org/obo/vo.owl
CMO: Clinical measurement ontology	Includes physiological and morphological concepts generated from clinical health programs	https://bioportal.bioontology.org/ontologies/CMO
EDAM: Bioscientific data analysis ontology	All familiar concepts that exist within the domain of computational biology and bioinformatics are included in EDMA ontology	https://bioportal.bioontology.org/ontologies/EDAM

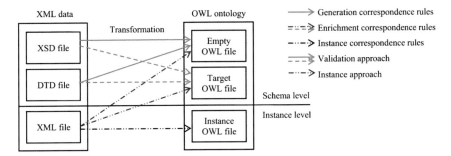

Fig. 3 Different rules and methods for transforming XML data to OWL ontologies

2. S-Trans are a method to transform healthcare data from XML form to OWL ontology proposed by Thuy et al. [5]. This study reveals that the approaches developed to transform XSD to the OWL suffer from the problem of duplicate elements. In some methods, the schema is given a unique identifier, but this solution results in data duplications in elements that can represent similar information. S-Trans method obtains data quality, by computing the similarity of duplicate elements. If the semantics of two elements is high then they are transformed into one ontology.

3. A method for generating OWL ontology automatically from XML data is proposed by Nora Yahia et al. [19]. This process is shown in Fig. 4.

The process can be explained in four steps.

Step1: A Java-based API called Trang is used to transform the XML document into XML Schema. This API takes the schema in XML syntax as input and produces XML Schema as output.

Step 2: This XML Schema is then analyzed using the XML Schema Object Model (XSOM).

Step 3: The derived XML Schema object will be fed as input to the Java Universal Network/Graph Framework (JUNG). The JUNG is used for graph-based manipulations which generate the XML- Schema Graph (XSG).

Step 4: This XSG is taken as input by the JENA API to generate OWL entities.

4. Many more approaches are proposed to transform XML into OWL ontology which includes Mapping XML to OWL by Hannes and Soren [20] where the relational data I serialized to XML and XSD format and relational structures are discovered which are later converted to owl format. A semantic approach for such transformation is suggested by Pham et al. [21]. Another approach for this transformation using a canonical data model is proposed in [22].

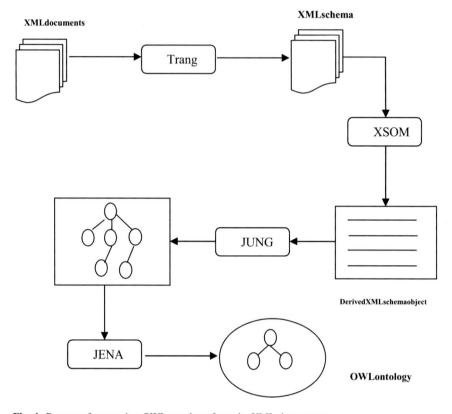

Fig. 4 Process of generating OWL ontology from the XML data source

8 Semantic Interoperability in Health Care: Benefits and Challenges

It is evident from Sects. 4 and 5 that data in the form of OWL ontology is best suitable for supporting semantic interoperability, whether the medical data ontology is constructed or transformed. This section highlights the major benefits and key challenges of semantic interoperability specifically in the healthcare domain. Table 3 summarizes the benefits followed by challenges [23].

9 Conclusions

Semantic interoperability in the domain of health care mainly aims at facilitating the seamless interchange of medical data among all healthcare providers and patients which helps in clinical decision making. However, the existing formats of data cause

Table 3 Benefits and challenges of interoperability

Benefits of interoperability	
1. Easy and safe access to patient's data	Even though data is stored in heterogeneous systems, interoperability facilitates easy access to patients' medical data. Patient scan also accesses their medical records completely which are maintained by several healthcare providers. The privacy of the patient's data is also secured
2. Accurate understanding of medical terms	The healthcare providers will be able to understand the medical terms transmitted from external systems without any difficulty Since the semantics of the content are preserved
3. Reduce medical errors	The primary concern in health care is medical errors since they are said to be the sixth most important cause of death in hospitals [24]. Interoperability in healthcare ensures medical data is formatted in such a way that machines can understand both structure and semantics of the information
4. Reduced healthcare cost	Increasing costs is a major challenge faced by the healthcare industry. Effective interoperability among various healthcare providers can be an essential factor for reducing the cost [24]
5. Data integration	Semantic interoperability allows healthcare systems to seamlessly Integrate data from different applications like laboratory systems, registration systems, healthcare providers, and vendors
Challenges of interoperability	
6. Complexity of the healthcare domain	The healthcare domain is complex as it involves many people like doctors, nurses, pharmacists, radiologists, laboratory technicians, and many more who together treat patients. Each of them generates a lot of information that could come in multiple formats which Makes the entire system complex
7. Legacy systems	The systems with limited interoperability capabilities are still in use today. These systems are designed for some particular tasks and become a challenge for data integration and exchange

a lack of understanding and confusion to different health applications. This problem can be addressed by representing medical data in form of OWL ontology which is rich in expressiveness and supports efficient semantic in-interoperability. The medical ontologies can be both constructed and transformed. This chapter gives a list of different medical ontologies and also gives insight into the different ways of transforming data into OWL format. Further, this chapter also discusses the benefits and challenges of semantic interoperability.

References

1. Shah T, Rabbi F, Ray P (2015) investigating an ontology-based approach for Big Data analysis of inter-dependent medical and oral health conditions. Cluster Compute 18:351–367. https://doi.org/10.1007/s10586-014-0406-8
2. Zenuni X, Raufi B, Ismaili F, Ajdari J (2015) State of the art of semantic web for healthcare. In: Procedia-social and behavioral sciences, vol 195, pp 1990–1998
3. Iftikhar S, Khan WA, Ahmad F, Fatima K (2012) Semantic interoperability in E-health for improved healthcare. Semant, Action-Applications Scenarios, vol 107
4. Taouli A, Djamel AB, Keskes N, Bencherif K, Hassan B (2018) Semantic for bigdata analysis: a survey. In: Proceedings of the INTIS2018: BigData and Internet of things IoT, Marrakech, Maroc, Dec 2018
5. Thuy PTT, Lee Y-T, Lee S (2012) S-trans: semantic transformation of XML healthcare data into OWL ontology. Knowl-Based Syst 35:349–356
6. Merelli I, P´erez-Sanchez H, Gesing S, D'Agostino D (2014) Managing, analysing, and integrating big data in medical bioinformatics: open problems and future perspectives. BioMed Res Int, ArticleID 134023, 13 pages
7. Chungoora T. Enter the realm of semantic web languages overview of markup languages in the Semantic WebStack. https://medium.com/swlh/enter-the-realm-of-semantic-web-languages-ee94ee68f544
8. Banu A, Fatima SS, Ur Rahman Khan K (2011) Semantic-based querying using ontology in relational database of library management system. Int J Web Semantic Technol (IJWesT) 2(4):21–32
9. Liyanage H, Krause P, de Lusignan S (2015) Using ontologies to improve semantic interoperability in health data. J Innov Health Inform 22(2):309–315
10. Semantic interoperability. https://en.wikipedia.org/wiki/Semantic_interoperability
11. Sure Y, Staab S, Studer R (2009) Ontology engineering methodology. In: Handbook on ontologies. Springer, Heidelberg, pp 135–152
12. https://en.wikipedia.org/wiki/Ontology_engineering
13. Spoladore D, Pessot E (2021) Collaborative ontology engineering methodologies for the development of decision support systems: case studies in the healthcare domain. Electronics 10(9):1060
14. Simperl EPB, Tempich C (2006) Ontology engineering: a reality check. In: On the move to meaningful internet systems. In: Proceedings of the OTM confederated international conferences, Montpellier, France, 29 Oct–3 Nov 2006. Springer, Cham, Switzerland, pp 836–854
15. Banu A, Fatima SS, Ur Rahman Khan K. Building OWL ontology: LMSO-library management system ontology. In: Meghanathan N, Naga-malai D, Chaki N (eds) Advances in computing and information technology. Advances in intelligent systems and computing, vol 178. Springer, Heidelberg. https://doi.org/10.1007/978-3-642-31600-5_51
16. Noy NF, McGuinness DL (2001) Ontology development 101: a guide to creating your first ontology. In: Stanford Knowledge Systems Laboratory Technical Report KSL-01-05 and Stanford Medical Informatics Technical Report SMI-2001-0880, Mar 2001
17. Farquhar A, Fikes R, Rice J (1996) The ontolingua server: a tool for collaborative ontology construction. In: Proceedings of the 10th knowledge acquisition for knowledge-based systems workshop, Banff, Canada, 9–14 Nov 1996

18. Hacherouf M, Bahloul SN, Cruz C (2015) Transforming XML documents to OWL ontologies: a survey. J Information Sci 41(2):242–259
19. Yahia N, Mokhtar SA, Ahmed AW (2012) Automatic generation of OWL ontology from XML data source. Int J Comput Sci Issues 9(2)
20. Bohring H, Auer S (2015) Mapping XML to OWL ontologies. In: Jantke KP, Fähnrich K-P, Wittig WS (Hrsg.), Marktplatz Internet: Von e-Learning bis e-Payment, 13. LeipzigerInformatik-Tage (LIT 2005). Gesellschaft für Informa-tik e. V., Bonn (S.147–156)
21. Thuy PTT, Lee YK, Lee S (2013) A semantic approach for transforming XML data into RDF ontology. Wireless Pers Commun 73:1387–1402. https://doi.org/10.1007/s11277-013-1256-z
22. Jounaidi A, Bahaj M (2018) Converting of an Xml schema to an Owl ontology using a canonical data model. J Theor Appl Information Technol 96(5)
23. Iroju O, Soriyan A, Gambo I, Olaleke J (2013) Interoperability in healthcare: benefits, challenges and resolutions. Int J Innov Appl Stud 3(1):262–270
24. Bock C, Carnahan L, Fenves S, Gruninger M, Kashyap V, Lide B, Nell J, Raman R, Sriram R (2005) Healthcare strategic focus area: clinical informatics. National Institute of Standards and Technology, Technology Administration, Department of Commerce, USA, pp 1–33, Sept 2005

A Novel Fusion Scheme for Face Recognition in Challenging Conditions

Shekhar Karanwal

Abstract Literature reveals that by the fusion of global and local descriptors more finer results are earned than either of them alone. Motivated by this, the proposed work introduces the novel fusion scheme by amalgamating the features of three effective descriptors, i.e., PCA, LBP and LPQ. PCA is availed as the global feature extractor and the local feature extraction are carried out by LBP and LPQ. This fusion scheme is termed as the PCA + LBP + LPQ. Prior to amalgamation, z-score normalization is carried out on the respective descriptor. The LBP and LPQ features are attained region wise from corresponding map images. The amalgamated size is on the bigger side; therefore, PCA services are exploited again for compact size. For matching SVMs are availed, and four datasets deployed are ORL, GT, JAFFE and Faces94. The PCA + LBP + LPQ pulls of superb recognition rates than either of PCA, LBP and LPQ. It also overshadow the numerous literature-based techniques.

Keywords Principal component analysis (PCA) · Local binary pattern (LBP) · Local phase quantization (LPQ) · Support vector machines (SVMs)

1 Introduction

Face recognition (FR) has accomplished noteworthy advancements over the last few years. This progress is because of the development of discriminant descriptors in changing conditions. The local ones turn out to be the dominant ones than the global ones and the amalgamation of both creates the robust descriptor than either of them. Sahan et al. [1] provide the fusion approach for FR in which local ones are extracted by the utilization of the radon transform, and the global ones are extracted by chebyshev moments (Fourier). Experiments on four datasets show the relevance of the fused approach. Wang et al. [2] introduced the emotion recognition by the fusion of principal component analysis (PCA) and pyramid Weber local descriptor (PWLD). Specifically, the entropy (information) is applied first for locating

S. Karanwal (✉)
Department of CSE, Graphic Era University (Deemed), Dehradun, UK 248002, India
e-mail: shekhar.karanwal@gmail.com

© The Author(s), under exclusive license to Springer Nature Singapore Pte Ltd. 2023   87
M. A. Chaurasia and C.-F. Juang (eds.), *Emerging IT/ICT and AI Technologies Affecting Society*, Lecture Notes in Networks and Systems 478,
https://doi.org/10.1007/978-981-19-2940-3_6

the occluded region. Second, PCA is utilized for the reconstruction of occluded region which is followed by employing the replace technique for image reconstruction, by the replacement of the region (occluded) with the premier matched image respective region in training size. Third, the PWLD feature extraction is done. The fusion method attains promising outcomes on JAFFE and CK datasets. Kisku et al. [3] presented the FR by amalgamating the KPCA and scale invariant feature transform (SIFT) features. Initially, image data is transfigured to the higher dimensional space, and then, PCA is utilized to procure the principal components for the computation of eigenvalues and eigenvectors. Then, SIFT is employed for extracting features from two distinct representation of eigenspace, which are merged next. On ORL face dataset, the FR method leads amazing results. Yang et al. [4] launched the FR method based on improved Weber local descriptor (IWLD), improved Weber binary coding (IWBC) and block-based Fisher linear discriminant (BFLD). In proposed IWLD, the local patterns are represented in more constructive manner by launching the Weber orientation and magnitude components. The more discriminativity is attained by the usage of IWBC. Finally, BFLD is employed for feature compression. On FERET, AR and PolyU-NIR, the method brings excellent results.

Chen et al. [5] proposed the sketch FR by the integration of pyramid histogram of oriented gradients (P-HOG), PCA and null spaced-based LDA (NLDA) features. First, the global and local P-HOG feature extractions are done from distinct image patches of face. Second, the size of the respective descriptors is reduced by the employment of PCA and NLDA, and third, the weighted features (sensitivity based) are merged. The recognition is carried by the nearest neighbor (NN), and on different datasets, the introduced FR validates its efficacy. Nordin et al. [6] discovered a new FR by joining the local binary pattern (LBP) and PCA features from T-Zone face. Precisely, the LBP features are extracted from distinct T-Zone face regions. Then, features are merged further for developing the full representation of the face. PCA is owned next for the dimension compression. The discovered method carries super rates on ORL dataset. Jayasimha et al. [7] invented the emotion recognition in which global and local feature extraction are carried out by HOG and LBP. Before imposing feature extraction procedure, the Viola Jones method and the hybrid Laplacian of Gaussian (HLOG) are availed for detection and preprocessing. Then, orthogonal local preserving projection (OLPP) is deployed for compressing the dimension, and matching algorithm adopted is support vector machines (SVMs). On CK+ dataset, the presented method fulfills amazing outcomes. In [8], the local feature extraction is done by the utilization of the novel descriptor termed as the distance ratio LBP (DR_LBP). DR_LBP overcomes the limitations of LBP by investigating the distances and using little in the ratio form. The DR_LBP feature size is formed from the upper and the lower regional halves of the face. The size is larger therefore LDA is bring into task for the compaction and then SVMs is deployed for classification. The fused method (of local and global) procures astounding rates than others on Yale, ORL and Faces94. Yaddaden et al. [9] developed the emotion recognition based on HOG, LBP and local linear embedding (LLE). From HOG and LBP, both global and local feature extraction are done. The image preprocessing is also deployed before extracting the features. To decrease the size of extracted feature, the LLE is adopted, and matching is covered up by SVMs. On three datasets, the global HOG performs better, and LBP is finer in local.

Jianxin et al. [10] launch the FR by usage of HOG-LBP, PCA and sparse representation. In beginning, block-wise LBP and HOG features are extracted, and then, all block features are joined for the developing of complete block HOG-LBP representation. The block-wise features extracted are huge in dimension, so PCA is applied for dimension depletion. Lastly, matching was conducted by SRC. Experiments obtained on AR and ORL datasets verify the potency of the launched FR. Zhou et al. [11] invented the new FR to solve lightning and misalignment issues by introducing the fusion of local, global descriptors and improved learning methods. LBP is operated first for extracting the local features. The working of global feature LDA is adopted second for size depletion followed by the division of fisher scores into sub-blocks as per the feature dimension. Finally, in third, the improved pairwise-constrained multiple metric learning (IPMML) is operated to derive the optimal Mahalanobis matrix, which is utilized for the computation of the robust distance for classification (by NN). Results illustrate that the invented method is effective in considered issues. Liu et al. [12] proposed the fusion scheme for recognition of emotion, in which initially the salient areas are obtained which is followed by the LBP and HOG feature extraction. Then, LBP and HOG features are merged for developing the LBP-HOG size. The merged dimension is made compacted by the deployment of PCA, and matching is performed by the numerous classifiers. In addition, the gamma correction method is also adopted for the evaluation. Both methods accomplish the good outcomes on JAFFE and CK+ datasets. Guo et al. [13] presented discriminant descriptor for emotion analysis so-called extended LBP on three orthogonal planes (ELBPTOP). ELBPTOP contains basically the three descriptors, i.e., LBPTOP, ADLBPTOP and RDLBPTOP. LBPTOP is the older one descriptor, and remaining two are the proposed ones. The latter two descriptors exploit the second-order information in angular and radial directions. The compressed dimension is gained by the utilization of whitened PCA (WPCA). The presented method pulls of finer results on many datasets.

Sun et al. [14] deployed the fusion of local and global methods to identify the seven emotions of the CK+ dataset. The local feature outcomes are acquired by the Gabor and LBP. Then, PCA is brought into task for the shortening of feature size to individual and the fused PCA features. Lastly, the optimization of feature is carried out by LDA with matching by SVMs. The fusion method gains the super results. Fang et al. [15] introduced the new descriptor for FR named as Weber local circle gradient pattern method (WLCGP). The WLCGP takes into consideration the connection among target pixel and the surrounded neighbors and the connection between the surrounded neighbors. The WLCGP full size is formed from the distinct image sub-blocks. Now, next is the turn of PCA to make the feature size small for classification (by 3-NN). On AR, Singapore infrared and ORL datasets, the amalgamated approach reflects the superb outcomes. Ameur et al. [16] discovered the local GLBSIF descriptor and used it with the PCA (global descriptor) for the production of discriminant FR. In GLBSIF, the G stands for Gabor, LB stands for LBP, and BSIF stands for binarized statistical image features. Therefore, GLBSIF feature presentation is carried out from the three descriptors. Prior to feature extracting the input image is normalized. The reduced size attained after PCA is fed into the KNN-SRC for matching. The discriminativitsy of invented FR is examined on six datasets, and it produces encouraging results. Patil et al. [17] proposed the FR based on LBP and Contourlet transform (CT). The block (multi) LBP is extracted from the original input image, and then, the block (multi) WLD is extracted from each enhanced sub-band, attained after CT decomposition. Then, all of them are joined at the feature level. The depletion in dimension compaction makes possible by LDA method. Finally, NN is deployed for matching. The astounding consequences are gained by the proposed method on six datasets.

Truong et al. [18] give the fusion method for FR by using bilateral line LBP (BL-LBP) and 2D-PCA. LLBP is deployed for extracting the features in vertical and the horizontal directions. Both of the directional features are shortened by 2D-PCA. Ultimately the SVM scores are fused for the learning methodology based on ensemble. On four datasets, tremendous results are accomplished by the fused method. Juneja et al. [19] developed its FR in following way. The noise affect is healed by the utilization of median filter which is followed by the operation of weighted LBP filter for generating the pattern region. Further the discriminant feature size is generated by the deployment of log Gabor filter (angle- and frequency-based). PCA is employed next for compaction. The developed FR brings amazing results on ORL dataset. Zhang et al. [20] invented the emotion recognition by adopting three local features, i.e., Gabor wavelet transform (GWT), LBP and local phase quantization (LPQ). PCA-LDA is used as the global feature. Gabor features are obtained from five scales and eight orientations. Next, LBP and LPQ are utilized for image (Gabor) encoding. Two-stage PCA-LDA used next for performing the size compaction. The launched method brings tremendous accuracy rates. Xu et al. [21] presented the polynomial contrast binary pattern (PCBPs) for the FR. In PCBPs, there is the efficient estimation of the local information (underlying) which is portrayed roughly as the projection coefficients (linear) in local regions (of entire pixels). The decomposition (polynomial-based) extracts discriminant information from various frequencies and orientations. FLDA is used for compacting the feature size. The outcomes on FERET and LFW illustrate the invented method. Karanwal et al. [22] deployed comparative analysis (study) among 14 descriptors for FR. The global-wise extraction of features is carried out from all the descriptors, and then, PCA, FLDA are applied for compaction. Among all, it is compound LBP (CLBP) which attains supreme outcomes. Karanwal et al. [23] give the fused method for FR by introducing the orthogonal difference-LBP (OD-LBP) descriptor. In OD-LBP, there is launching of the new comparison function which makes him superior than the previous local descriptors. The compression in size and then matching are carried out by the PCA and SVMs. On five datasets, OD-LBP proves its value.

In recent times, the deep learning-based methods have gain huge popularity due to its successful utilization in different applications, especially in FR. Some of deep convolutional neural network (CNN)-based methods which are successfully utilized in literature are [24, 24]. Some more examples of such networks are AlexNet, VGG,

ResNet-50 and LetNet. Despite proving relevance in difference applications, there are various short comings observed in these deep CNN-based methods, and these are requirement of training data in huge amount, huge computational complexity and lack of adapting the parameters with respect to specific method. In recent times, the local descriptors are proven better than some of these deep learning methods in the uncontrolled conditions. Some such work can be explored in [26–28]. Motivating with that the proposed work introduces the hybrid descriptor by integrating one global and two local methods termed as PCA + LBP + LPQ. The PCA + LBP + LPQ attains stupendous outcomes compared to the individual and from many techniques in literature. More detailed elaboration on the proposed method is given in the next paragraph.

Literature reveals that by the fusion of global and local descriptors more finer results are earned than either of them alone. Motivated by this, the proposed work introduces the novel fusion scheme by amalgamating the features of three effective descriptors, i.e., PCA [29], LBP [30] and LPQ [31]. PCA is availed as the global feature extractor and the local feature extraction are carried out by the LBP and LPQ. This fusion scheme is termed as the PCA + LBP + LPQ. Prior to amalgamation, z-score normalization is carried out on the respective descriptor. The LBP and LPQ features are attained region wise from corresponding map images. The amalgamated size is on the bigger side; therefore, PCA services are exploited again for compact size. For matching, SVMs [32] are availed, and four datasets deployed are ORL [33], GT [34], JAFFE [35] and Faces94 [36]. The PCA + LBP + LPQ pulls of superb recognition rates than either of PCA, LBP and LPQ. It also overshadow the numerous literature-based techniques.

Road map of remaining work is organized as: Illustration of PCA, LBP and LPQ are discussed in Sects. 2, and 3 discusses about fused scheme. Results with conclusions and future scope are placed in Sects. 4 and 5.

2 Illustration of PCA, LBP and LPQ

2.1 Principal Component Analysis (PCA)

The vital PCA [29] steps are defined below.

A. All the images in the gray form are transformed into column vectors for the computation of the mean value. Suppose E is the variable used for the storage of the mean value, and it is defined as

$$E = \frac{1}{t} \sum_{i=1}^{t} L_i, \tag{1}$$

L_i signifies the column vectors.

B.   Then, there is computation of the difference between L_i and E. The variable T is availed for the difference value storing, given as

$$T = L_i - E \tag{2}$$

C.   Next eigenvalues and eigenvectors are calculated from $T.T^T$. But $T.T^T$ takes memory in huge amount, so they are computed from $T^T.T$. The eigenvectors of z largest eigenvalues are availed for the making of eigenface space. Suppose N is utilized for storage of Y largest eigenvectors.

$$N = [Y_1, \ldots, Y_n] \tag{3}$$

D.   Finally eigenfaces are derived by the usage of N and T which is stated as $A = N.T$, where A is the variable used for the storage of eigenfaces. Eventually there is the projection of images in the face space specified as S by the utilization of T and A, given as

$$S = A^T T \tag{4}$$

When global feature extraction is done by PCA, then 125 top most eigenvectors are selected for creating eigenface space. Therefore, after all images are projected in face space, the feature size produced is 125. When PCA is implemented separately for comparison, then size of projected features is those features which are produced from all descriptors after dimension reduction, feature size used by classifier. Rest details are in results section.

2.2   Local Binary Pattern (LBP)

In LBP [30], the neighboring pixels are collated to pixel placed at the center. 1 is issued to those places (neighbors) which have > or = value to the pixel placed at the center, else 0 is issued. After collation, there is emergence of the 8 bit pattern whose decimal value is deduced by the weights apportion. Same criteria are deployed in all places which form the LBP image. The regional (3×3) merged LBP histogram makes the wholly size of 2304. The single region size is 256. Equation 5 constructs the LBP code for one place. $I, J, V_{J,i}$ and V_c are the neighboring size, radius, distinct spots (places) and center spot. Figure 1 shows the LBP concept for single place.

$$\mathrm{LBP}_{I,J}(x_c) = \sum_{i=0}^{I-1} k\left(V_{J,i} - V_c\right)2^i, \quad k(p) = \begin{pmatrix} 1 & p \geq 0 \\ 0 & p < 0 \end{pmatrix} \tag{5}$$

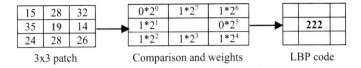

Fig. 1 LBP concept for single place

2.3 Local Phase Quantization (LPQ)

In [31], Ojansivu et al. invented the LPQ descriptor which shows the property of blur invariance. In LPQ, there is the extraction of STFT information from N × M neighbors h_x for each single position x of original image $f(x)$. STFT is defined as

$$K(y, x) = \sum_{v \in h_x} f(x - v)e^{-j2\pi y^T v} \tag{6}$$

For extracting the real and imaginary parts, it should be enough to pick the four frequency parts. Ojansivu in [31] picks four frequency parts q_0, q_1, q_2, q_3 where $q_0 = [b, 0]^T$, $q_1 = [0, b]^T$, $q_2 = [b, b]^T$ and $q_3 = [b, -b]^T$. Further, there is formation of vector which carries low frequency complex parts as mentioned in Eq. 7.

$$K(x) = [K(q_0, x), K(q_1, x), K(q_2, x), K(q_3, x)] \tag{7}$$

So, the extraction of feature is brought from $K(x)$, as specified as

$$F(x) = \left[\text{Real}(K(x)) \text{Imag}(K(x)) \right] \tag{8}$$

Finally, by the consideration of the value (i.e., sign of $F(x)$, with +ve or 0 are taken as 1 else 0), there is the evolution of the binary pattern, reserved in U_i. Then, weights granting and values addition take place for forming the decimal code for one place, as defined in Eq. 9.

$$\text{LPQ}_{I,J}(x_c) = \sum_{i=0}^{I-1} (U_{J,i}) 2^i \tag{9}$$

I and J signify patch size and radius, and $U_{J,i}$ signifies distinct locations in patch. Same procedure is adopted for every position. There is the emergence of LPQ image after applying this earlier process for all positions. The regional (3 × 3) merged LPQ histogram makes the wholly size of 2304. The single region size is 256. Figure 2 shows the LPQ concept. From STFT patch, the four parts (frequency) selected are (1, 2), (1, 3), (2, 1) and (2, 2).

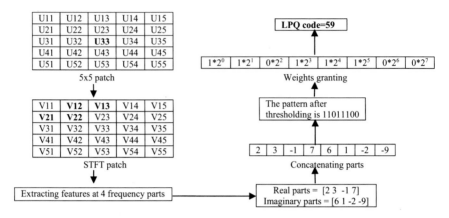

Fig. 2 LPQ concept for single place

3 The Fusion Scheme (PCA + LBP + LPQ)

The proposed fusion scheme is created by amalgamating the features of three effective descriptors, i.e., PCA, LBP and LPQ, as described earlier. Prior to amalgamation, z-score normalization is carried out on respective descriptor. It is found during the experimentation that there is huge improvement in recognition rate after the deployment of z-score normalization. Therefore, after amalgamation, the PCA + LBP + LPQ builds the size of [125 2304 2304] = 4733. PCA services are exploited again for compacting huge size, and then, radial basis function (RBF), the SVMs-based function are deployed for classification. Figure 3 shows complete flow diagram.

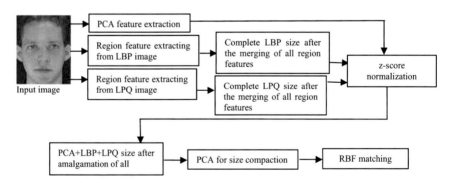

Fig. 3 Complete flow diagram

4 Results

4.1 Datasets Description

The **ORL** which is earlier known as AT and T dataset is used several times in literature for testing. This database is composed from 40 subjects, and subject-wise images (gray) are 10. Therefore, the full database has 400 images. The intrapersonal conditions in these images are lightning, emotion and pose. All are captured in similar image size of 112×92. Some of them are conveyed in Fig. 4a.

The **GT** is also used several times in literature for testing. This database is composed from 50 subjects, and subject-wise images (color) are 15. Therefore, the full database has 750 images. The intrapersonal conditions in these images are lightning, scale, emotion and pose. All are captured in varied image sizes. Some of them are conveyed in Fig. 4b.

The **JAFFE** database is composed from ten subjects, and the subject-wise images are nearly 21. Therefore, the full database has 213 images (gray). Precisely there are 2–4 images of seven distinct emotions which are anger, disgusting, fear, happiness, neutral, sad and surprise. In cluttered scene the image capturing is done as the result the size of the images are 256×256, which are on larger side. The image cropping is performed for the removal of the cluttered scene, and after that, the size of the images is set to 141×131. The crop images of first subject are conveyed in Fig. 4c.

The **Faces94** database is composed from 152 subjects, and the subject-wise images are 20. Therefore, the full database has 3040 images (color). The intrapersonal conditions in these images are emotions. All are captured in similar image size of 200×180. In this work, there is the utilization of only 30 subject images, i.e., 600. Some of them are conveyed in Fig. 4d.

4.2 Details with Respect to Feature Size

The high-dimensional original image size will definitely impact the computational cost; therefore, the original image is reduced to the size of 49×46. For gray images, this is carried out directly, for color images, first gray conversion is carried out, and then, reduction is performed. From reduced size image, the PCA, LBP, LPQ and PCA + LBP + LPQ are deployed for extracting the features. When PCA is separately implemented, then input to PCA is the column vectors of the raw input reduced image which produces the size of 2254, and after applying PCA, the classifier size is [29 40 20 8] on [ORL GT JAFFE Faces94] databases. LBP and LPQ build the size of 2304, and after PCA, the classifier size is same as earlier defined. PCA + LBP + LPQ feature size is 4733, as PCA is 125, LBP and LPQ have the size of 2304. For PCA feature size creation, the 125 top most eigenvectors are selected. Prior to fusion of features z-score normalization is performed separately on the feature size of all three descriptors. Finally, PCA + LBP + LPQ size is reduced further by PCA,

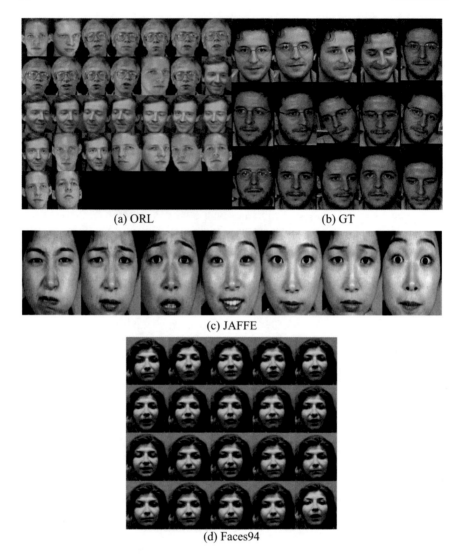

(a) ORL   (b) GT

(c) JAFFE

(d) Faces94

Fig. 4 Some of the images of four datasets

as employed to other descriptors. The feature size used by the classifier is same as defined earlier, i.e., [29 40 20 8] on [ORL GT JAFFE Faces94] datasets. For fair comparison, the compact size is similar for all descriptors. MATLAB R2018a is considered for all the testing.

4.3 Recognition Rate Calculation

The recognition rate (in %) is generated by formula as shown in Eq. 10. In Eq. 10, RR, C_{ts} and F_c are recognition rate, test set and false counts. The remaining parameter C_{tst} not included in Eq. 10 indicates the training set. The RR is estimated on C_{ts} based on the F_c values. The parameter F_c indicates the total samples which are not matched correctly to the respective class sample. For example, if C_{ts} represents 360 samples and F_c are 10, then RR $= ((360–10)/360)*100 = 97.22\%$. In similar way, the RR is calculated on every subset.

$$[RR = \frac{C_{ts} - F_c}{C_{ts}} * 100] \tag{10}$$

On **ORL**, C_{tst} varies from **1:4, and** C_{ts} varies from **9:6**. The finer outcome is achieved after 30 executions. On all subsets, **PCA + LBP + LPQ** exceeds the results of the solitary ones. **PCA + LBP + LPQ** achieves the astonishing accuracy of [86.11% 94.37% 99.28% 100%] on training size of 1:4. All these obtained accuracies are much better than PCA, LBP and LPQ. Table 1 conveys it all, and the graph illustration is revealed in Fig. 5a.

On **GT**, C_{tst} varies from **5:8, and** C_{ts} varies from **10:7**. The finer outcome is achieved after 25 executions. On all subsets, the **PCA + LBP + LPQ** exceeds the results of the solitary ones. **PCA + LBP + LPQ** achieves the amazing accuracy of [87.20% 88.00% 91.75% 92.28%] on training size of 5:8. All these obtained accuracies are much better than PCA, LBP and LPQ. Table 1 conveys it all, and the graph illustration is revealed in Fig. 5b.

On **JAFFE**, C_{tst} varies from **2:4, and** C_{ts} varies from **19:17**. The finer outcome is achieved after 28 executions. On all subsets, **PCA + LBP + LPQ** exceeds the results of the solitary ones. **PCA + LBP + LPQ** achieves the accuracy of [91.19% 97.26% 99.42%] on training size of 2:4. All these obtained accuracies are much better than the PCA, LBP and LPQ. Table 2 conveys it all, and the graph illustration is revealed in Fig. 5c.

Table 1 RR on ORL and GT datasets

	ORL results				GT results			
	C_{tst} attributes				C_{tst} attributes			
	$C_{tst} = 1$	$C_{tst} = 2$	$C_{tst} = 3$	$C_{tst} = 4$	$C_{tst} = 5$	$C_{tst} = 6$	$C_{tst} = 7$	$C_{tst} = 8$
Descriptors	RR in %				RR in %			
PCA	73.61	85.00	92.50	95.00	71.40	74.00	77.25	79.42
LBP	78.61	90.93	96.07	98.75	82.20	83.55	85.75	87.14
LPQ	78.05	90.31	96.07	98.33	80.20	83.33	85.75	86.57
PCA + LBP + LPQ	**86.11**	**94.37**	**99.28**	**100**	**87.20**	**88.00**	**91.75**	**92.28**

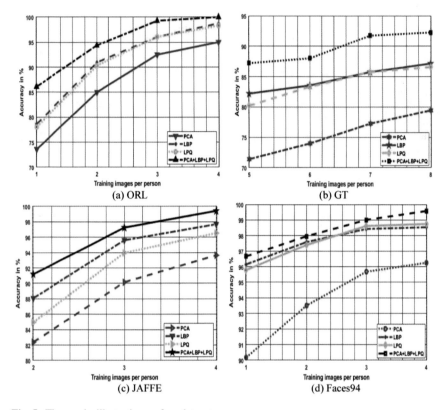

Fig. 5 The graphs illustration on four datasets

Table 2 RR on JAFFE and Faces94 datasets

	JAFFE results			Faces94 results			
	C_{tst} attributes			C_{tst} attributes			
	$C_{tst} = 2$	$C_{tst} = 3$	$C_{tst} = 4$	$C_{tst} = 1$	$C_{tst} = 2$	$C_{tst} = 3$	$C_{tst} = 4$
Descriptors	RR in %			RR in %			
PCA	82.38	90.16	93.64	90.17	93.51	95.68	96.25
LBP	88.08	95.62	97.68	96.14	97.59	98.43	98.54
LPQ	84.97	93.98	96.53	95.78	97.40	98.62	98.75
PCA + LBP + LPQ	**91.19**	**97.26**	**99.42**	**96.66**	**97.96**	**99.01**	**99.58**

On **Faces94**, C_{tst} varies from **1:4, and C_{ts}** varies from **19:16**. The finer outcome is achieved after the 20 executions. On all subsets, the **PCA + LBP + LPQ** exceeds the results of the solitary ones. **PCA + LBP + LPQ** achieves the accuracy of [96.66% 97.96% 99.01% 99.58%] on training size of 1:4. All these obtained accuracies are much better than PCA, LBP and LPQ. Table 2 conveys it all, and the graph illustration is revealed in Fig. 5d.

4.4 Recognition Rate Comparison with Literature Techniques

4.4.1 ORL Dataset

On ORL, total 14 techniques are compared with the hybrid method **PCA + LBP + LPQ**. These 14 techniques fall into types of local, classification, global, dimension reduction and sparse representation. Their identifications with accuracy achieved are defined as: LC-LBP [22], VELBP [22], NCDB-LBPac [37], NCDB-LBPc [37], HWLRR [38] and ASWNLRR [38] gains the RR of [52.22% 64.68% 73.92% 78.33%], [61.94% 77.81% 83.21% 86.66%], [78.88% 91.25% 95.35% 98.75%], [78.33% 91.87% 95.71% 97.91%], [66.20% 74.50% 81.70% 84.70%] and [58.10% 66.90% 70.40% 76.20%] on $C_{tst} = 1{:}4$. OD-LBP [23] and HOG [23] succeeds in leading the RR of [82.77% 94.37% 98.21%] and [70.83% 89.68% 91.07%] when $C_{tst} = 1{:}3$. $(2D)^2PCA + (2D)^2LDA$ [39], $(2D)^2MMC + LDA$ [39] and $(2D)^2PCA + LDA$ [39] pulls of the RR of 97.61%, 95.28% and 97.22% when $C_{tst} = 4$. Ratio sum for compaction [40] and trace ratio for compaction [40] achieves the RR of [91.43% 95.00%] and [91.07% 93.75%] on $C_{tst} = 3{:}4$. DLSR [41] attains the RR of [88.12% 89.64% 94.58%] on $C_{tst} = 2{:}4$. The **PCA + LBP + LPQ** technique comprehensively outruns the RR of all 14 techniques. This shows the effectiveness of hybrid method **PCA + LBP + LPQ** against the other 14 compared methods. Table 3 gives the all RR.

4.4.2 GT Dataset

On GT also total 14 techniques are compared with the hybrid method **PCA + LBP + LPQ**. These 14 techniques fall into types of local, regression and learning and classification Their identifications with accuracy achieved are defined as: LC-LBP [22] and VELBP [22] reach the RR of [59.00% 59.71%] and [65.50% 67.14%] on $C_{tst} = 7{:}8$. OD-LBP [23] and HOG [23] succeeds in leading the RR of [85.20% 87.33%] and [71.60% 74.00%] when $C_{tst} = 5{:}6$. NCDB-LBPac [37], NCDB-LBPc [37], DLRR_AG [43], DLSR [43] and JGLDA [43] obtain the RR of [80.60% 82.66% 84.50% 86.00%], [81.80% 82.66% 84.50% 86.57%], [63.60% 68.22% 73.00% 75.42%], [54.40% 65.33% 69.88% 72.10%] and 59.32% 66.79% 68.95% 72.84%] when $C_{tst} = 5{:}8$. CZZBP [42] and CMBZZBP [42] attain RR of [76.00% 79.00% 80.57%] and [82.88% 85.50% 86.85%] on $C_{tst} = 6{:}8$. SAPFR-1 [44] and NUPDWT [44] lead in succeeding the RR of [84.82% 89.45%] and [73.34% 79.05%] when $C_{tst} = 5$ and 7. HOG + SVM [45] pulls the RR of 84.80% on $C_{tst} = 8$. The PCA + LBP + LPQ technique comprehensively outruns the RR of all 14 techniques. This shows

Table 3 RR comparison on ORL dataset

All techniques	C_{tst} attributes			
	$C_{tst} = 1$	$C_{tst} = 2$	$C_{tst} = 3$	$C_{tst} = 4$
All techniques	RR in %			
LC-LBP [22]	52.22	64.68	73.92	78.33
VELBP [22]	61.94	77.81	83.21	86.66
OD-LBP [23]	82.77	94.37	98.21	N/F
HOG [23]	70.83	89.68	91.07	N/F
NCDB-LBPac [37]	78.88	91.25	95.35	98.75
NCDB-LBPc [37]	78.33	91.87	95.71	97.91
HWLRR [38]	66.20	74.50	81.70	84.70
ASWNLRR [38]	58.10	66.90	70.40	76.20
$(2D)^2PCA + (2D)^2LDA$ [39]	N/F	N/F	N/F	97.61
$(2D)^2MMC + LDA$ [39]	N/F	N/F	N/F	95.28
$(2D)^2PCA + LDA$ [39]	N/F	N/F	N/F	97.22
Ratio sum for compaction [40]	N/F	N/F	91.43	95.00
Trace ratio for compaction [40]	N/F	N/F	91.07	93.75
DLSR [41]	N/F	88.12	89.64	94.58
PCA + LBP + LPQ	**86.11**	**94.37**	**99.28**	**100**
NF-Not found				

the effectiveness of the hybrid method **PCA + LBP + LPQ** against the other 14 compared methods. Table 4 gives the all RR.

4.4.3 JAFFE Dataset

On JAFFE total 12 techniques are compared with the hybrid method **PCA + LBP + LPQ**. These 12 techniques fall into types of matching, local, dimension reduction, global and manifold. Their identifications with accuracy achieved are defined as: RNNC [46], NCC [46], Gradient + BLDA [50] and HE + BLDA [50] lead in attaining the RR of 95.65, 90.76, 94.12 and 92.35% on $C_{tst} = 3$. DWT + Statistical features [47] carry the RR of 92.00% on $C_{tst} = 4$. Contourlet transform [48] succeeds in getting the RR of [94.62% 98.75%] when $C_{tst} = 3{:}4$. HDR [49], LGCF [49] and SLFT [49] have the RR of 71.00, 91.00 and 78.50% when $C_{tst} = 2$. CDFA [51], PCA [51] and LPP [51] bring off the RR of [93.04% 94.50% 96.17%], [85.84% 89.10% 91.62%] and [83.84% 89.32% 91.33%] when $C_{tst} = 2{:}4$. The PCA + LBP + LPQ technique outruns the RR of 11 techniques wholly. The RR of CDFA [51] is slight better than PCA + LBP + LPQ on $C_{tst} = 2$. On $C_{tst} = 3{:}4$, the PCA + LBP + LPQ pulls better RR than CDFA [51]. This shows the effectiveness of the hybrid method **PCA + LBP + LPQ** against other 11 compared methods. Table 5 gives all RR.

Table 4 RR comparison on GT dataset

All techniques	C_{tst} attributes			
	$C_{tst} = 5$	$C_{tst} = 6$	$C_{tst} = 7$	$C_{tst} = 8$
	RR in %			
LC-LBP [22]	N/F	N/F	59.00	59.71
VELBP [22]	N/F	N/F	65.50	67.14
OD-LBP [23]	85.20	87.33	N/F	N/F
HOG [23]	71.60	74.00	N/F	N/F
NCDB-LBPac [37]	80.60	82.66	84.50	86.00
NCDB-LBPc [37]	81.80	82.66	84.50	86.57
CZZBP [42]	N/F	76.00	79.00	80.57
CMBZZBP [42]	N/F	82.88	85.50	86.85
DLRR_AG [43]	63.60	68.22	73.00	75.42
DLSR [43]	54.40	65.33	69.88	72.10
JGLDA[43]	59.32	66.79	68.95	72.84
SAPFR-1 [44]	84.82	N/F	89.45	N/F
NUPDWT [44]	73.34	N/F	79.05	N/F
HOG + SVM [45]	N/A	N/F	N/F	84.80
PCA + LBP + LPQ	**87.20**	**88.00**	**91.75**	**92.28**

N/F-Not found

Table 5 RR comparison on JAFFE dataset

All techniques	C_{tst} attributes		
	$C_{tst} = 2$	$C_{tst} = 3$	$C_{tst} = 4$
	RR in %		
RNCC [46]	N/F	95.65	N/F
NCC [46]	N/F	90.76	N/F
DWT + Statistical features [47]	N/F	N/F	92.00
Contourlet Transform [48]	N/F	94.62	98.75
HDR [49]	71.00	N/F	N/F
LGCF [49]	91.00	N/F	N/F
SLFT [49]	78.50	N/F	N/F
Gradient + BLDA [50]	N/F	94.12	N/F
HE + BLDA [50]	N/F	92.35	N/F
CDFA [51]	**93.04**	94.50	96.17
PCA [51]	85.84	89.10	91.62
LPP [51]	83.84	89.32	91.33
PCA + LBP + LPQ	**91.19**	**97.26**	**99.42**

N/F-Not found

4.4.4 Faces94 Dataset

On Faces94 total 11 techniques are compared with the hybrid method **PCA + LBP + LPQ**. These 11 techniques fall into types of local, transform domain, classification, normalization and global. Their identifications with accuracy achieved are defined as: LC-LBP [22], MBP [22], HELBP [22], genetic algorithm (Gen:0) [56] and genetic algorithm (Gen:1) [56] gains RR of [86.31% 94.07% 96.47% 97.50%], [86.31% 93.51% 96.27% 97.08%], [91.22% 96.66% 98.62% 98.95%], [76.32% 80.92% 92.11% 92.76%] and [76.97% 82.24% 96.05% 94.74%] on C_{tst} = 1:4. Ridgelet transforms [52] pull of the RR of [96.66% 97.33% 97.14%] when C_{tst} = 1:3. DGLELM [53], GEELM [53] and GELM [53] attain the RR of [96.98% 98.07% 98.37%], [96.53% 97.62% 97.73%] and [96.79% 97.64% 97.85%] when C_{tst} = 2:4. DWT E-CLAHE [54] succeeds in bringing the RR of 98.84% on C_{tst} = 2. FPGA-based FR [55] achieve the RR of 95.00% when C_{tst} = 1. The PCA + LBP + LPQ technique outruns the RR of ten techniques wholly. The RR of DWT E-CLAHE [54] is slightly superior than PCA + LBP + LPQ. This shows the effectiveness of the hybrid method **PCA + LBP + LPQ** against the other ten compared methods. Table 6 gives all the RR.

Table 6 RR comparison on Faces94 dataset

| | C_{tst} attributes | | | |
	C_{tst} = 1	C_{tst} = 2	C_{tst} = 3	C_{tst} = 4
All techniques	RR in %			
LC-LBP [22]	86.31	94.07	96.47	97.50
MBP [22]	86.31	93.51	96.27	97.08
HELBP [22]	91.22	96.66	98.62	98.95
Ridgelet Transforms [52]	96.66	97.33	97.14	N/F
DGLELM [53]	N/F	96.98	98.07	98.37
GEELM [53]	N/F	96.53	97.62	97.73
GELM [53]	N/F	96.79	97.64	97.85
DWT E-CLAHE [54]	N/F	**98.84**	N/F	N/F
FPGA based FR [55]	95.00	N/F	N/F	N/F
Genetic Algorithm (Gen:0) [56]	76.32	80.92	92.11	92.76
Genetic Algorithm (Gen:1) [56]	76.97	82.24	96.05	94.74
PCA + LBP + LPQ	**96.66**	**97.96**	**99.01**	**99.58**

N/F-Not found

5 Conclusions and Future Scope

This work launched the novel fusion scheme by the amalgamation of three effective descriptors, i.e., PCA, LBP and LPQ. PCA is availed as the global feature extractor, and LBP, LPQ are utilized as the local features. The fusion scheme is called as PCA + LBP + LPQ. Before amalgamating the features, the normalization (z-score) is carried out on respective descriptor. The LBP and LPQ features are formed region wise from corresponding map images. The amalgamated size is on the bigger side; therefore, PCA services are exploited for compact size. For matching, SVMs are availed, and four datasets deployed are ORL, GT, JAFFE and Faces94. The PCA + LBP + LPQ pulls of superb recognition rates than either of PCA, LBP and LPQ. It also overshadow the numerous literature-based techniques. From literature, 51 techniques are taken into consideration for comparison, and the fusion scheme outstrip completely 49 techniques on all sets. So at last it is concluded easily the potency of the fusion scheme on numerous challenging conditions.

The future direction of this work is to deploy proposed fusion scheme in other application also. So, its efficacy can be tested on other applications also.

Beside this, the launching of the novel descriptor is proposal for the upcoming research.

References

1. Sahan AM, Itbi ASA (2021) The fusion of local and global descriptors in face recognition application. In: ACCT, pp 1397–1408. https://doi.org/10.1007/978-981-15-5341-7
2. Wang X, Xia, C, Hu, M, Ren F (2018) Facial expression recognition under partial occlusion based on fusion of global and local features. In: Proceedings of ICGIP
3. Kisku DR, Tistarelli M, Gupta P, Sing JK (2015) SIFT fusion of kernel eigenfaces for face recognition. In: Proceedings of SPIE, pp 965200-1–965200-8
4. Yang BQ, Zhang T, Gu CC, Wu KJ, Guan XP (2016) A novel face recognition method based on IWLD and IWBC. Mult Too App 75:6979–7002
5. Chen Z, Yao S, Liu C, Cai L (2019) Sketch face recognition: P-HOG multi-features fusion. Int J Patt Recog Art Intell 33(4):1956003-1–1956003-17
6. Nordin MJ, Hamid AAKA (2011) Combining LBP and PCA on T-zone face area for face recognition. In: Proceedings of ICPAIR
7. Jayasimha Y, Reddy RVS (2021) A facial expression recognition model using hybrid feature selection and SVMs. Int J Inf Com Sec 14(1)
8. Khanbebin SN, Mehrdad V (2020) Local improvement approach and LDA-based LBP for face recognition. Neu Comp App
9. Yaddaden Y, Adda M, Bouzouane A (2021) FER using LLE with LBP and HOG descriptors. In: IWHCSEHW, pp 221–226
10. Jianxin Z, Junyong W (2020) Local occluded face recognition based on HOG-LBP and SR. In: ICAICA, pp 808–813
11. Zhou L, Wang H, Lin S, Hao S, Lu ZM (2020) Face recognition based on LBP and IPC MML. Mult Tool Appl 79:675–691
12. Liu Y, Li Y, Ma X, Song R (2017) FER with fusion features extracted from salient facial areas. 17(4):1–18

13. Guo C, Liang J, Zhan G, Liu Z, Pietkainen M, Liu L (2019) Extended LBP for efficient and robust spontaneous facial micro-expression recognition. IEEE Acc 7:174517–174530
14. Sun Y, Yu J (2017) Facial expression recognition by fusing Gabor and LBP features. In: ICMC, pp 209–220
15. Fang S, Yang J, Liu N, Sun W, Zhao T (2018) Face recognition using weber local circle gradient pattern method. Mult Too App 77:2807–2822
16. Ameur B, Belahcene M, Masmoudi S, Derbel AG, Hamida AB (2017) A new GLBSIF descriptor for face recognition in the uncontrolled environments. In: ICATSIP, pp 1–6
17. Patil HY, Kothari AG, Bhurchandi KM (2016) Expression invariant face recognition using LBP and Contourlet transform. Opt 127(5):2670–2678
18. Truong HP, Khoa MBN, Kim YK (2020) Bilateral line LBP for face recognition. Prep
19. Juneja K, Rana C (2019) LBP pattern-processed Log-Gabor filter for expression and illumination robust facial recognition. In: ACAWSEE, pp 795–803. https://doi.org/10.1007/978-981-13-6772-4
20. Zhang B, Liu G, Xie G (2016) Facial expression recognition using LBP and LPQ based on GWT. In: ICCC, pp 365–369
21. Xu Z, Jiang Y, Wang Y, Zhou Y, Li W, Liao Q (2019) LP-CBP for face recognition. Neuroco 355:1–12
22. Karanwal S (2021) A comparative study of 14 state of art descriptors for face recognition. Mult Too App
23. Karanwal S, Diwakar M (2021) OD-LBP: orthogonal difference LBP for face recognition. Dig Sig Proc 110
24. Wu W, Tao D, Li H, Yang Z, Cheng J (2021) Deep features for person re-identification on metric learning. Patt Recogn 110
25. Krizhevsky A, Sutskever I, Hinton GE (2017) Imagenet classification with deep convolutional neural networks. Comm ACM 60(6):84–90
26. Sharma RP, Dey S (2021) A comparative study of handcrafted local texture descriptors for fingerprint liveness detection under real world scenarios. Mult Too App 80:9993–10012
27. Chen Z, Zhang L, Cao Z, Guo J (2018) Distilling the knowledge from handcrafted features for human activity recognition. IEEE Trans Ind Infor 14(10):4334–4342
28. Makhmudkhujaev F, Wadud MAA, Iqbal MTB, Ryu B, Chae OL (2019) Facial expression recognition with local prominent directional pattern. Sig Proc Img Comm 74:1–12
29. Jia Y, Liu H, Hou J, Kwong S, Zhang Q (2021) SAML via dual-channel information recovery. IEEE Trans Cyb 1–12
30. Ojala T, Pietikainen M, Harwood D (1996) A comparative study of texture measures with classification based on featured distributions. Patt Recog 29(1):51–59
31. Ojansivu V, Heikkila J (2008) Blur insensitive texture classification using LPQ. In: ICISP, pp 236–243. https://link.springer.com/conference/icisp
32. Ma Z, Li B (2020) A DDoS attack detection method based on SVM and K-nearest neighbour in SDN environment. Int J Comp Sci Eng 23(3):224–234
33. http://www.cl.cam.ac.uk/research/dtg/attarchive/facedatabase.html
34. http://www.anefian.com/research/face_reco.htm
35. http://www.kasrl.org/jaffe.html
36. http://cswww.essex.ac.uk/mv/allfaces/faces94.html
37. Karanwal S, Diwakar M (2021) Neighborhood and center difference based LBP for face recognition. Patt Anal App
38. Fu Z, Zhao Y, Chang D, Wang Y (2021) A hierarchical weighted low-rank representation for image clustering and classification. Patt Recog 112:1–12
39. Huang G (2011) $(2D)^2$ PCA plus $(2D)^2$ LDA: a new feature extraction for face recognition. In: ICDIP, pp 800934-1–800934-4
40. Liang K, Yang X, Xu Y, Wang R, Nie F (2021) Ratio sum formula for dimensionality reduction. Mult Too App 80:4367–4382
41. Huang M, Shao G, Wang K, Liu T, Lu H (2021) Discriminative locality-constrained sparse representation for robust face recognition. J Phys Conf Ser 1780:1–6

42. Karanwal S, Diwakar M (2021) Two novel color local descriptors for face recognition. Opt 226(2):1–15
43. Wang J, Liu Z, Lu W, Zhang K (2020) Discriminative label relaxed regression with adaptive graph learning. Comp Intell Neuro 1–10
44. Li ZM, Li WJ, Wang J (2020) SAPS for face recognition. Int J Patt Recog Art Intell 34(2):1–19
45. Mandal B, Wang Z, Li L, Kassim AA (2016) Performance evaluation of local descriptors and distance measures on benchmarks and first-person-view videos for face identification. Neuroco 184:107–116
46. Zafar A, Nawaz R, Iqbal J (2013) Face recognition with expression variation via robust NCC. In: ICET, pp 1–5
47. Divya A, Raja KB, Venugopal KR (2019) Sorting pixels based face recognition using DWT and statistical features. In: ICISPC, pp 150–154
48. Patil HY, Kothari AG, Bhurchandi KM (2014) Expression invariant face recognition using CT. In: ICIPTTP, pp 1–5
49. Fernandes S, Bala J (2017) A comparative study on various state of the art face recognition techniques under varying facial expressions. IAJIT 14(2):254–259
50. Chang CY, Chang XW, Hsieh CY (2011) Applications of block LDA for face recognition. 2(3):259–269
51. Tunc B, Dagli V, Gokmen M (2012) Class dependent factor analysis and its application to face recognition. Patt Recog 45(12):4092–4102
52. Kautkar S, Koche R, Keskar T, Pande A, Rane M, Atkinson GA (2010) Face recognition based on RT. Procedia Comput Sci 2:35–43
53. Chu Y, Lin H, Yang L, Diao Y, Zhang D, Zhang S, Fan X, Shen C, Xu B, Yan D (2020) DGLP-ELM for image classification. Neuroco 387:13–21
54. Ayyavoo T, Suseela JJ (2018) Illumination pre-processing method for face recognition using 2D DWT and CLAHE. IET Biometrics 7(4):380–390
55. Schaffer L, Kincses Z, Pletl S (2017) FPGA-based low-cost real-time face recognition. In: ISISI, pp 35–38
56. Subban R, Mankame D, Nayeem S, Pasupathi P, Muthukumar S (2014) Genetic algorithm based human face recognition. In: Proceedings of ICAICNN

High-Performance Computing with Artificial Intelligence Benefits for the Civilization Impacted by the COVID-19 Pandemic

B. N. Chandrashekhar and H. A. Sanjay

Abstract A deadly virus that creates diseases in all living things on earth [animals/humans] is known as coronavirus. According to World Health Organization, it is named as COVID-19 (CoV-19). Areas like bioinformatics, epidemiology, and molecular modeling did extensive research and came up with strategies to fight against COVID-19. In this repute, the main aim of this chapter is to propose how powerful high-performance computing (HPC) with artificial intelligence (AI) techniques benefitted to the society impacted by the COVID-19 pandemic. The COVID-19 helps to aggregate a variety of geographically distributed computational resources, such as supercomputers, computer clusters, data sources, storage systems, scientific instruments, and presents them as a unified, dependable resource for solving large-scale computations, and data-intensive computing applications from HPC to help COVID-19 scientists execute complex computational study to assist with battling the infection.

1 Introduction

A huge arrangement of infections known as COVID is accessible that cause illness in a moderate to outrageous style among individuals and animals. Different such infections are found to make illnesses related to breathing conflicting from the typical infection to more fundamental sicknesses. CoV-19 has a spot with the gathering of coronaviruses that has been found lately. In Wuhan, China, it impoverished out after December 2019. As of now, it is causing unimaginable ruin all through the globe.

B. N. Chandrashekhar (✉)
Department of Information Science & Engineering, Nitte Meenakshi Institute of Technology, Yelahanka, Bangalore, Karnataka 560064, India
e-mail: chandrashekar.bn@nmit.ac.in

H. A. Sanjay
Department of Information Science & Engineering, M.S. Ramaiah Institute of Technology, Bengaluru, Karnataka 560054, India
e-mail: sanjay.ha@msrit.edu

© The Author(s), under exclusive license to Springer Nature Singapore Pte Ltd. 2023
M. A. Chaurasia and C.-F. Juang (eds.), *Emerging IT/ICT and AI Technologies Affecting Society*, Lecture Notes in Networks and Systems 478,
https://doi.org/10.1007/978-981-19-2940-3_7

Depletion, fever, and dry hack are a part of the ordinary aftereffects achieved by CoV-19 that are as of now found [1].

Also, a portion of the less incessant CoV-19 indications contains torments and throbs, blockage in the nose, migraine, throat torment, not having the option to smell or taste skin rashes, and loose bowels alongside shading misfortune in toes and fingers. A portion of these indications become hard to identify since they might be exceptionally gentle and truly begin to become serious throughout a process of things working out.

One more basic issue in identifying COVID-19 is that sure individuals might show exceptionally gentle side effects or nothing. Individuals can fill in as transporters of this infection and prompted its uncontrolled feast. People who not showed any indications are known as asymptotic [2]. Moreover, the larger part of the people (around 80%) is found to recuperate from CoV-19 without going through any medicines in emergency clinics. Be that as it may, 1 of 5 people who have procured this ailment can turn out to be fundamentally debilitated and foster serious breathing trouble. Such people should go through hospitalization right away. The people who are viewed as in danger of CoV-19 are senior people and people having intrinsic clinical issues, for example, high BP, issues identified with lungs and heart, diabetes, or disease. Yet, a review has uncovered that even a typical individual might become basically ill as a result of CoV-19. In any case, the possibilities are viewed as uncommon. Research likewise shows that any individual regardless of old enough/sexual orientation encountering breathing trouble with fever/hack needs to go through clinical treatment very quickly without anything to do. WHO has proposed prescribing CoV-19 impacted people to call the medical services suppliers (HCP) who can give appropriate bearings to the persistent [3, 4]

The spread of COVID-19 has been characterized by four phases liable upon the type of spread and the associated period of time. Every nation practical a few methods to avoid this spread involving lockdown (stay at home), cover mouth and nose with masks, avoiding unnecessary traveling, staying away from enormous get-togethers, cleaning hands utilizing sanitizers, alongside disinfection of areas. Pointless development of residents was confined in certain nations. This has prompted a genuine effect on the economy as little and intermediate-scale businesses unfavorably affected. Presently, the GDP has dropped all throughout the world out of nowhere. These days, the main treatment accessible is confinement of the contaminated individual who is given sure normal medications that can help him/her in recuperation. Since the antibody isn't yet accessible for CoV-19, the passing rate has flooded radically all throughout the planet. The single separation of people has prompted serious wretchedness among touchy people prompting even self-destruction who can give appropriate headings to the patient [4, 5].

In growing nations like India, daily life has been totally affected due to the lockdown. At first, most of the confirmed cases were expected of the people getting back from abroad. Subsequently, it was started spreading in the local area. The standard stage of the sick people in India is 39. Sensibly, people in age among 21 and 40 are viewed as contaminated more. Henceforth, an optimized forecast of CoV-19 is a lot of fundamental for decreasing the blowout of this twenty-first-century epidemic [6, 7].

In order to process and compute the huge amount of real-time data generated during era of CoV-19 was not possible with the single workstations and limited resources that was a big hurdle. To address this issue, scientist and researchers utilized power of HPC to process huge amount real-time data CoV-19 data into simulations. Using HPC-based simulations, it became feasible for the scientists and researchers to study and improve the new strategies to precautionary measures. By merging modeling and simulation capacities with new methods in artificial intelligence (AI) and machine learning, these reproductions are currently turning out to be considerably more precise. That is the reason through our joint efforts with overall driving examination habitats that are utilizing our HPC and artificial intelligence solutions.

The remaining section is arranged as follows. Section 2 reflects on the HPC exploited for COVID-19. Section 3 depicts the heterogeneous clusters utilization for data gathering and dispensation in COVID-19. Section 4 describes some AI techniques for CoV-19 using HPC. The chapter ends with conclusion.

2 HPC for AI

There will be heavy demand for high-end configuration-based hardware in the HPC market. In 2000, the most impressive supercomputer bragged max execution 3.2 teraflops (TF), while it is 187.7 petaflops (PF), an increment in core by multiple times. Furthermore, the current trends of HPC demand are to provide sufficient memory ability to deal with data is a serious responsibility. Henceforth, having high-end processors, memory parts, and different parts in HPC frameworks [8] is a crucial challenge. HPC administrations are additionally getting a recent trend. Organizations, like HPE, have referenced zeroing in on giving HPC administrations like warning administrations and expert administrations for combination and establishment. In order to process the real-time COVID-19 applications in a reliable, quickly and efficiently HPC is utilized. Newer workloads such as COVID-19 applications, storm detection, machine learning, weather modeling are being executed on HPC systems. Due to this, HPC infrastructure contributes a lot to IT market and day-to-day excellence of life.

Scientists and researchers used HPC power to decide how the investigated object might act in genuine real-life circumstances by using a simulations environment. Due to this, scientists and researcher while developing more subjective and analogous items will save cost and execution time in research and development. This would be an exceptionally overwhelming assignment to manage without model simulation, which for the most part comprises large frameworks, numerous mathematical details, and complex physical science. This section examines the famous HPC framework applied for COVID-19, and Fig. 1 shows the need of HPC and AI to deal with the spreading of the infection.

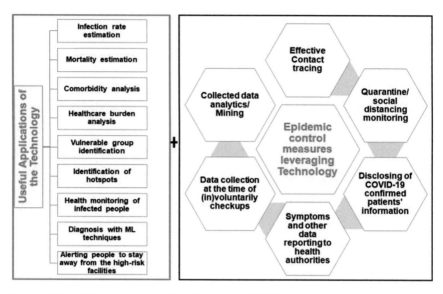

Fig. 1 HPC and AI techniques-based applications for COVID-19

Utilizing valuable applications and advanced technologies through the HPC with AI assists society with controlling the pandemic circumstance. Due to privacy issues [6], use of digital solutions is less even though advanced technologies in the era of COIVD-19 are available. Moreover, a few nations have embraced digital solutions by disparaging security issues, and huge outcomes have been acquired as far as infection prevention. Additionally, individuals' nervousness and stress stay higher in such nations. For example, in order to contact the infected people in Korea they were adopted CCTV, arbitrary calls and versatile signs, etc. Likewise, to control the infection in China they were utilized decision support systems (DSS) networks. In certain nations, mobile applications were utilized to track down strongly infected patients. Through a definite combination of the writing and examination of different accessible applications and reports, measurements of ten nations as far as higher reception of AI/ML throughout a period of close to two years are shown Fig. 2. Besides, bits of

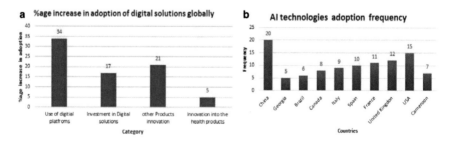

Fig. 2 Statistics **a** digital solution and **b** the adoption of AI around the world

knowledge into the interest in innovation by organizations all throughout the planet to battle the emergency brought about by the continuous pandemic [7].

Until now many literature works were carried out in the COVID-19 situation. However, these literatures have for the most part centered around the overall uses of the innovation in the academic sector, industry regions, the travel industry area, and so on, in any case, a substantial framework of HPC and AI/ML utilization in the CoV-19 setting, and their value in the alleviation and managing the epidemic, is not discussed in past. To address this challenges trend utilized ML and HPC thinking about plague control measures, information life cycle in plague frameworks, and multipurpose use of ML and HPC. All the more explicitly, we highlight the job, need, and utility of ML and HPC in the CoV-19 period. To this end, a reduced outline of the devotion of ML and HPC procedures in the time of CoV-19 is shown practically. Through this succinct viewpoint, we desire to give a strong establishment to future examination in the COVID-19 section.

Until now, many literatures have given inclusion of AI/ML methods in the era of CoV-19. These literatures have principally concentrated around the overall solving the dangers of CoV-19 requirements in these areas, for example, bioinformatics, the study of disease transmission, and atomic displaying, which require stages with the colossal computational capacity to determine complex intelligent issues and inter-face with enormous datasets in more restricted time intervals. To facilitate the US Division of Energy and IBM shaped the COVID-19 events, White House Office of Science and Technology Policy utilized the HPC Association comprising of the central government [9]. A portion of the affiliation accomplices incorporate Google Cloud, Microsoft, Amazon Web Services, Hewlett Packard Enterprise, NASA and NVIDIA, Scientists can present their recommendations to the association to get to little groups and the absolute biggest supercomputers on the planet. Genome sequencing utilized the HPC to know the precise natural construction of the infection and demonstrating different medicines. Computer-based intelligence-driven HPC stages can be utilized to find fitting enemy of viral medications and immunizations for CoV-19. The documents [9, 10] have talked about the utilization of HPC to handle COVID-19.

As a brilliant illustration, headways in computer vision (CV) keep on driving numerous advanced AI and ML frameworks. CV is speeding up pretty much every area in the business empowering associations to change the manner in which machines and business frameworks work—producing, independent driving, and medical care. Practically all CV frameworks have moved on from conventional rule-based programming worldview to huge scope, information-driven ML worldview. Furthermore, subsequently, GPU-based equipment assumes a basic part in guaran-teeing top-notch expectations and characterization by aiding crunching monstrous measures of preparing information (frequently in the scope of petabytes).

3 Data Gathering and Dispensation in COVID-19 from Heterogeneous Clusters

In a current trend, to help the countries in innumerable ways, with human conduct examination, suggest, strategy development, and working on the expectations for everyday comforts of individuals, to give some examples. The most recent advancements are truly adept at finding bits of knowledge from enormous scope information [11]. We highlight the data gathered and handled in the COVID-19 time and compare analytics presented by them in Fig. 3.

Aside from these strategies, big analytics on existing information, like products information, have significant advantages in isolating the weak population [9].

Regardless of the far-reaching investigation given in Fig. 3, we give the accompanying examination works that can be taken advantage of to check the sickness spread.

- Tracking down the most uncovered individuals through area investigation (e.g., temples/cinema hall).
- Removal of individuals where actual existence used for a task is compulsory.
- Collecting necessary documents from the people were doing a job in high-hazard zones for identification purpose.
- Replacing the people whenever they are about infected by virus.
- Investigating contact among individuals and anticipating the danger scored dependent on multi-criteria.
- Detecting people having close connection in a security-saving way.
- Investigating the elements of the illness regarding age gatherings.
- Identifying deadly blends of the sicknesses to bring down mortality possibilities

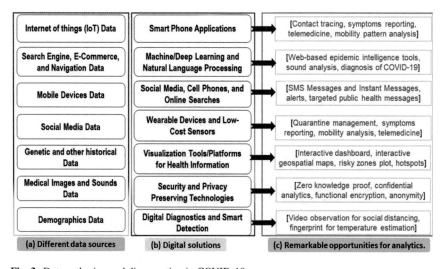

Fig. 3 Data gathering and dispensation in COVID-19

- Providing the likelihood of disease score dependent on numerous boundaries.
- Finding the relationship of the diseases and economics to all the almost certain helper people to work on their prosperity.
- Evaluating the relationship among ICU and signs to move for resource composing.

By using HPC/AI above-mentioned activities will be accomplished with adequate precision to further develop medical services. By using the above capacities, the weight of medical care can be altogether diminished. Moreover, robotized analysis and remedies can be given simplicity.

4   AI Techniques for CoV-19 Using HPC

The consolidated use of ML and HPC can make ready for further developed medical care and the compelling moderation and manage CoV-19. The methods ML and HPC have exhibited adequacy in monitoring sickness. Power full HPC is exceptionally useful in handling large-scale information about people. In this chapter, we present the task of AI/ML and HPC in the era of CoV-19 period.

4.1   ML/AI and HPC Techniques

The data life cycle is a huge construction square of information-based applications. It has seven phases, and every stage has uncommon abilities to be performed on the data. Table 1 shows data life cycle framework.

The different stages given in Table 1 are conventional and taken for COVID-19 applications/structures. The usage of ML and HPC in the recently referenced times of the data life cycle in Fig. 4. AI/ML techniques are generally used for extricating needed data from information. Alternately, HPC strategies can be significant for taking care of temporary results or taking care of tremendous degree data.

Table 1   Brief framework of the DLC to battle the COVID-19

Stage number	Stage name	Roles
1	Gathering	Info is gathered from respective personalities
2	Loading	Info is kept in catalogues for future practice
3	Processing	Info is prepared for future practice
4	Analyze	Info is managed with AI/ML and HPC for understandings
5	Custom	Info is utilized for preferred conclusions
6	Scattering	Info is collective for study resolves
7	Documentation	Info is documented for required purpose

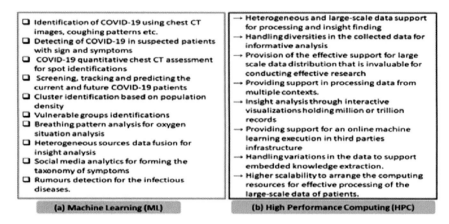

Fig. 4 COVID-19 data analyzation using ML and HPC

Information has turned into a critical piece of applications. They open the information bounded in the basic information. The most recent investigation strategies with hyperparameter tuning functionalities have demonstrated fruitful in numerous applications. We present the job of ML and HPC in the investigation of the COVID-19 period in Fig. 5. ML strategies can aid information extraction. For example, they can be generally utilized in distinguishing patients with and without COVID-19.

ML strategies can likewise be utilized to rank the comorbidities that can lead to deaths, ICU admission, oxygen need, and so on. They can likewise be utilized to isolate people that can without much of a stretch become focuses of COVID-19 because of their work nature or cleanliness rehearses. What's more, they can aid patterns investigation, security safeguarding examination, and information dissemination across associations.

Interestingly, HPC strategies have greater utility according to an organization.

Perspective; for instance, local area bunching includes enormous scope information handling. In such a manner, HPC methods are helpful to stack, process, hold middle outcomes, and convey results to invested individuals. In this way, both these procedures assume a basic part in the COVID-19 time. At times, the two methods are together used to perform applicable errands. The possible conversation for more helpful highlights of the two advancements is still up in the air from the most recent investigations [12, 13].

4.2 Multipurpose Applications with ML and HPC Techniques

ML and HPC strategies not only used for special purpose [diagnostic and analytics] but even used for multipurpose also, sum up the potential multipurpose uses of AI/ML and HPC as follows.

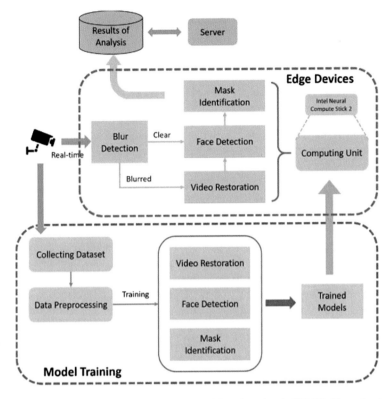

Fig. 5 Architecture of mask detection using VR and face detection in COVID-19 pandemic

- Educating about COVID-19 through sentiment analysis and suggesting wellbeing tips.
- Providing clinical measurements for COVID-19 side effects since genuine incident information.
- Extended the area of medical clinics and facilities.
- Growing information effectiveness by suggesting/division consequences on cells.
- Extracting and melding multi-style information for the exact investigation.
- Combining multimode results to eradicate individuals' nervousness on outcomes' validity.
- Analyzing and sharing human direct for feeble associations unmistakable evidence locally.
- Analyzing the innovation suggestions on society through pertinent elements.

To help society in more than one way [13–15], both HPC and ML have different general applications [16]. However, both HPC and ML were utilized for estimating and investigating the information of numerous establishments for better examination of the COVID-19 pandemic.

4.3 HPC for AI in Edge Computing

HPC is required to improve the efficiency of edge computing and that has been advanced and furthermore became famous because of its capacity of low organization inertness, what's more, security assurance with minimal expense. In this chapter, we have considered the edge computing framework for mask detection to recognize mask-wearing progressively during COVID-19. To begin with, in light of genuine video information of transport driver checking, we use obscure recognition techniques and VR to further develop identification precision for the video information with obscure issues from minimal expense cameras, which could be viewed as a piece of the video preprocessing [17]. Then, at that point, the face finder is prepared and checked by the public informational indexes and transport drive observing informational index. Subsequent to getting and trimming face regions of the transport drive observing informational index, the detection of mask can be prepared.

Figure 5 mostly incorporates two models, i.e., training and live streaming investigation. By using VR and face detection models, we can detect the face masks in the real-time video. Subsequent to gathering and preprocessing our transport drive checking informational gathering and the collecting public information. This can be done with edge devices. Along these lines, in the system, observing information will be sent to elite execution hardware for helper model preparing, and afterward communicated to the edge gear for constant assessment. To be explicit, in the part of continuous video investigation, we will execute the video obscure identification with the Laplacian administrator to decide if VR is required, which can diminish the high computational overhead brought by VR. Then, at that point, the ongoing video information is inputted on the ensuing models to get the consequences of discovery also distinguishing proof and is sent back to show to the board.

4.3.1 Facemask Detection Using Edge Computing for COVID Patients

Considering the architecture of mask detection system of our face detection is done by using HPC with edge computing devices [18], the framework is displayed in Fig. 6. The frameworks are segregated into two phases exact and effective on the edge gadgets with the contracting of the spatial size of the information, advancing the responsive fields. In pragmatic application, the casings of video as info pictures have a high goal. Accordingly, to diminish the computational cost, we really want to speed up downsampling as soon as could be expected.

It gives various branches unique portions, which are distinctive responsive fields. Then, at that point, through the few convolutional layers, the last two down samplings are finished. Like SSD, multiscale include maps are created for identification.

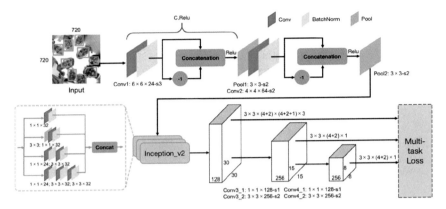

Fig. 6 Architecture of mask detection using edge computing in COVID-19 pandemic

5 Conclusions and Future

The CoV-19 pandemic had issues severely affecting the health, the economy of a few nations internationally. The most recent advances in the battle are in contradiction of the unexpected test of CoV-19. In particular, presented a framework of the pestilence includes ML and HPC procedures to attend mankind in a viable manner by relieving the epidemic thru innovation. Depicted the huge number of heterogeneous kinds of information gathered in the COVID-19 time and the surprising chances proposed when investigated with cutting-edge AI/ML and HPC strategies. Featured the adequacy of AI/ML and HPC procedures in the alleviation and management of the CoV-19 epidemic concluded their special use in an assortment of uses.

We gave possible exploration headings and difficulties in the specialized reception of ML and HPC because of information challenges such as inaccessibility, sparseness, uncertain, information harming, and so on. We acknowledge that this survey gives a solid foundation to future assessments around here tantamount to pandemic components and related data. The HPC and AI/ML methods were utilized in the CoV-19 period through separate information in a circle. Notwithstanding the key commitments given over, an energetic space of examination is creating adaptable anonymization strategies that can without much of a stretch be tuned, in light of the first information qualities or conditions to encourage information reusability. What's more, utilizing ML and HPC methods to deal with huge scope information in a protection-saving way is of fundamental significance for future endeavors.

Research headings and difficulties in the specialized reception of ML and HPC because of information problematic such as inaccessibility, sparseness, uncertainty, information harming, and so on We accept that this remarkable review gives a strong establishment to future investigations around here according to pandemic elements and comparing information.

References

1. Chamola V, Hassija V, Gupta V, Guizani M (2020) A comprehensive review of the COVID-19 pandemic and the role of IoT, drones, AI, blockchain, and 5G in managing its impact. IEEE Access 8:90225–90265
2. Saxena SK, Kumar S, Maurya VK, Sharma R, Dandu HR, Bhatt MLB (2020) Current insight into the novel coronavirus disease 2019 (COVID-19). In: Saxena S (eds) Coronavirus disease 2019 (COVID-19). Medical virology: from pathogenesis to disease control, pp 1–8
3. Saxena SK, Kumar S, Maurya VK, Sharma R, Dandu HR, Bhatt M (2020) Current insight into the novel coronavirus disease 2019 (COVID-19). In: Coronavirus disease 2019 (COVID-19): epidemiology, pathogenesis, diagnosis, and therapeutics, pp 1–8
4. World Health Organization (2020). https://www.who.int/emergencies/diseases/novel-corona virus-2019/question-and-answers-hub/q-a-detail/q-a-coronaviruses. Accessed 5 July 2020
5. Kaushik S, Kaushik S, Sharma Y, Kumar R, Yadav JP (2020) The Indian perspective of COVID-19 outbreak. Virusdisease 31(2):1–8
6. Sujath R, Chatterjee JM, Hassanien AE (2020) A machine learning forecasting model for COVID-19 pandemic in India. Stoch Env Res Risk Assess 34:959–972
7. Keller JM, Liu D, Fogel DB (2016) Introduction to computational intelligence. In: Fundamentals of computational intelligence: neural networks, fuzzy systems, and evolutionary computation. Wiley-IEEE, pp 1–4
8. Chandrashekhar BN, Sanjay HA, Lakshmi H (2020) Prediction model for scheduling an irregular graph algorithms on CPU GPU hybrid cluster framework. In: Fifth IEEE international conference on inventive computation technologies (ICICT-2020), India, on 26–28 Feb 2020. ISBN:978-1-7281-4685-0/20/2020 IEEE
9. COVID-19 HPC Consortium (2020) The COVID-19 high performance computing consortium. Retrieved from https://covid19-hpcconsortium.org/. Accessed on 10 Apr 2020
10. Smith M, Smith JC (2020) Repurposing therapeutics for COVID-19: supercomputer-based docking to the SARS-CoV-2 viral spike protein and viral spike protein-human ACE2 interface. ChemRxiv. Preprint. https://doi.org/10.26434/chemrxiv.11871402.v3
11. Wieringa J, Kannan PK, Ma X, Reutterer T, Risselada H, Skiera B (2021) Data analytics in a privacy-concerned world. J Bus Res 122:915–925
12. Swapnarekha H, Behera HS, Nayak J, Naik B (2020) Role of intelligent computing in COVID-19 prognosis: a state-of-the-art review. Chaos Solitons Fractals 138:109947
13. Zivkovic M, Bacanin N, Venkatachalam K, Nayyar A, Djordjevic A, Strumberger I, Al-Turjman F (2021) COVID-19 cases prediction by using hybrid machine learning and beetle antennae search approach. Sustain Cities Soc 66:102669
14. Arora A, Chakraborty P, Bhatia MPS (2021) Problematic use of digital technologies and its impact on mental health during COVID-19 pandemic: assessment using machine learning. In: Emerging technologies during the era of COVID-19 pandemic, vol 348, pp 197–221
15. Chandrashekhar BN, Sanjay HA (2020) Prediction model of an HPC application on CPU-GPU cluster using machine learning techniques. In: 2nd IEEE Scopus international conference on innovative mechanisms for industry applications (ICIMIA), India. IEE, , 5–7 Mar 2020. https://doi.org/10.1109/ICIMIA48430.2020.9074866
16. Majeed A, Lee S (2021) Applications of machine learning and high-performance computing in the era of COVID-19. Appl Syst Innov 4:40. https://doi.org/10.3390/asi4030040
17. Wang X, Ning Z, Guo S (Feb 2021) Multi-agent imitation learning for pervasive edge computing: a decentralized computation offloading algorithm. IEEE Trans Parallel Distrib Syst 32(2):411–425
18. Zhang S, Zhu X, Lei Z, Shi H, Wang X, Li SZ (2017) Faceboxes: a CPU real-time face detector with high accuracy. In: Proceedings of IEEE international joint conference on biometrics (IJCB), pp 1–9

Quality of Life Estimation Using a Convolutional Neural Network Technique

B. A. Manjunatha and K. Aditya Shastry

Abstract Life quality is a well-being metric that considers physical and mental health as well as social activities. A novel method for assessing a user's life quality score is presented, which employs a deep learning architecture. A convolutional neural network and a support vector machine (SVM) were coupled to develop this solution for multimodal data. Three tests were conducted to determine the accuracy of the estimation method. In the research on living standard estimation, eight life factors were chosen to make up life quality. Next, the regression of life quality score experiment gives scores of each parameter where the distribution of the difference between the actual and predicted outcomes is known as error of neural network. Except for the mental health category, these findings suggest that the attributes required for life quality assessment may be retrieved from data. One of the reasons why estimating the mental health scale was challenging might be because the learning framework was unable to extract an adequate feature for estimation. As a result, we used eye movement to determine mental health. It was shown that estimate is achievable, and the suggested approach employing different modes of activity data proved efficacy for estimating makeup life quality, as well as for extracting high-dimensional information about a human's life quality, such as their pleasure level toward social activities. Lastly, recommendations and arguments concerning the likely behavior of the estimation. Discussions on improving comprehensive life quality by applying this quality of life estimation approach to robots in the elderly care business.

Keywords Quality of life · Convolutional neural network · Robots · Mental health · Support vector machine

B. A. Manjunatha (✉) · K. Aditya Shastry
Nitte Meenakshi Institute of Technology, Bangaluru, India
e-mail: Manjunatha.ba@nmit.ac.in

K. Aditya Shastry
e-mail: Adityashastry.k@nmit.ac.in

M. A. Chaurasia and C.-F. Juang (eds.), *Emerging IT/ICT and AI Technologies Affecting Society*, Lecture Notes in Networks and Systems 478,
https://doi.org/10.1007/978-981-19-2940-3_8

1   Introduction

1.1   Scope and Purpose

This global population of people aged 60 and more was expected to be 962 million in 2017, more than double what it was in 1980 [1]. According to statistics, one out of every five women and one out of every 10 males live alone. It is difficult to avoid mishaps like sickness, fates, and damage among senior people. Do not have frequent healthcare facilities due to issues like a lack of medical professionals. Furthermore, taking prompt action and suitable actions in the event of an occurrence may be challenging. As a result, both local governments and the commercial sector have created monitoring systems for older people in recent years. This tendency may be seen in the introduction of monitoring robots. Human–agent interaction (HAI) using robots for the well-being of old people necessitates estimating the subjects' well-being. Life quality is an indicator that may be used not only to measure physical discomfort, but also to manage mental and social activities holistically. Many research in the field of healthcare strive to increase life quality nonetheless, there are a number of issues with life quality. It is frequently stated that introducing nursing routines that are robots enhance excellence life quality, which cannot be measured numerically. Improvements in quality of life, for example, might be attributed to increased frequency and/or duration of conversations or improvements in the instrument's operability. Despite the fact that there is a quantitative method for assessing quality of life, this is the case. Despite its relevance, the relationship between HAI and each component of quality of life has received little attention thus far.

As a result, the main objective of this research is to develop a quality of life estimation system based on natural HAI. The assumption that the HAI specification document contains: initial step, personal satisfaction assessment should be performed by means of noninvasive and normal questioning, and second, information on personal satisfaction should be extricated and afterward used to adjust according to agent behavior dependent on the human's personal satisfaction state. The SF-36v2 [2] quality of life indicator was employed in this investigation. This measure distinguishes between actual working (AW), job physical (JP), body torment (BT), general well-being (GW), vitality essentialness (VE), social working (SW), job enthusiastic (JE), and psychological well-being (PW). Each of the 36 items on the SF-36v2 questionnaire can be replied by picking one of three to five replies. As shown in Fig. 1, the replies to the survey are taken care of into a scoring system, which produces the score for each scale. The scores may all be calculated using the data from these measurements (RSCS) "role/social component summary score." It is also possible to compare the generated life quality score to the national average.

To incorporate normal language collaboration in HAI, scientists make a correspondence framework with a conversational specialist that utilizations voice data and incorporate morphological, Chat API, and Open JTalk examination. The specialist takes video and sound of the client's chest area because of the client's answer. The information obtained through natural language conversion is used to estimate quality

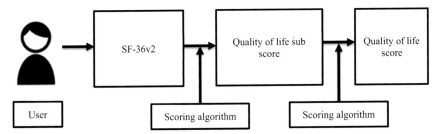

Fig. 1 Method of measuring quality of life using SF-36v2

of life. In order to measure life quality, deep learning technique is applied to the convolutional neural network (CNN) [3], videos data use as an input. The data mining technique such as SVM, audios data use as an input. The leave-one-out approach [4] is used to assess the model's accuracy. The video, audio data samples are classified into two types: training data and testing data. One sample is chosen from all of the data to be used as test data. In the initial trial, a score of life estimate experiment on the eight subscales, and the second is an estimation of each quality of life questionnaire response. Each experiment is subjected to an accuracy check. These two experiments reveal that assessing mental health, one of the scales that make up quality of life, is difficult, also, the explanations behind this are on the grounds that factors connected to eye movement are not recovered by the system. As a consequence, focused on eye movement in the last trial and improved the CNN architecture to evaluate mental wellness.

In this study, the authors propose that natural language dialogue be utilized to assess life quality as a technique for understanding the client through collaboration. Accepting that these outcomes are adequate for passing judgment on the improvement of associations and conduct choice dependent on client understanding, conversations on significant realities, for example, offering suitable help dependent on the user's life quality, will start interestingly from this assessment result. This study is unique in that we use quality of service estimation rather than emotion assessment to estimate human state. When compared to emotion detection by facial expression analysis, quality of service is a complete list that arrangements with a scope of components, and in light of the fact that the many scales can be measured, quality of life can deliver more data. Furthermore, because we perform experiments with real-world social activities in mind, data is not acquired consciously rather, data is gathered through natural events like discussions with the specialist, making our strategy more down to earth. As a result, our concept for a realistic quality of life estimation method is also novel. In the domain of senior consideration, various watching robots have been assembled, tried, and utilized. Be that as it may, just a subjective improvement in personal satisfaction has been accounted for. Human-agent interaction relates to other quality of life measures. It becomes feasible to discuss the relationship between individual scales and the association of testing agents. It serves as a starting point for discussions on improving overall quality of life by applying this quality of life estimation approach to robots in the elderly care business.

This paper is organized as follows. Section 2 presents our related work, Sect. 3 discusses the proposed approach based on convolutional neural network, Sect. 4 results and discussion, Sect. 5 concludes and suggests future work.

2 Related Work

- This section discusses the relevant works on quality of life estimation using neural network can precisely measure the human's status. Yao and Fei-Fei [5] introduced a shared setting model with depict articles and human postures in human–object association exercises and exhibited that their model was fruitful in identifying objects, assessing human postures, and sorting human–object connection exercises in the field of HAI. Cao et al. [6] proposed utilizing part affinity fields, and direction of appendages in the information picture, to estimate ongoing multi-individual 2D human posture assessment.
- There are vision, audio, text, and EEG-based [7–11], several multimodal techniques to emotion estimate [12]. Numerous vision-put together exploration center with respect to look investigation on the grounds that the face assumes such a crucial part in both passionate demeanor and discernment during correspondence. Face analysis is used to estimate emotions largely using two approaches. 2D spatiotemporal face characteristics are used in one technique, while 3D spatiotemporal facial features are used in the other. Geometric feature-based techniques for 2D spatiotemporal face features have been proposed by Feris et al. [13], Pantic and Bartlett [14], and Kotsia and Pitas [15]. Ren and Huang [16] proposed a method for extracting feature points on the face to automatically learn facial emotions. For reading emotions resulting from context-aware facial expressions, Gorbunov et al. [17] developed a genetic algorithm (GA) method. 3D expression data was used by Vieira et al. [18] and Yin et al. [19] for face appearance detection utilizing 3D spatiotemporal facial features. Bourdev et al. [3] recently discovered that 3D ConvNets are better at learning spatiotemporal facial features than 2D ConvNets and suggested a new form of 3D convolutional network called convolutional 3D ConvNets (C3D). C3D is a method for doing three-dimensional convolution. Lu et al. [20] introduced a C3D and recurrent neural network-based video-based emotion identification system (RNN).
- Scherer [21] and Chan et al. [22] proposed the speech synthesis that examined the effect of feeling. The proposed method can analyze the particular emotion, pace of the audio, and pitch in change of moods. The connection among feeling and verbal dis-fluencies like quiet stops was demonstrated by Vidrascu and Devillers [23].
- Furthermore, it is critical for agents who engage with users to evaluate emotions that arise spontaneously during conversation rather than emotions that are stated consciously. Most previous research on facial expression recognition, according to Tian et al. [24], depends on arranged and habitually overstated facial emotions. Cohn and Schmidt [25] looked on the differences between natural and deliberate

facial emotions. They estimate utilizing spontaneous manifestations of emotion in research.

- Bartlett et al. [26] and Ioannou et al. [27] are two further types of person-to-person communication focused at gathering audio and video data. In HAI, however, robots must communicate independently. As a result, this research simulates an autonomous communication scenario by simulating a whole human–agent discussion using a chat API.

3 Proposed Approach

In this study, we propose a novel method for evaluating quality of life. As a result, before estimating quality of life, we create a discussion agent utilizing a characteristic language exchange framework and perform an interpersonal experiment to collect data for machine learning. The agent records a procedure in which a person conducts a natural language discussion with the agent and results in the elements gathered from the chest area movements as an .avi document and the sound information as a .wav record in the experiment. In the end, we want agents to be able to estimate the link between the aforementioned information and the eight variables that make up quality life. The responses from the life of quality indicator SF-36v2, which are received immediately before the user engages with the natural language chat bot, are used as learning data. In quality of life measurement, a scoring method that uses the Japanese national standard value can be used to generate the score. The entire data collection configuration. The interpersonal experiment and natural language discussion system, both of which are important components of the proposed technique, are described in depth in the following paragraphs. Figure 2 depicts a flowchart of our research's general frame design.

3.1 Natural Language System

On the computer, we first develop a conversational agent. An audio component captures the user's voice an audio-recognition component translates the recorded audio into a character a conversations component provides appropriate responses for the user and a voice output component outputs the response through speech synthesis.

3.1.1 Input Voice Data

The voice input component's recording time is set to start toward the finish of the agent's discourse and finish toward the finish of the client's discourse, assuming that the agent and the user have a one-to-one interaction. As a result, the recorded content contains entirely of the user's natural conversation speech. As a result, the moment

Fig. 2 Designing of a frame

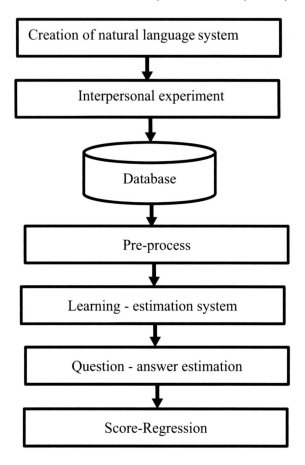

of utterance as well as the stillness at the time of utterance may be recorded. The tone of voice derived from regularity analysis of the speech data supplied here, as well as the tone of voice feature for silence in speech analysis, can both be used.

3.1.2 Voice Recognition Part

For speech recognition, the Tata telecommunications developers provide a voice recognition API [28]. Using a large amount of audio data and a future coming voice recognition algorithm, a high rate of speech recognition may be achieved. The user's speech is converted into text in real time in this part.

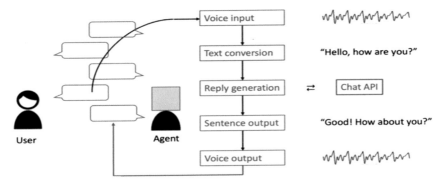

Fig. 3 Conversation agent

3.1.3 Conversation Part

The agent must not only maintain a continuous interaction with the user, but also respond appropriately to the user's query, but also continue the discussion. As a result, we've added a chat dialogue API [28] to the agent this time, allowing us to incorporate the most up-to-date information in the discussion.

3.1.4 Sound Output Part

A technique shouting out the agent's replies is utilized to provide interaction comparable to that of a conversation between persons. In addition to deliver the character's sequence as voice, it is necessary to synthesize the speech. The voice output is created using Open JTalk, a hidden Markov chain text-to-speech synthesis technique. Based on certain Japanese words, this technology can modify the voice's quality and pitch at will. We use an Open JTalk-based framework to produce chat-dialogue-API responses to user utterances in this study. The detailed integration of the conversational agent created in this work is shown in Fig. 3.

3.2 Experiment with Interpersonal Relationships

3.2.1 Setting

The experiment includes 14 individuals as participants. After obtaining prior details about the experiment and verifying their free will agreement, all subjects participate. To start the trial, subsequent to reacting to SF-36v2, the member sits before the PC where the normal language visit framework is introduced. At the point when we execute the normal language cooperation framework, the specialist welcomes the subject with a Hello. The subject and the agent then take turns speaking. What do you have planned for today? After this, I'm going to have lunch with a buddy. There are no constraints on the substance of the discussion or the overall conversation length during the experiment, with the exception that the number of talks must be 30 or more, with each chat being restricted to around 10 s.

3.2.2   Collecting Data

The agent collects data from the subject through free contact with the user. We expect to be able to notice the changes inside the data displayed here on user's face over time, that is impossible to see in a single still shot, as well as learn from the data by collecting and studying time information from video instead of photographs.

The following scoring system is applied to the gathered SF-36v2 response results. Each of the eight scales is given a score. To initiate, add together the scores for selected responses to questions from each level.

The aforementioned technique yields scores for the eight scales, which are referred to as PF, RP, BP, GH, VT, SF, RE, and MH. The SF-36v2 allows for scoring based on national standard values. As a result, the scale of each score is transformed to a 50-point average indicating national standard values, with a 10-point standard deviation. This is accomplished by doing the following calculation. The standard deviation is divided by the mean of the national standard value, which is subtracted from the score values, which ranges from 0 to 100 points.

3.3   Preprocessing of Audio Data

Subjects who submit the same answers are grouped together for each question, and this group is given an answer number. While approaches based on neural networks and methods based on support vector machines (SVM) exist for voice recognition and estimation, established methods do not. As a result, we use SVM to classify speech data collected with Talkbox Scikit. The simplest linear threshold element is used as a neuron model in SVM to generate two-class pattern classification shown in Eq. (1). To optimize the margin, the nonlinear function objects parameters are learned from the training data.

$$\min_{w,\xi} J(w,\xi) = \frac{1}{2} w^T w + \frac{1}{2} C \sum_{i=1}^{n} \xi_i^2$$

subject to

$$y_i(w^T w\varphi(x_i) + b) \geq 1 - \xi_i \text{ and } \xi_i \geq 0 \text{ for } i = 1, \ldots n$$

$$(w^T x_i + b) \geq 1 - \xi_i \text{ for } y_i = +1;$$

$$(w^T x_i + b) \geq -1 + \xi_i \ \text{ for } \ y_i = -1 \tag{1}$$

3.3.1 Input Video Data Preprocessing

In this, utilized convolutional neural network for multidimensional data is found as machine learning algorithm. As follows, it translates the captured video data into a suitable form for the input data. The moving image is divided into number of frames, and it follows that the time sequence is used for input. Furthermore, just the subject's face portion is retrieved when the data is divided into frames to observe the relevance of the user's facial emotions. We utilize a 50 by 50-pixel frame size since number of frame size is related to learning time. Furthermore, the .avi file lengths for all subjects are standardized to 100 frames, and each pixel is assigned three RGB values. After splitting of frames, next is to the detection of a face.

3.3.2 Face Detection

Face detection recognizes several faces in a picture, as well as crucial facial traits such as emotional reaction and whether or not they are wearing headgear. The responsibilities of facial extracting features are common in face detection techniques. The use of machine learning algorithms like fuzzy classifiers [29–31] and neural networks is to create classification models. As a consequence, for each image, we prepare and construct 100, 50, 50, 3 array data as input data. In the learning phase, convolutional 3D is employed.

Convolutional 3D is nothing but three-dimensional convolutional neural network that combines vertical, horizontal, and depth dimensions to turn three-dimensional input data into a three-dimensional convolution. C3D contains knowledge about views, and movements in a video, allowing them to be utilized for a wide range of applications without need to fine-tune the model for each activity. All of the qualities of a good descriptor: It is broad, small, straightforward, and effective.

Among the restricted set of architectures investigated, scientists discover that the 3 * 3 * 3 convolution kernel for all layers makes the most sense. The size of all convolution kernel is d, where d is kernel momentary profundity. Since these convolution layers are totally applied with appropriate padding and stride 1 * 1 * 1, the size of the contribution to result of these layers doesn't shift. All pooling layers, except for the first, are max pooling, with kernel size 2 * 2 * 2 and stride 1 * 1 * 1, advising that the result signal is 8 times less than the input signal. As a result of employing this method, three-dimensional data containing time-series information from the input data is produced as output.

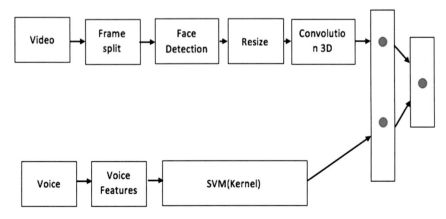

Fig. 4 Learning and estimating system

Fig. 5 Implemented architecture for the learning part, using C3D

3.4 Learning Estimation System

As illustrated in Fig. 4, the learning and estimating system is constructed. After preprocessing the optical and audio input, each anticipated result is integrated into a distinct production. Figure 4 depicts the convolutional-3D architecture that has been implemented. A total of eight layers are used to produce three-dimensional convolution. The maximum pooling layer is provided after a total of five convolutional layers as shown in Fig. 5. The layers fc6 and fc7 are entirely linked. The output layer is then added. The sizes of the C3D kernels are now 3 * 3 * 3 that has been stated to have the highest correctness in a C3D deep learning experiment [3].

The initial layer of the pool kernel has a size of 122, though the others are 222. The FC7 layer may create a total of 4096 units; however, the final layer has five dimensions. That is, the number of alternatives for each question on the SF-36v2 is the score outputs for the eight quality of life scales. To assess whether a neuron should be activated or not, the activation function produces a weighted sum and then adds bias to it. The activation function's purpose is to make a neuron's output nonlinear.

3.4.1 Activation Function

A neural network is nothing more than a linear regression model without an activation function. The activation function alters the input in a nonlinear manner, allowing it to learn and do more difficult tasks. Rectified linear unit is abbreviated as RELU.

It is the most common technique of activation. Convolutional neural networks with hidden layers are employed.

3.5 Questionnaire–Answer Estimation

In the SF-36v2, experiments are put up for each question to estimate questionnaire responses. Each pair of input data is labeled with a five-dimensional vector in order to distinguish which of the five alternatives is a response. The label for the user's input data is [0, 1, 0, 0, 0], with 1 replacing the next element of the vector and 0 replacing the remaining components, when the client picks the second choice as the answer. It creates 36 label, these correspond to the entire series of queries in the SF-36v2 questionnaire and then complete learning process. After learning the relationship between labels and video and audio data, the estimating system will attempt to generate vectors comprising five items against the test data, such as [0, 0.92, 0.08, 0, 0]. The likelihood that the linked choice is the best option is represented by each element. For example, [0, 0.92, 0.08, 0, 0] indicates that the second choice would be likely right.

3.6 Score Regression

The approach is designed to produce an eight-dimensional vector that correlates to the 8 scores obtained on the quality of life scale. This enables us to estimate the eight subscale scores immediately. For example, [58.5, 55.7, 61.7, 62.8, 40.2, 57.0, 56.1, 43.8] = [PF, RP, BP, GH, VT, SF, RE, MH] is data from a single user's output. Such labels are added to every piece of data collected from the user when it comes to the training data. The system attempts to create scores for the eight subscales in the form of vectors with eight components each in this experiment. A categorization issue is the difficulty in computing the questionnaire response. The scale score, on the other hand, is explicitly learned and created in this experiment, and therefore, it is simplified to a regression problem. To test the model's estimating accuracy, a leave-one-out approach [4] is applied (for both responses and subscale scores).

3.7 Questionnaire–Answer Estimation

In case of a characterization issue, where the assessment depends on which things are effectively sorted or erroneously arranged, it is feasible to figure the appraisal's precise answer rate to check the model's accuracy. The SF-36v2 contains all 36 questions that were subjected to the same procedure. Each of the 36 questions was assigned to one of the eight subscales [32]. Since question 2 was excluded from the calculation of the eight subscale scores, it was inspected separately. The correct answer rate is represented on the vertical axis (percent). The question number correlates to the horizontal axis, and there are a total of 36 questions. When the average response from the questionnaire is used as the estimation result.

Given the existence of research on emotion estimation using audio [33, 34], video [35], and both [20], it is probable that knowledge of unhappiness has an influence on the user's speech and emotions, which may be observed by others. The psychosomatic state of an individual can be measured using physical characteristics, as per studies on pulse estimation from looks utilizing photoelectric volumetric heartbeat wave recording [36, 37], the connection between head vibration and heartbeat [38, 39], and the connection among heartbeat and feeling [40, 41]. Indeed, Microsoft Azure's Emotion API is now extensively used to categorize facial emotions into eight fundamental feelings. Moreover, research [42–44] has been done on using vocal characteristics to undertake emotional assessments, and therefore, these aspects are thought to be beneficial for categorizing or assessing human emotions.

4 Results and Discussion

4.1 Results

Figure 6 depicts a violin graph of the distribution of errors (loss), which depicts the difference between the system's estimation and the actual result determined using the SF-36v2 scoring algorithm. The mental health (MH) scale stands out when contrasted with the other scales. It is discovered that estimating mental health is the most challenging.

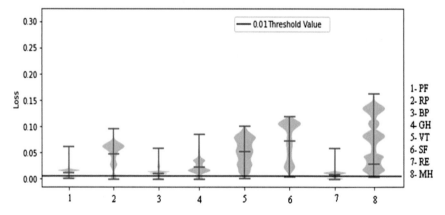

Fig. 6 Error distribution (loss) between the actual and projected scores

4.2 Discussion

Because of the existence quality score relapse test, seven out of eight scales might be assessed with minor mistakes. Due to the employed time-series data in our learning estimation architecture, it had greater features than two-dimensional convolution with static images. This time, we concentrated on the face. As a result, unnecessary evidence that is head movement was discarded, and rather they were collected characteristics on the face surface. This led to the inclusion of features in the face that are useful for quality of life estimation, and applied learning estimation system utilizing C3D was successful in reducing the high-dimensional information about life quality.

Additionally, despite the fact that the gathered movies were reduced to 50×50 pixels this time, estimate was demonstrated to be still achievable. This experiment showed that using three-dimensional convolution to time-series data for calculating quality of life, which is one of the measures of human status, is successful. This method may also be used to estimate facial expression detection and has shown to be successful (FER). As a consequence, the experimental findings show that our innovative telecare strategy, it is successful in analyzing quality of life using 3D convolutional neural networks.

As plausible causes, the following four can be mentioned.

- The framework made neglected to separate components that were valuable for assessing the mental health scale.
- Difference among abstract feeling demonstrated in the poll preceding the examination.
- Feeling communicated all through the regular language contact with the conversational AI.
- Emotional changes.

For the first reason, human expressions are supposed to carry information about emotions like despair, as evidenced by research that estimates depression from facial expressions. The authors suggest that visual attention patterns are one of the most useful indicators of human health. As a result, while this video had eye movement data, it also included a lot of other information, so it is conceivable that it did not extract the qualities needed to evaluate mental health. In regard to the second hypothesis, question addressed the individual for his or her current feeling; so, it is feasible to believe that the fault was simply with this component and not with emotion estimate. In general, there are two types of evaluation methods in emotional analysis: Self-assessment that is a subjective judgment made by the user, and objective analysis made by a third party. The primary emotions were defined both by self-assessment and by objective assessment of the voice integrating emotion, according to the findings with the percentages of discrepancy among both evaluated.

5 Conclusion and Feature Work

We applied natural language communication in conjunction with a computer-based conversational agent to measure user's quality of life in this study. Via a hybrid C3D and SVM architecture, the conversational using moving pictures and words, AI gathered spatiotemporal characteristics. We compared real questionnaire responses to system-calculated questionnaire replies, as well as the scores for eight subscales as directly assessed by the system, in the accuracy evaluation. Developed a technique to estimate quality in this study, life quality is a useful metric for assessing an individual's or a society's physical and mental well-being. We concluded from the findings that multimodal data may be used to derive high-dimensional information on a human's life, such as their degree of happiness with everyday life and social activities.

The strategy for estimating quality of life utilized in this study is adequate, and this is the only way to compare how different welfare-related machines (robots) perform on each of the characteristics that make up quality of life via human interaction. This issue, not simply monitoring robots, is expected to do general assessment technique for any robots that develop cooperative contacts through people. Researchers hope that their findings will serve as a springboard for talks aimed at enhancing all quality of life measures at the same time. This study uses a 3D convolutional neural network, SVM, to assess quality of life to better understand senior health care. In the future, we expect HAI approaches to take into consideration situations and optimal behavior. Looking forward to hearing presentations and conducting research on how to better understand individuals and change their behavior depending on their personalities in the future.

References

1. World population ageing 2017, United Nations, 2017. www.un.org/en/development/desa/population/publications/pdf/ageing/WPA2017_Highlights.pdf. Accessed 22 Aug 2019
2. Fukuhara S, Bito S, Green J, Hsiao A, Kurokawa K (1998) Translation, adaptation, and validation of the SF-36 health survey for use in Japan. J Clin Epidemiol 51(11):1037–1044. https://doi.org/10.1016/S0895-4356(98)00095-X
3. Tran D, Bourdev L, Fergus R, Torresani L, Paluri M (2015) Learning spatiotemporal features with 3d convolutional networks. In: Proceedings of the IEEE conference on computer vision, pp 4489–4497
4. Kocaguneli E, Menzies T (2013) Software effort models should be assessed via leave-one-out validation. J Sys Softw 86(7):1879–1890. https://doi.org/10.1016/j.jss.2013.02.053
5. Yao B, Fei-Fei L (2012) Recognizing human-object interactions in still images by modeling the mutual context of objects and human poses. IEEE Trans Pattern Anal 34(9):1691–1703. https://doi.org/10.1109/TPAMI.2012.67
6. Cao Z, Simon T, Wei SE, Sheikh Y (2017) Realtime multi-person 2d pose estimation using part affinity fields. In: Proceedings of the IEEE computer society conference on computer vision and pattern recognition, pp 7291–7299

7. Ren F, Xin K, Quan C (2016) Examining accumulated emotional traits in suicide blogs with an emotion topic model. IEEE J Biomed Health 20(5):1384–1396. https://doi.org/10.1109/JBHI.2015.2459683

8. Quan C, Ren F (2010) Sentence emotion analysis and recognition based on emotion words using REN-CECPS. Int J Adv Intell 2(1):105–117

9. Lin YP, Wang CH, Jung TP, Wu TL et al (2010) EEG-based emotion recognition in music listening. IEEE Trans Biomed Eng 57(7):1798–1806. https://doi.org/10.1109/TBME.2010.2048568

10. Jenke R, Peer A, Buss M (2014) Feature extraction and selection for emotion recognition from EEG. IEEE Trans Affect Comput 5(3):327–339. https://doi.org/10.1109/TAFFC.2014.2339834

11. Ren F, Dong Y, Wang W (2018) Emotion recognition based on physiological signals using brain asymmetry index and echo state network. Neural Comput Appl 1–11. https://doi.org/10.1007/s00521-018-3664-1

12. Bänziger T, Grandjean D, Scherer KR (2009) Emotion recognition from expressions in face, voice, and body: the multimodal emotion recognition test (MERT). Emotion 9(5):691–704. https://doi.org/10.1037/a0017088

13. Chang Y, Hu C, Feris R, Turk M (2006) Manifold based analysis of facial expression. Image Vision Comput 24(6):605–614. https://doi.org/10.1016/j.imavis.2005.08.006

14. Pantic M, Bartlett MS (2007) Machine analysis of facial expressions. In: Delac K, Grgic M (eds) Face recognition. IntechOpen, Vienna, Austria, pp 377–416

15. Kotsia I, Pitas I (2006) Facial expression recognition in image sequences using geometric deformation features and support vector machines. IEEE Trans Image Process 16(1):172–187. https://doi.org/10.1109/TIP.2006.884954

16. Ren F, Huang Z (2016) Automatic facial expression learning method based on humanoid robot XIN-REN. IEEE Trans Hum Mach Syst 46(6):810–821. https://doi.org/10.1109/THMS.2016.2599495

17. Barakova EI, Gorbunov R, Rauterberg M (2015) Automatic interpretation of affective facial expressions in the context of interpersonal interaction. IEEE Trans Hum Mach Syst 45(4):409–418. https://doi.org/10.1109/THMS.2015.2419259

18. Chang Y, Vieira M, Turk M, Velho L (2005) Automatic 3D facial expression analysis in videos. In: Zhao W, Gong S, Tang X (eds) International workshop on analysis and modeling of faces and gestures. Springer-Verlag, Berlin Heidelberg, Heidelberg, Germany, pp 293–307

19. Wang J, Yin L, Wei X, Sun Y (2006) 3D facial expression recognition based on primitive surface feature distribution. In: Proceedings of the IEEE computer society conference on computer vision and pattern recognition, pp 1399–1406. https://doi.org/10.1109/CVPR.2006.14

20. Fan Y, Lu X, Li D, Liu Y (2016) Video-based emotion recognition using CNN-RNN and C3D hybrid networks. In: Proceedings of the 18th ACM international conference on multimodal interaction. Association for Computing Machinery, New York, pp 445–450

21. Scherer KR (2003) Vocal communication of emotion: a review of research paradigms. Speech Commun 40(1–2):227–256. https://doi.org/10.1016/S0167-6393(02)00084-5

22. Kwon OW, Chan K, Hao J, Lee TW (2003) Emotion recognition by speech signals. In: Proceedings of the eighth European conference on speech communication and technology. International Speech Communication Association, Baixas, France, pp 125–128

23. Vidrascu L, Devillers L (2005) Detection of real-life emotions in call centers. In: Proceedings of the ninth European conference on speech communication and technology. International Speech Communication Association, Baixas, France, pp 1841–1844

24. Tian YL, Kanade T, Cohn JF (2005) Facial expression analysis. In: Li SZ, Jain AK (eds) Handbook of face recognition. Springer, New York, pp 247–275

25. Cohn JF, Schmidt K (2003) The timing of facial motion in posed and spontaneous smiles. In: Li JP, Zhao J, Liu J, Zhong N, Yen J (eds) Active media technology. World Scientific Publishing, Singapore, pp 57–69

26. Bartlett MS, Littlewort G, Frank M, Lainscsek C, Fasel I, Movellan J (2005) Recognizing facial expression: machine learning and application to spontaneous behavior. In: Proceedings of the

IEEE computer society conference on computer vision and pattern recognition, pp 568–573. https://doi.org/10.1109/CVPR.2005.297

27. Ioannou S, Raouzaiou A, Tzouvaras V, Mailis T, Karpouzis K, Kollias S (2005) Emotion recognition through facial expression analysis based on a neurofuzzy method. Neural Netw 18(4):423–435. https://doi.org/10.1016/j.neunet.2005.03.004

28. DOCOMO developer support, NTT DOCOMO (n.d.). https://dev.smt.docomo.ne.jp. Accessed 15 May 2019

29. Juang C-F, Shiu S-J (2007) Fuzzy system learned through fuzzy clustering and support vector machine for human skin color segmentation. IEEE Trans Syst, Man Cyber Part A: Syst Hum 37(6):1077–1087

30. Juang C-F, Shiu S-J (2008) Using self-organizing fuzzy network with support vector learning for face detection in color images. Neurocomputing 71(16–18):3409–3420

31. Guan CN (2012) Face localization using fuzzy classifier with wavelet-localized focus color features and shape features. Digital Sig Process 22(6):961–970

32. Shuzo M, Yamamoto T, Shimura M, Monma F, Mitsuyoshi S, Yamada I (2011) Construction of natural voice database for analysis of emotion and feeling. Trans Inf Proc Soc Jpn 52(3):1185–1194

33. Shuzo M, Yamamoto T, Shimura M, Monma F, Mitsuyoshi S, Yamada I (2011) Construction of natural voice database for analysis of emotion and feeling. Trans Inf Proc Soc Jpn 52(3):1185–1194. http://id.nii.ac.jp/1001/00073614/

34. Ren F, Matsumoto K (2015) Semi-automatic creation of youth slang corpus and its application to affective computing. IEEE Trans Affect Comput 7(2):176–189. https://doi.org/10.1109/TAFFC.2015.2457915

35. Emotion API, Microsoft Azure (n.d.). https://azure.microsoft.com/ja-jp/services/cognitive-services/emotion/. Accessed 15 May 2019

36. Cennini G, Arguel J, Akşit K, van Leest A (2010) Heart rate monitoring via remote photoplethysmography with motion artifacts reduction. Opt Express 18(5):4867–4875. https://doi.org/10.1364/OE.18.004867

37. Li X, Chen J, Zhao G, Pietikainen M (2014) Remote heart rate measurement from face videos under realistic situations. In: Proceedings of the IEEE computer society conference on computer vision and pattern recognition, pp 4264–4271

38. Balakrishnan G, Durand F, Guttag J (2013) Detecting pulse from head motions in video. In: Proceedings of the IEEE computer society conference on computer vision and pattern recognition, pp 3430–3437

39. Cohn JF, Reed LI, Ambadar Z, Xiao J, Moriyama T (2004) Automatic analysis and recognition of brow actions and head motion in spontaneous facial behavior. In: Proceedings of IEEE international conference on systems, man and cybernetics, pp 610–616. https://doi.org/10.1109/ICSMC.2004.1398367

40. Cannon WB, de la Paz D (1911) Emotional stimulation of adrenal secretion. Am J Physiol 28(1):64–70. https://doi.org/10.1152/ajplegacy.1911.28.1.64

41. Appelhans BM, Luecken LJ (2006) Heart rate variability as an index of regulated emotional responding. Rev Gen Psychol 10(3):229–240. https://doi.org/10.1037/1089-2680.10.3.229

42. Shibasaki K, Mitsuyoshi S (2005) Evaluation of emotion recognition from intonation: evaluation of sensibility technology and human emotion recognition. IE-ICE Tech Rep 105(291):45–50

43. Kwon OW, Chan K, Hao J, Lee TW (2003) Emotion recognition by speech signals. In: Proceedings of the eighth European conference on speech communication and technology, Baixas, France, International Speech Communication Association, pp 125–128

44. Calvo MG, Lang PJ (2004) Gaze patterns when looking at emotional pictures: motivationally biased attention. Motiv Emot 28(3):221–243. https://doi.org/10.1023/B:MOEM.0000040153.26156.ed

E-learning During COVID-19—Challenges and Opportunities of the Education Institutions

B. Anusha, Ginu George, and Anson Kangirathingal Joy

Abstract As part of the COVID-19 lockdown, educational institutions were closed and adopted e-learning to keep the learning process going. Due to the COVID-19 pandemic, e-learning has become a required component of all educational institutions such as schools, colleges, and universities worldwide. This pandemic has thrown the offline teaching process into chaos. This chapter discusses the concept and role of e-learning during the pandemic and various challenges and opportunities of e-learning encountered by educational institutions. Three broad challenges identified in e-learning are inaccessibility, self-inefficacy, and technical incompetency. E-learning opportunities are no geographic barriers, flexibility, creativity, and critical learning incorporation increased utilization of online resources and reinforced distance learning.

Keywords E-learning · COVID-19 · Educational institutions · Challenges · Opportunities

1 Introduction

COVID-19, which led to the pandemic, influenced the everyday lives of several million people since 2019 and counting. "Change is the only constant" is strikingly accurate in this context. This period brought many changes, learning, and experiences in people's lives. The sphere of education underwent significant modifications; classroom teaching became virtual, books replaced with e-material, and the lines between

B. Anusha
Department of Commerce (PG), Krupanidhi Degree College, Bengaluru, India

G. George (✉) · A. K. Joy
School of Commerce, Finance and Accountancy, CHRIST (Deemed to be University, Bengaluru, India
e-mail: ginu.george@christuniversity.in

A. K. Joy
e-mail: anson.kj@christuniversity.in

homework and classwork blurred. Academic institutions that were once hesitant to change their traditional pedagogical approach left with no option but to adopt online teaching as a new normal to continue the learning process. Education institutions, educators, students, and parents strode forward on the path of revolutionizing a new way of education.

This new learning model has its own set of advantages and challenges. The significant merit of this new model of instruction is that it allows students access to a wide range of sources to refer from within their discipline and other disciplines, thus widening their understanding of the subjects. It has facilitated a necessary aspect of flexibility in recording the sessions and learning at a convenient pace. Activities and fests went online whereby students and faculty members could interact with diverse individuals from different parts of the world and introduce diverse cultures in learning. The inherent belief that examinations are the ultimate ends of education changed from mere examinations to a more cohesive evaluation pattern. The flip side was financial constraints; the students from weaker sections of the society could not purchase the necessary devices to cope with this changed circumstance. The shift led to high reliance on technology, challenging many teachers, especially those unfamiliar and comfortable using technological devices. Tutors and educators adapted to teaching online quickly with multiple barriers such as limited laptops or computers, poor Internet, and other technical-related problems [18].

Online education also has other problems: for instance, sitting in front of computer screens for a long-time harms students' and teachers' physical and mental well-being. Physical education classes are a widely accepted and integrated part of every student's educational journey, focusing on a healthy and active lifestyle [13] affected during the online classes. Despite all these challenges and problems, the pandemic has shown that anything is possible with education, for learning never ends. As the world proceeds toward normalcy, it is up to us to take our lockdown learnings and spread it across, for education is like lighting a candle and spreading the light.

This chapter discusses the concept of e-learning, various e-learning tools and platforms used in the education sector, the difference between e-learning and online learning, the role of e-learning during COVID-19, and its opportunities and challenges of e-learning, and the chapter ends with a conclusion on how e-learning is effectively incorporated by education sector especially during these pandemic times.

2   E-learning

E-learning is an extensively used mode of education during the COVID times. It is a learning framework dependent on formalized instruction and linked with learning objects such as PCs, tabs, and other technological gadgets, and Internet utilization shapes a significant part of e-learning. E-learning tools can be classified based on student interaction and remote learning models. Many universities adopted remote learning platforms as a transition filler to implement interactive models. For instance, students had to enroll for MOOC courses of their choice as part of curriculum credits in the initial phases of lockdown. Lockdown further gave time to universities to conduct training along the lines of MOOC courses to start interactive models of teaching, learning, and evaluation. Some of the tools used for interactive engagements were Blackboard, Institution Learning Management Systems, Google Classroom, Webex, Zoom, Google Meets, and Google Dashboard. This phase also led to many e-learning platforms such as Coursera, BIJUS, Udemy, Khan Academy, upGrad, and many more.

2.1 E-learning Platforms

These platforms are a webspace where academic content and other related resources are available to students. They can access lectures and materials and interact with educators' students across the globe. These provide an opportunity for students to learn at their convenience. Thus, it is an integrated set that offers students and tutors a different learning experience with fun, interactive, and complete engagement. It is observed that most colleges and universities have started integrating e-learning platforms into their traditional teaching as it provides an element of global teaching exposure to the students.

Other reasons why e-learning platforms like Coursera, BIJUS, Udemy, Khan Academy, and UpGrad are gaining much importance is because they provide a robust learning experience for students that almost simulates a classroom experience. All features such as course instructor and student interaction, creative assignments, evaluation, and students also get an opportunity to clarify their doubts over chats.

Thus, the convenience students get through e-learning without losing the essence of classroom learning makes the e-learning platform more attractive. Another exciting feature is that e-learning platforms provide varied content formats to cater to students learning styles. The most observed content formats are videos, PowerPoint presentations, case discussions, webinars, and interviews. All the recorded sessions are accompanied by transcripts and learning materials, which helps students refer back if they have not understood a specific concept. Most courses offered in these e-learning platforms comprised activities, quizzes, and assignments after every session, and the final score is determined by taking an average score from all the assignments. This continuous evaluation makes students more attentive, increases knowledge retention level, higher performance, and more involvement in the learning, which is sometimes absent in traditional learning.

2.2 E-learning Tools

Apart from various platforms mentioned in the section above, many new tools are available in the open platforms for educators to enhance their teaching effectiveness. These tools are synchronized with most of the e-learning platforms.

Plagiarism checker: One of the most critical issues educators face in online education is academic dishonesty displayed by students by submitting plagiarized work. To control this academic dishonesty, the educators must try to create an environment of honesty by clearly outlining the consequences of such dishonesty. Different assessment methods are recommended, such as discussion forums, presentations, projects, and research-based assignments, to reduce instances of academic dishonesty.

Further to assist educators to assess such assessments and give constructive feedback for student improvement, some of the tools that could be of use are DupliChecker, Grammarly, PaperRater, Plagiarisma, Search Engine Reports, Plag-Tracker, Plagium, Copyleaks, PlagScan, Unplug Checker, and similarity report in google classroom.

Assessments: Another difficulty that educators face is assessing the online submissions. It becomes essential to understand that assessment without giving clarity/feedback to students would not result in learning, and therefore, educators must develop rubrics for various forms of assessments. Some platforms that could help educators design rubrics are Rcampus and SCU.EDU. These rubrics are mapped in the google classroom assignments for better and straightforward assessments.

Class engagement tools: Online education cannot be purely driven by lecture mode since this would only result in decreased interest and attention in the learning process, and hence, it is essential to ensure that students participate in the learning process. In this regard, one of the tools that could be of use is Mentimeter. Mentimeter can conduct polls, quizzes, word clouds, scales, and open-ended questions. Mentimeter might not be the only tool that helps conduct such engagements; most video conferencing applications have also evolved to create options for polling and conducting quizzes; such features are also available now in Webex and Zoom.

Other tools such as Book creator could help educators design creative assignments where students can use a combination of text, images, and videos to create interactive stories, digital portfolios, research journals, poetry books, and many more.

Screencastify is another tool that could be utilized for blended learning. This tool helps educators create learning video content. The software tools record the system screen as well as the webcam. Therefore, the end product is similar to the videos made available in MOOC courses.

A visual representation of data in the correct format could convey more meaning than literature content. Thus, these following tools could help educators create creative student assessments through infographics. These tools are Infogram, Venngage, Piktochart, Canva, and Easel.ly.

3   Difference Between E-learning and Online Learning

E-learning has been emphasized a lot due to COVID lockdown and the need for continuing education. The concept of technology-assisted learning dates back to the 1960s, called computer-assisted learning [3]. E-learning, first introduced in a research article titled "Synthesis of Research on Electronic Learning." Wherein e-learning meant "learning via electronic sources, such as television, computer, videodisk, Tele-text, videotext." [12]. While [2] states that the broader field is online learning. Many other terminologies have evolved, defining the same as "e-learning, Internet learning, distributed learning, networked learning, tele-learning, virtual learning, computer-assisted learning, Web-based learning, and distance learning." Therefore, considering the overlapping usage of various terminologies, it becomes essential to differentiate e-learning and online learning. Online learning defined as accessing online materials, interacting with in—instructors and classmates, and acquiring knowledge and growing from ones' own experience [2]. At the same time, e-learning is defined as a "web-based system that makes information or knowledge available to users or learners and disregards time restrictions or geographic proximity" [8].

The online learning definition clearly articulates the learner's interaction with the instructor and support during the learning process. Further, even though the learner can choose to engage with the content available on the Internet directly, they would choose a process of engagement that involves a teacher's presence and a complete cycle of learning, assessment, and feedback, thus further motivating learners to continue their learning engagement. But this is not the case with e-learning; the learner has the advantage of time and location, which implies that the learning is at the pace that the learner chooses.

Further e-learning could also imply the massive knowledge distribution to all Internet users, in which case terminologies such as MOOC could also form a part of the broader term e-learning. Thus, catering to a global audience is not possible considering the physical space limitations of universities.

4 E-learning During COVID-19

COVID-19 is a humanitarian crisis that led to disruption in educational opportunities. Lack of education for the long term can have adverse effects on children's well-being and the entire nation's development [9]. Psychological-related problems, social isolation, and lack of physical exercise [12] are some problems students face during the pandemic. For the nations, it was an increased skill gap among individuals. So, providing education in a crisis like this has a much broader objective than just controlling the effect of school/college closures. Reopening schools/colleges is unthinkable due to the fear of increased disease transmission. Thus, it becomes essential to continue education through alternative modes. UNESCO surveyed 59 countries globally to understand how nations have addressed the challenges of school/college closure. Most nations use mixed media and interactive classes to ensure continued education from the survey. Mixed media involved the usage of television and radio broadcasting as sources for information dissemination. On average, 75% of the countries used existing national online learning platforms, 62% used free open-access learning platforms, 41% encouraged private players to provide e-learning material and platforms, and 60.75% provided interactive online classes [12].

5 Challenges of E-learning

Even though governments, administrators, and educators worldwide ensured continued education. But on a large-scale assessment, only a few privileged received online or e-learning opportunities, while it did deny opportunities for approximately 258 million children who dropped out from school (UIS 2019). According to the World Bank report in 2019, "53% of all children in low- and middle-income countries suffered from learning poverty." Thus, the data highlights the pre-COVID crisis in the educational space, especially for the vulnerable population.

The challenges identified are accessibility, self-efficacy, and technical competency. Accessibility implies the provision of good Internet and compatible devices [1]. The UNESCO survey highlights that approximately 87% of the educational systems across a survey sample of 59 countries, consisting of middle-income countries, reported unequal access to ICT facilities at home due to insufficient Internet and power infrastructure [12, 13].

Self-efficacy refers to self-belief or an expectation of one's own ability to complete a definitive task [14]. The lack of self-efficacy among students and teachers becomes a barrier toward e-learning utilization. Therefore, it can be noted in many research studies that even though the e-learning platforms have a wider reach, sustaining learner motivation has always been a difficulty [7].

Technical competency deals with discomfort in adopting technological usage for providing e-learning content and its usage by the learner. From the instructor's perspective, the focus should be on creating exciting and engaging content while ensuring intrinsic motivation to continue the course. It also implies that the student, by the end of the course, realizes the course's intended learning outcomes through his engagement in the course activities. The challenge is to ensure all this happens through a digital platform at the pace that the learner chooses. In developed economies, the challenge would be from the provider's end, while in most developing economies, the challenges are on both the instructor and learner.

6 Opportunities of E-learning

The educational institutions, mostly in developing countries, have often faced the challenge of providing Quality education even before the pandemic. They have always thrived hard to bring in various facilities and approaches to enhance their teaching-learning process. Considering Quality as their top priority, educational institutions can use the current situation to enhance it by considering a few opportunities.

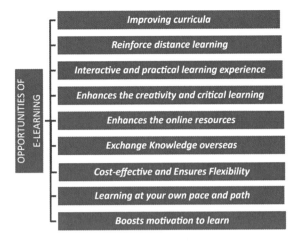

6.1 Improving Curricula

During the pandemic, the shift from offline classes to online classes was a significant change in the paradigm of educational institutions. Both the education providers and seekers had challenges adapting to the new methodologies.

Curricular priorities proposed are concerning the academic skills and knowledge students needed to maintain depending on their age, and grade level in subjects was discounted as non-essential during COVID-19. E-learning prioritized global competencies within curricula that can be leveraged in a wide variety of situations.

6.2 Reinforce Distance Learning

Educational institutions provided different possibilities for students who returned to their hometown to continue their education. Students who either moved back to their home or hometown did not have to compromise on their learning process. Students could enroll and complete distance education courses related to their ongoing courses.

Moreover, educational institutions could offer options for students who had barriers of locational distance and hesitated to choose to continue their education. Thus, educational institutions could provide courses accessible to students from other parts of the country and world.

6.3   Interactive and Practical Learning Experience

Many students at their educational institutions have become more engaged with their classmates and teachers due to the rise of e-learning technologies [4]. A more linked learning community that employs interactive learning materials more regularly has resulted from the mix of e-learning technologies and traditional print-based approaches. Students said they had better access to instructors, materials, and educational experiences [6], which is one of the fundamental goals of all educational institutions.

6.4   Enhances the Creativity and Critical Learning

Globally, new teaching and learning approaches have emerged in educational institutions, the most important of which is e-learning, driving this change from teacher-centered to student-centered learning [17]. E-learning provides students the situations that allow them to understand the materials better.

According to studies, effective education is a combination of the learner's memory and how well they retain material, how engaged they are in classroom activities, and prior understanding of the issue. As a result, students who attend an educational institution with E-learning facilities benefit from better analytical skills, creative thinking, and comprehension.

6.5   Enhances the Online Resources

Many educational institutions and universities focus on building online databases to facilitate their students and faculty members to enable learning abilities and encourage their research activities. Learner empowerment has achieved new heights thanks to the evolution of e-learning opportunities and technologies combined with massive online resources. YouTube, TED Talks, the Khan Academy, and digital sites like Noodle and Udemy have grown in popularity to make teaching and learning relevant and practical.

Subscription to magazines, journals, and books in the library and research culture in an educational institution contributes to the institution's reputation. The educational institutions that provide online resources can grow their popularity. With the pandemic situation around, educational institutions that have switched from offline to online mode of teaching-learning process contributed toward more enhancements of online resources, hence creating a brand image in the market.

6.6 Exchange Knowledge Overseas

Many higher educational institutions, especially during the COVID-19 pandemic, have implemented good e-learning strategies to benefit all stakeholders. The primary goal is to provide possibilities for knowledge construction among student peer groups from all around the world [10]. Researchers have recently discovered that social media platforms such as Facebook are useful for exchanging ideas, sharing instructional materials, and interacting with instructors, resulting in greater involvement and effectiveness.

Various online platforms like Zoom, Cisco WebEx, Google Meet, and Microsoft Teams app have helped learners and instructors to connect overseas from the comfort of their homes. Conferences, webinars, professional development courses, workshops, and training can be organized at an international level that can be made readily available for participants to attend and exchange knowledge with just a click.

6.7 Cost-Effective and Ensures Flexibility

Educational institutions charged higher fees to cater to the needs of physical schooling. The face-to-face form is less cost-effective than the online mode when considering the considerable costs of physical and educational institutions [5]. Online training courses were cost-effective due to increased enrollment, increased student access to quality programs and resources, and other benefits. Educational institutions can provide access to recorded lectures that students may see and absorb at their leisure, making scheduling more flexible. According to [5], higher enrolments increased student access to quality programs, resources, and other benefits.

6.8 Facilitate Learning at Your Own Pace and Path

Students can use the digital education system to examine what they need to know to seek and use online resources. Through digitization, educational institutions can assist each student in learning at their own pace and on their path.

Educational institutions can offer more courses, especially for working professionals. They usually look for some new technical courses as they must be up-to-date with the current skill set as new technologies enter the market. E-learning has grown in popularity as the Internet has gotten more affordable over time. Most working professionals prefer e-learning because it allows them to study at their own pace. They are not cut off from the workplace, and they may put their new abilities to use right away.

6.9   Boosts Motivation to Learn

Better teaching aids and methods help educational institutions create an elevated learning environment [8], most experienced during the pandemic. A professor can use e-learning to make lectures more engaging through visual presentations.

Lectures can be more successful with digital teaching aids like videos, photos, graphs, and animations because they improve their reading comprehension skills, illustrate or reinforce a skill or concept, differentiate instruction, and relieve anxiety or boredom by presenting information in a new and exciting way.

7   Conclusion

All educational institutions have emphasized E-learning during the pandemic and the need for a continuous learning process. While the lockdown caused inconveniences, e-learning gained a lot of importance and attention from the educational institutions and their stakeholders. Institutions could continue to cater to their students those courses stopped in the offline sessions. Though the sudden shift posed the institutions' many challenges, some good learning and practices can be adopted even post-pandemic.

Some of them are incorporating a hybrid teaching model, using online platforms for conducting and submitting assignments, using online resources, and providing exclusive online modules to achieve creativity and critical learning. To conclude, while classroom learning is more personalized, structured, and systematic, e-learning provides various advantages and opportunities that otherwise could not be achieved through traditional teaching.

References

1. Aboagye E, Yawson JA, Appiah KN (2021) COVID-19 and e-learning: the challenges of students in tertiary institutions. Soc Educ Res 1–8. https://doi.org/10.37256/ser.212021422
2. Anderson T (ed) (2004) Theory and practice of online learning. Athabasca University
3. Bernhardt KS (1960) Review of the undirected society and automatic teaching: the state of the art. Can J Psychol/Revue Can Psychol 14(1):60–61. https://doi.org/10.1037/h0083460
4. Gregson J, Jordaan D (2009) Exploring the challenges and opportunities of m-learning within an international distance education programme. In: Mobile learning transforming the delivery of education and training. http://tvoccd.oerp.ir/sites/tvoccd.oerp.ir/files/articles/mobile%2C%20learning.pdf#page=203
5. Jung I (2005) Cost-effectiveness of online teacher training. Open Learning: J Open Distance e-Learn 20(2):131–146. https://doi.org/10.1080/02680510500094140
6. Jung I, Rha I (2000) Effectiveness and cost-effectiveness of online education: a review of the literature. Educ Technol 40(4):57–60

7. Khalil H, Ebner M (2014) MOOCs completion rates and possible methods to improve retention—a literature review. In: World conference on educational multimedia, hypermedia and telecommunications, pp 1305–1313
8. Mathew NG, Alidmat AOH (2013) A study on the usefulness of audio-visual aids in EFL classroom: implications for effective instruction. Int J Higher Educ 2(2):p86. https://doi.org/10.5430/ijhe.v2n2p86
9. Reuge N, Jenkins R, Brossard M, Soobrayan B, Mizunoya S, Ackers J, Jones L, Grace Taulo W (2021) Education response to COVID 19 pandemic, a special issue proposed by UNICEF: editorial review|Elsevier Enhanced Reader. https://doi.org/10.1016/j.ijedudev.2021.102485
10. Sarker MFH, Mahmud RA, Islam MS, Islam MK (2019) Use of e-learning at higher educational institutions in Bangladesh: opportunities and challenges. J Appl Res Higher Educ 11(2):210–223. https://doi.org/10.1108/JARHE-06-2018-0099
11. Sun P-C, Tsai RJ, Finger G, Chen Y-Y, Yeh D (2008) What drives a successful e-Learning? An empirical investigation of the critical factors influencing learner satisfaction. Comput Educ 50(4):1183–1202. https://doi.org/10.1016/j.compedu.2006.11.007
12. UNESCO (2020) National education responses to COVID-19—summary report of UNESCO online survey. https://unesdoc.unesco.org/ark:/48223/pf0000373322/PDF/373322eng.pdf.multi
13. UNICEF (2021) Effectiveness of digital learning solutions to improve educational outcomes. A review of the evidence. Working paper. https://www.unicef.org/documents/effectiveness-digital-learning-solutions-improve-educational-out-comes
14. Wang AY, Newlin MH (2002) Predictors of web-student performance: the role of self-efficacy and reasons for taking an online class. Comput Hum Behav 18(2):151–163. https://doi.org/10.1016/S0747-5632(01)00042-5
15. White MA (1983) Synthesis of research on electronic learning. Educ Leadersh 40(8):13–15. Retrieved 9 Dec 2021 from https://www.learntech-lib.org/p/134860/
16. Welch R, Alfrey L, Harris A (2021) Creativity in Australian health and physical education curriculum and pedagogy. Sport Educ Soc 26(5):471–485. https://doi.org/10.1080/13573322.2020.1763943
17. Yusuf N, Al-Banawi N (2013) The impact of changing technology: the case of e-learning. Contemp Issues Educ Res 6(2):173–180. https://doi.org/10.19030/cier.v6i2.7726
18. Zalat MM, Hamed MS, Bolbol SA (2021) The experiences, challenges, and acceptance of e-learning as a tool for teaching during the COVID-19 pandemic among university medical staff. PLoS ONE 16(3):e0248758. https://doi.org/10.1371/journal.pone.0248758

Real-Time Human–Machine Interaction Through Voice Augmentation Using Artificial Intelligence

M. N. Sumaiya, B. V. Sreekanth, U. S. Akash, Aravind Sharma Kala, and G. M. Dharanendra Gowda

Abstract In real time, there is a huge demand to access dynamic, personalized, and adaptive information. Interacting through voice augmented systems yields efficient and faithful access to great deal of knowledge. And the obtained information becomes more meaningful and understandable because it is appropriately customized in the aspect of real-time physical, digital, and virtual interactions. The objective of this project is to display the interactions of voice augmentation functions to achieve human to machine interaction (HMI). The quality of the services can be improved with the help of artificial intelligence through deep learning models to illustrate the possibilities of its potential. There exist a good range of applications which include vending machines, billing counters, home assistant, etc. This work includes audio processing, artificial intelligence, text-to-speech conversion, speech to text, audio conversion, various machine learning algorithms and models, implementation using the Internet of things, and applications in various fields, to evaluate the model.

Keywords Human–machine interaction · Voice augmentation · Artificial intelligence · Voice assistant · Recurrent neural network

1 Introduction

The need for acquiring knowledge and information is increasing day by day. A couple of decades ago, few of the primitive methods to accumulate information were through books, scriptures, and manuscripts. As technology has risen to a level where human can access every data in this world within fraction of seconds, the need to access information has also increased. In this real world, human to machine interactions (HMI) through voice are essential for aged, physically challenged people and people involved in multiple tasks. Instead of using physical movements of their body to interact with machines, easily, they can interact through their voice to get their work

M. N. Sumaiya (✉) · B. V. Sreekanth · U. S. Akash · A. S. Kala · G. M. Dharanendra Gowda
Department of Electronics and Communication Engineering, Dayananda Sagar Academy of Technology and Management, Bangalore 560082, India
e-mail: drsumaiyamn@dsatm.edu.in

© The Author(s), under exclusive license to Springer Nature Singapore Pte Ltd. 2023 147
M. A. Chaurasia and C.-F. Juang (eds.), *Emerging IT/ICT and AI Technologies Affecting Society*, Lecture Notes in Networks and Systems 478, https://doi.org/10.1007/978-981-19-2940-3_10

done. It is a very challenging task to bring the machines to communicate on the same level as us. So, this is a step exploring toward that domain, which is the real-time interactive voice augmentation. A device capable of communicating on the same level as a human makes it easily accessible to everyone irrespective of their knowledge of language, age and abilities. The objectives are to build a voice augment mainframe device, capable of carrying real-time conversation with the user using audio processing, artificial intelligence and machine learning algorithm, to establish a database for voice recognition and audio synthesis module, capable of speech-to-text and text-to-speech conversion. To build a machine learning model, by training it with the database and producing the optimum model capable of choosing the optimum solution and to deploy and demonstrate the wide range of potential applications in the field of banking, kiosks, reservation for restaurants, ticket booking vending machines for transportations and public vending machines. For example, Amazon offers Transcribe, an automatic speech recognition (ASR) service that permits developers to feature speech-to-text capability to their applications. Once the voice capability is integrated into the application, users can analyze audio files and, in return, receive a text file of the transcribed speech. Google has made moves in making assistant more universal by opening the software development kit through actions, which allows developers to create voice into their own products that support AI. Another one of Google's speech recognition products is the AI-driven cloud speech-to-text tool which enables developers to convert audio to text through deep learning neural network algorithms.

In [1], a sequence-to-sequence-based voice conversion (VC) method is proposed, which is capable enough of converting voice characteristics and the pitch contour along with duration of the input speech. It enables the use of batch normalization in all the hidden layers. A few drawbacks are restricted only to speaker to identity conversion tasks, and the framework of S2S learning approach has a lot of room to improve in the aspects of accuracy. In [2], recently, neural networks are applied to tackle audio pattern recognition problems. Audio pattern recognition is a vital research topic within the machine learning area and includes several tasks like audio tagging, acoustic scene classification, music classification, speech emotion classification, and sound event detection. It is studied that the system is inspired by conventional cognitive models for memory. A few drawbacks are that the trained data is confined to the audio set's dataset, and PANNs are susceptible to multiple pattern recognition tasks, which is time consuming. Recently, there has been increasing progress in end-to-end automatic speech recognition (ASR) architecture, which transcribes speech to text with none pre-trained alignments [3]. These online systems have the advantages over the offline baselines in both decoding latency and decoding speed. In the application of low latency encoders, the recognition accuracy was observed to be low. The connectionist temporal classification (CTC) architecture is primitive at best, which has to be tuned perfectly to form hybrid CTC. Speech separation is a method to extract the speech data from the ambience noise and background distortions [4]. The latest method in speech separation illustrates as a problem of supervised learning, by training lots of datasets on differential patterns of speech, speakers, and background noise. Implementing deep learning models on supervised speech separation methods yields effective results. Learning machines, training targets, and acoustic features are the three significant constituents of deep learning-based supervised speech separation. It provides a general overview on the various steps involved in speech separation and provides a comprehensive overview of deep neural network (DNN) based on supervised speech separation. The main drawback is that the DNN-based speech enhancement as described has met the criterion in limited conditions, but not in all conditions, and the DNN-based speech enhancement has many flaws in the aspect of background speech and noise separation. The introduction of smart mobile devices is of great advantage for user interaction as these devices are equipped with numerous sensors, making applications context aware [5, 6].

To further improve user experience, the most mobile operating systems and repair providers are gradually shipping smart devices with voice-controlled intelligent personal assistants, reaching a replacement level of human and technology convergence. It is observed that, to provide defense mechanisms against many of these attacks, the underlying operating system, in this case Android, first needs to decouple voice input and output. Data is only accessible through appropriate authenticated channel, without which the data is secure from malicious treats. Data is susceptible to malware attacks and is open to spyware attacks as the connectivity is branched throughout. Identity security is another aspect where improvement is immediately needed [7]. The literature review also covers differing types of deep architectures, like deep convolution networks, deep residual networks, recurrent neural networks, reinforcement learning, variational auto-encoders, etc. [8–10]. Convolution neural network (CNN) can progressively extract higher representations of the image after each layer and finally recognize the image [11–13]. Numerous research works are based on machine learning which is a method that learns from past experiences and uses gained knowledge to do better in the future [14–16]. Machine learning emphasizes on automatically learning and adapting when exposed to data without the need of human intervention. The reason for that is based on the fact that no system can be described as intelligent if it does not have the ability to learn and adapt [17–19]. In order to exemplify applications of supervised and unsupervised learning, we will offer annotated tips that could be the literature on machine learning for communication systems. Tasks are administered at the sting of the network, that is, at the bottom stations or access points and at the associated computing platforms, from tasks that are instead responsibility of a centralized cloud processor connected to the core network [20–22]. Indeed, for many tasks in communication networks, it is possible to gather or generate training datasets and there is no need to apply sense or to supply detailed explanations for how a decision was made [23–25]. Alternatively, under an algorithm deficit, a physics-based model, if available, can be possibly used to carry out computer simulations and obtain numerical performance guarantees [26–28]. As a solution to unstable gradient values problem, a novel RNN architecture was developed which avoids vanishing and exploding gradients, while it can be trained with conventional RNN learning algorithms. The improved RNN architecture is referred to as long short-term memory neural network (LSTM) [29–31]. The context length exploited when applying the neural network can be longer than the context length considered in training [32]. Supervised speech separation has also been shown to generalize well given sufficient training data [33–35]. A huge amount of research has gone into finding ways of constraining GMMs to extend their evaluation speed and to optimize the trade-off between their flexibility and therefore the amount of training data required to avoid serious overfitting [36, 37]. Other types of models may work better than GMMs for acoustic modeling if they can more effectively exploit information embedded in a large window of frames [38]. In fact, two decades ago, researchers achieved some success using artificial neural networks with a single layer of nonlinear hidden units to predict HMM states from windows of acoustic coefficients [39]. The application of recurrent networks to speech detection and recognition in clean and noisy environments have been proposed in some early studies [40–42]. As we have come across abundant journal papers which have covered every possible aspect in audio processing, speech-to-text conversions, machine learning models and training them, different ways of communication methods and the IoT aspect of it have better connectivity and accessibility. It is evident that there are a lot of hurdles to overcome to achieve this project. So, the objective of the project is to bring the machines to communicate on the same level as us. Then the purpose of the algorithm is to map the audio signal into textual inputs with a proper conversion method which is capable to discriminate background noise with higher accuracy. Then, the machine learning algorithm must be designed, developed, built and trained, the model must be re-trained with huge number of iterations, since it will be deployed in various applicative environments. To choose an optimum algorithm with the highest accuracy to run the model, which can access the database and interpret the input accurately and with the help of the trained dataset, recognize the situation and analyze the inputs and other constraints which restricts or governs the situation, and then to be capable of arriving at the most appropriate output is the biggest achievement of all. So, this is a small step exploring toward that domain which is the real-time interactive voice augmentation. A device capable of communicating on the same level as a human makes it easily accessible to everyone irrespective of their knowledge of language, age, and abilities.

The chapter is organized as follows: Sect. 2 describes the methodology and materials used for this work. The experimental quantitative and qualitative results are discussed, and results are tabulated in Sect. 3. Finally, conclusion is drawn in Sect. 4.

2 Methodology and Materials

A voice augment mainframe device is built which is capable of carrying real-time conversation with the user, using audio processing, artificial intelligence, and machine learning algorithm. Also, a database for voice recognition and audio synthesis module is established, which is capable of speech-to-text and text-to-speech conversion. Now, the main objective of this chapter is to design and build a machine learning model and training it with the database and producing the optimum model capable of choosing the optimum solution. To choose an optimum algorithm with the highest accuracy to run the model, which can access the database and interpret the input accurately and with the help of the trained dataset, recognize the situation and analyze the inputs and other constraints which restricts or governs the situation, and then to be capable of arriving at the most appropriate output is the biggest achievement of all. The block diagram of the proposed model is shown in Fig. 1.

The most widely used algorithm for natural language processing (NLP) and speech recognition is recurrent neural network (RNN) that can been trained to ingest speech data into small frames. The RNN has three layers: input layer, hidden layer, and the output layer; the hidden layers are the computation layers. At each time step, the non-recurrent layers work on independent data. The hidden layer may be a bidirectional recurrent layer with two hidden unit sets. One set has forward recurrence, while the other has backward recurrence, which requires some memory to perform the recurrence as illustrated in Fig. 2.

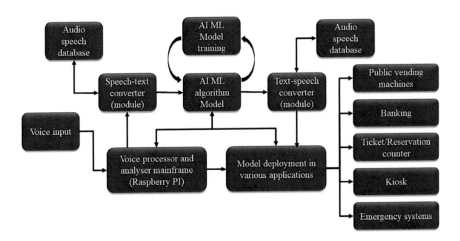

Fig. 1 Block diagram of the proposed system

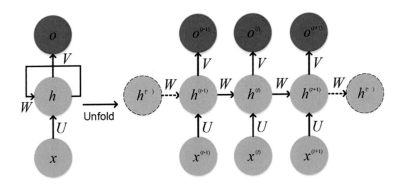

Fig. 2 General block diagram of RNN

One of the prominently used recurrent neural networks for the applications that involve sequential input data processing, such as speech recognition, image recognition, music composition, and handwriting recognition, is long short-term memory recurrent neural network (LSTM-RNN). First, considering the input speech data which are nothing but sequence of words, every word will be encrypted to unique binary vectors, then the neurons are initialized with random values coined as weights. During the training, when the binary vector of the input sequence is applied to the input layer, the nodes of the layer perform the respective logical function or sigma function on the weights to get an arbitrary value, and with the help of activation function, this value is fed to the successive layers. In application such as speech recognition, recognizing only the current input word from the speech data is not the aim, but to generate a value by co-relating the weight and the binary vector along with the previous word vector of the input sequence and then storing the value arbitrarily to co-relate with the successive vectors of the input sequence.

At the output layer, the difference of the predicted vector value to the actual vector value is calculated using a loss function; according to the obtained loss value, using a gradient descent function, the weights of the neurons are calibrated. This process is iterated as many times as required to train the model with minimum loss. Gradient descent function calculates the next optimum value from the current value of the weight and reduces by one step after scaling to the learning rate, and it is subtracted from the present weight because we want to minimize the loss function value. The learning rate controls the step size of the gradient descent which effects strongly on the performance of the model. Larger the value of learning rate, larger will be the step size, which will usually miss the optimum weight value. Smaller the value of learning rate, higher will be the possibility of attaining the optimum weight value, in larger number of steps of small size.

One more major task is to determine optimum parameters for the neural network. In speech recognition, since the number of words in each sentence might vary for every instance, determining the number of nodes in input layer is quite difficult. The hyperparameters determined for this system are determined as follows:

- The number of nodes in input and output layer or also called as batch size is equal to 10.
- The number of hidden layers is equal to 3.
- Learning rate for the gradient descent function is equal to 0.001.
- The number of epochs to determine the number of iterations to train the dataset is equal to 1000.

The sigma function or the logical function of the nodes is a simple linear transformation function $Y = a_1 * T * (v_1) + a_2 * T * (v_2)...$, where Y is the arbitrary value of the node, T is the transformation function, a_1 and a_2 are the weights of the nodes, and v_1 and v_2 are the input vectors. The activation function used is called as rectified linear unit function (RELU), which is a piecewise linear function that will output the input data. For loss detection of the predicted value, cross-entropy loss function is implemented for calibrating the weights.

PyTorch library is used to build and deploy the proposed work as it provides the required framework to design the machine learning algorithm. After designing all the mentioned modules, this project will be implemented to perform some of the real-time machines like ATM, vending machine, ticket reservation machine, and kiosk machine, where all these machines can be controlled using voice rather than using buttons, with the help of software development toolkit to design the graphical user interface.

For the mainframe device, we have used Raspberry Pi 4 which features a quad-core ARM Cortex-A72 processor, dual video output, and a good selection of other interfaces. It also requires a microphone module compatible with Raspberry Pi for audio recognition, a speaker module for the output, and a Bluetooth module to enable the Raspberry Pi to be capable of conveying output via Bluetooth speakers. The main software, Raspbian OS, is used to operate and simulate Raspberry Pi.

Since the system is being deployed in different environment, the dataset required to train the model in various environment will vary. Hence, the datasets were obtained separately for personal home assistant, vending machine, and digital assistant. About ten thousand words in each domain of application were collected to train the model.

First, the dataset required to train the model is preprocessed before using it to train the data. For example, for any given sentence, the very first step is to perform segmentation of the sentence into smaller words, and this process is called as stemming. Then, the stemmed words will be of any form according to the grammar of the sentence in which it is used; hence, it is essential to filter down to its root form. This process is called as tokenization. These tokenized words are mapped into its corresponding binary symbols with unique value into a binary array.

This preprocessed dataset is used for the model training. Now, we have to build an appropriate neural network model with appropriate batch size, training parameters, and learning parameters. The model based on recurrent neural network algorithm is built, since it is the most suitable algorithm to develop natural language processing. The dataset is divided into small batches, and each batch is called as epoch. This is done so that after each epoch training, the model adjusts the weight of the neuron by calculating the cross-entropy loss. Cross-entropy loss is the measurement of the amount of difference from the predicted values to the actual value. By rigorous training, the model predicts the exact words, with loss ranging maximum up to 0.09%.

Now, the trained model saves the data into a path file, which consists of the trained data, and this is used imported in the functionality program, where we have to map each and every functionality to the predicted words and sentence. Various library functions along with the path file are imported, in order to accommodate as many as functionalities as possible. Then, in order to present the voice augmented system, a graphical user interface program is devised which enables the user to operate by audio and visual feedbacks.

3 Experimental Results

The experimental results, at the progressing stage of the project, yielded successful outcomes in several implemented applications such as voice operated interactive personal assistant in digital systems, voice operated interactive personal assistant in home automation, voice operated vending machine, and voice operated interactive e-commerce systems, as illustrated in Fig. 3a–c.

3.1 Quantitative Measures

Dataset consists of 1000 words which was taken from Google Speech Command dataset. As the model is trained gradually with considerable number of datasets, the initial accuracy rate of the recurrent neural network algorithm is achieved 88.54%. In order to have a comprehensive understanding of this result, as given in Table 1, the same number of datasets was trained in different algorithms such as linear regression and polynomial regression algorithms which yielded accuracy rate of 53.61% and 72.89%, respectively. Hence, we could set a benchmark for the accuracy rate of the model, trained in recurrent neural network algorithm. After the model training was carried out further with some more datasets, the accuracy rate of 99.56% was achieved, and these results are only due to the primitive datasets trained to the model.

From Table 1, it is evident that recurrent neural network is the optimal algorithm for training the datasets with maximum accuracy. The major difference between the recurrent neural network and the rest of the supervised algorithms is that it is more suitable for natural language processing and audio processing, which is essential in human to machine interaction (HMI). In the case of regression algorithms, linear or polynomial, it determines an arbitrary equation through which it can determine the features in the datasets and produce or predict the optimum result during deployment. This is highly incompatible for language processing, because there are many situations in the datasets, where the equations that determine the features are incapable to recognize multiple possibilities out of a single input data. Since we have discussed the process of tokenization of datasets in the methodology section, where the input word is filtered to its root form, there are many homophones present in English that are read the same but have different meanings, this feature is highly difficult to be recognized using the arbitrary equations of regression algorithms, hence the accuracy level reduces, and we might find a little improvement in the accuracy if the epoch number is increased, but not as good as expected.

In the case of recurrent neural network, after the preprocessing of dataset, each word which is converted into its binary counterpart with unique binary value is fed into different neurons of different layers. A neuron can be any complicated mathematical function developed according to the application. Basically, a neural network consists of three layers: input, hidden, and output layers. The batch size or the number of input neuron, number of hidden layers, and number of output neurons depend

(a) e-commerce systems

(b) vending machine

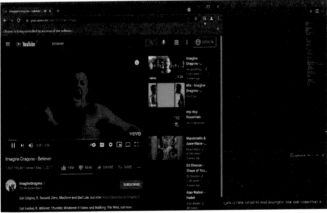

(c) Personal Assistant

Fig. 3 Voice operated interactive systems **a** e-commerce systems, **b** vending machine, **c** personal assistant

Table 1 Accuracy of various algorithms

Algorithms	Accuracy (%)
Recurrent neural network	88.54
Linear regression	53.61
Polynomial regression	72.89

on the application, for a given input sentence of minimum number of words in the sentence, each word with its unique binary value is fed into each neuron, and after the mathematical function is performed, an activation function passes the output to the next neuron in the hidden layer. In this layer according to the neural network developed, it performs forward recurrence or backward recurrence, i.e., when a particular binary value of a homophonic word is obtained, it stores the obtained mathematical output of the function in a temporary memory, and this algorithm is called as recurrent neural network long short-term memory (RNN-LSTM). Backward recurrence is performed with the homophonic binary value with the next binary value of the successive word in the input sentence. Hence, the predicted output value will be more accurate as every possibility of the homophonic binary value is calculated by performing recurrence. Therefore, the accuracy is considerably more, with more epoch numbers, the accuracy increases much more.

3.2 Qualitative Measures

The performance of the proposed algorithm is measured by using the error rate, training accuracy and comprehensive accuracy at different stages of the proposed model. Quality metrics formulas are listed below in Eqs. (1)–(2).

$$\text{Accuracy} = \left(\left(\sum V_i / N\right)\right) * 100 \tag{1}$$

where V_i is the predicted value of input words in each instance and N is the number of words in that epoch.

$$\text{Cross entropy loss} = H(P, Q) = H(P) + \Delta(P||Q) \tag{2}$$

where P is the predicted value, Q is the actual value, $\Delta(P||Q)$ is the divergence from Q to P, and $H(P)$ is entropy of P. In the accuracy of recognizing homophones in every situation, there will be similarly sounding words whose meaning will depend on the context it is being used. It was calculated by a built-in function of the PyTorch library.

Cross-entropy loss, model training accuracy, and the comprehension accuracy at the various stages of the proposed algorithm are given in Table 2. It is observed that the mode training accuracy approaches 99.56% at the final stage due to recurrent neural

Table 2 Quality metrics of the proposed algorithm

Quality metrics	Initial stage (%)	Intermediate stage (%)	Final stage (%)
Cross-entropy loss	10.63	01.80	00.03
Model training accuracy	88.54	91.28	99.56
Comprehension accuracy	85.54	81.36	83.56

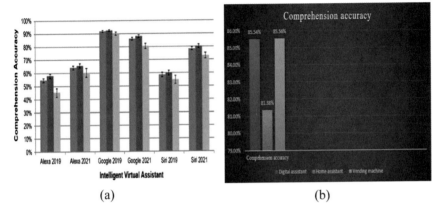

(a)   (b)

Fig. 4 Comprehension accuracy **a** intelligent virtual assistants **b** proposed system

network model. The comprehension accuracy of other intelligent virtual systems and the proposed system, which were trained on various datasets which included thousands of words in many languages and class of demographic each year, is shown below in Fig. 4.

The results show that the average comprehension accuracy is 57.3%, 89.64%, and 68.39% for Alexa, Google, and Siri, respectively. These results are due to the huge number of datasets monitored in various regions around the world, taking accents and language into consideration. The proposed algorithm yields an average accuracy about 83%, which is attained by considerably less and primitive dataset from English language alone, as shown in Fig. 4b.

The interfacing program successfully could recognize the voice input and appropriately open the respective windows and applications. When the user wishes to open YouTube, the model asks the user's desired video to watch in YouTube and open appropriate video in YouTube browser. If the user wishes to open Amazon shopping website, it will revert back to user with what does the user wish to buy and open appropriate section in the Amazon Web site. Similarly, it can access every application installed in the system and open them at the request of the user. If the user wishes to buy any given condiment from the vending machine, it will display the available stock of the condiments. When the user wishes to buy, it will enquire the number of items the user wishes to buy and display the final amount to be paid. After the transaction, it will update the stock and display the updated stock.

4 Conclusion

Voice-controlled digital assistants are becoming more natural, as it is integrated into everyday devices. Also, the emulations of human conversations will become much more natural. The main purpose of this chapter is achieved one step closer towards having a very intelligent system. In this chapter, a successful description of the terms used to describe actions makes the devices user friendly which is described in detail.

This chapter has its implementation in various domains, right from being personal assistant in a household, to being a tool to deploy sophisticated operations and functions in various industries. Moreover, this chapter has the advantage of being flexible in the aspect of implementation, it is not only restricted in industries and household, but it has the scope to be implemented in field of public, government, and defense. In other words, both scope and scale of this project are vast. In the future, a much more interactive experience environment through every digital channel is possible. Voice technology is becoming increasingly accessible to developers. And as consumers are getting increasingly easier and reliant upon using voice to speak to their phones, cars, smart home devices, etc., voice technology will become a primary interface to the digital world and with it, expertise for voice interface design and voice app development are going to be in greater demand. The future possibilities of advancements in the field of voice augmented systems are enormous. To build a strong speech recognition experience, the AI behind it is to become better at handling challenges like accents and ambience noise.

Acknowledgements This work is funded and supported by VGST-KFIST L1, Karnataka, India, with the grant number GRD 786.

References

1. Kameoka H, Tanaka K, Kwaśny D, Kaneko T, Hojo N (2020) ConvS2S-VC: fully convolutional sequence-to-sequence voice conversion. In: IEEE/ACM Transactions on audio, speech, and language processing, vol 28, pp 1849–1863
2. Kong Q, Cao Y, Iqbal T, Wang Y, Wang W, Plumbley MD (2020) PANNs: large-scale pretrained audio neural networks for audio pattern recognition. In: IEEE/ACM Transactions on audio, speech, and language processing, vol 28, pp 2880–2894
3. Miao H, Cheng G, Zhang P, Yan Y (2020) Online hybrid CTC/attention end-to-end automatic speech recognition architecture. IEEE/ACM Trans Audio Speech Lang Process 28:1452–1465
4. Wang D, Chen J (2018) Supervised speech separation based on deep learning: an overview. IEEE/ACM Trans Audio Speech Lang Process 26(10):1702–1726
5. Alepis E, Patsakis C (2017) Monkey says, Monkey does: security and privacy on voice assistants. IEEE Access 5:17841–17851
6. Rubio-Drosdov E, Díaz-Sánchez D, Almenárez F, Arias-Cabarcos P, Marín A (2017) Seamless human-device interaction in the internet of things. IEEE Trans Consum Electron 63(4):490–498
7. Teo JH, Cheng S, Alioto M (2020) Low-energy voice activity detection via energy-quality scaling from data conversion to machine learning. IEEE Trans Circ Syst I Regul Pap 67(4):1378–1388

8. Shrestha A, Mahmood A (2019) Review of deep learning algorithms and architectures. IEEE Access 7:53040–53065
9. Nassif AB, Shahin I, Attili I, Azzeh M, Shaalan K (2019) Speech recognition using deep neural networks: a systematic review. IEEE Access 7:19143–19165
10. Simeone O (2018) A very brief introduction to machine learning with applications to communication systems. IEEE Trans Cogn Commun Network 4(4):648–664
11. Delang K, Todtermuschke M, Schmidt PA, Bdiwi M, Putz M (2019) Enhanced service modelling for flexible demand-driven implementation of human–robot interaction in manufacturing. IET Collab Intell Manuf 1(1):20–27
12. Saini S, Sahula V (2020) Cognitive architecture for natural language comprehension. Cogn Comput Syst 2(1):23–31
13. L'Heureux A, Grolinger K, Elyamany HF, Capretz MAM (2017) Machine learning with big data: challenges and approaches. IEEE Access 5:7776–7797. https://doi.org/10.1109/ACCESS.2017.2696365
14. Yang C, Zeng C, Liang P, Li Z, Li R, Su C (2018) Interface design of a physical human-robot interaction system for human impedance adaptive skill transfer. IEEE Trans Autom Sci Eng 15(1):329–340. https://doi.org/10.1109/TASE.2017.2743000
15. Amarú L, Gaillardon P, De Micheli G (2014) Biconditional binary decision diagrams: a novel canonical logic representation form. IEEE J Emerg Sel Top Circ Syst 4(4):487–500. https://doi.org/10.1109/JETCAS.2014.2361058
16. Vorm ES (2020) Computer-centered humans: why human-AI interaction research will be critical to successful AI integration in the DoD. IEEE Intell Syst 35(4):112–116. https://doi.org/10.1109/MIS.2020.3013133
17. du Boulay B (2016) Artificial intelligence as an effective classroom assistant. IEEE Intell Syst 31(6):76–81. https://doi.org/10.1109/MIS.2016.93
18. Toda T, Black AW, Tokuda K (2007) Voice conversion based on maximum-likelihood estimation of spectral parameter trajectory. IEEE Trans Audio Speech Lang Process 15(8):2222–2235. https://doi.org/10.1109/TASL.2007.907344
19. Deng L, Li X (2013) Machine learning paradigms for speech recognition: an overview. IEEE Trans Audio Speech Lang Process 21(5):1060–1089. https://doi.org/10.1109/TASL.2013.2244083
20. Le H, Oparin I, Allauzen A, Gauvain J, Yvon F (2013) Structured output layer neural network language models for speech recognition. IEEE Trans Audio Speech Lang Process 21(1):197–206. https://doi.org/10.1109/TASL.2012.2215599
21. Erro D, Moreno A, Bonafonte A (2010) INCA algorithm for training voice conversion systems from nonparallel corpora. IEEE Trans Audio Speech Lang Process 18(5):944–953. https://doi.org/10.1109/TASL.2009.2038669
22. Dahl GE, Yu D, Deng L, Acero A (2012) Context-dependent pre-trained deep neural networks for large-vocabulary speech recognition. IEEE Trans Audio Speech Lang Process 20(1):30–42. https://doi.org/10.1109/TASL.2011.2134090
23. Louridas P, Ebert C (2016) Machine learning. IEEE Softw 33(5):110–115. https://doi.org/10.1109/MS.2016.114
24. Sheth A, Yip HY, Shekarpour S (2019) Extending patient-Chatbot experience with internet-of-things and background knowledge: case studies with healthcare applications. IEEE Intell Syst 34(4):24–30. https://doi.org/10.1109/MIS.2019.2905748
25. Makishima N et al (2019) Independent deeply learned matrix analysis for determined audio source separation. IEEE/ACM Trans Audio Speech Lang Process 27(10):1601–1615. https://doi.org/10.1109/TASLP.2019.2925450
26. Tu Y, Du J, Lee C (2019) Speech enhancement based on teacher-student deep learning using improved speech presence probability for noise-robust speech recognition. IEEE/ACM Trans Audio Speech Lang Process 27(12):2080–2091. https://doi.org/10.1109/TASLP.2019.2940662
27. Cui X, Goel V, Kingsbury B (2015) Data augmentation for deep neural network acoustic modeling. IEEE/ACM Trans Audio Speech Lang Process 23(9):1469–1477. https://doi.org/10.1109/TASLP.2015.2438544

28. Nakashika T, Takiguchi T, Ariki Y (2015) Voice conversion using RNN pre-trained by recurrent temporal restricted Boltzmann machines. IEEE/ACM Trans Audio Speech Lang Process 23(3):580–587. https://doi.org/10.1109/TASLP.2014.2379589
29. Sundermeyer M, Ney H, Schlüter R (2015) From feedforward to recurrent LSTM neural networks for language modeling. IEEE/ACM Trans Audio Speech Lang Process 23(3):517–529. https://doi.org/10.1109/TASLP.2015.2400218
30. Receveur S, Weiß R, Fingscheidt T (2016) Turbo automatic speech recognition. IEEE/ACM Trans Audio Speech Lang Process 24(5):846–862. https://doi.org/10.1109/TASLP.2016.2520364
31. Nakashika T, Takiguchi T, Minami Y (2016) Non-parallel training in voice conversion using an adaptive restricted Boltzmann machine. IEEE/ACM Trans Audio Speech Lang Process 24(11):2032–2045. https://doi.org/10.1109/TASLP.2016.2593263
32. Gannot S, Vincent E, Markovich-Golan S, Ozerov A (2017) A consolidated perspective on multimicrophone speech enhancement and source separation. IEEE/ACM Trans Audio Speech Lang Process 25(4):692–730. https://doi.org/10.1109/TASLP.2016.2647702
33. Wang Y, Narayanan A, Wang D (2014) On training targets for supervised speech separation. IEEE/ACM Trans Audio Speech Lang Process 22(12):1849–1858. https://doi.org/10.1109/TASLP.2014.2352935
34. Abdel-Hamid O, Mohamed A, Jiang H, Deng L, Penn G, Yu D (2014) Convolutional neural networks for speech recognition. IEEE/ACM Trans Audio Speech Lang Process 22(10):1533–1545. https://doi.org/10.1109/TASLP.2014.2339736
35. Truong H, Dustdar S (2015) Principles for engineering IoT cloud systems. IEEE Cloud Comput 2(2):68–76. https://doi.org/10.1109/MCC.2015.23
36. Guo Y, Stolyar AL, Walid A (2020) Online VM auto-scaling algorithms for application hosting in a cloud. IEEE Trans Cloud Comput 8(3):889–898. https://doi.org/10.1109/TCC.2018.2830793
37. Akata Z et al (2020) A research agenda for hybrid intelligence: augmenting human intellect with collaborative, adaptive, responsible, and explainable artificial intelligence. Computer 53(8):18–28. https://doi.org/10.1109/MC.2020.2996587
38. Kucherbaev P, Bozzon A, Houben G (2018) Human-aided bots. IEEE Internet Comput 22(6):36–43. https://doi.org/10.1109/MIC.2018.252095348
39. Hinton G et al (2012) Deep neural networks for acoustic modeling in speech recognition: the shared views of four research groups. IEEE Signal Process Mag 29(6):82–97. https://doi.org/10.1109/MSP.2012.2205597
40. Juang CF, Cheng CN, Chen TM (2009) Speech detection in noisy environments by wavelet energy-based recurrent neural fuzzy network. Expert Syst Appl 36(1):321–332
41. Juang CF, Lai CL, Tu CC (2009) Dynamic programming prediction errors of recurrent neural fuzzy networks for speech recognition. Expert Syst Appl 36(3P2):6368–6374
42. Tu CC et al (2012) Recurrent type-2 fuzzy neural network using Haar wavelet energy and entropy features for speech detection in noisy environments. Expert Syst Appl 39(3):2479–2488

Gamification in Education and Its Impact on Student Motivation—A Critical Review

Mifzala Ansar and Ginu George

Abstract In education, gamification refers to including game characteristics and design ideas in the classroom setting. Over the previous five years, gamification has increased student motivation and academic performance. This study will examine the previous literature to see how gamification will disseminate over time, educational level (from nursery to college), causes, and the most frequently used game elements. A systematic literature review will search interdisciplinary databases for quantitative experimental studies examining educational gamification and providing information on current research lines. According to the findings of a comprehensive research study, gamification can be advantageous at all academic levels, from elementary school to college. Following systematic research, gamified learning can increase students' motivation and intellectual accomplishment. Student learning may be made more pleasurable via gamification, which is the first advantage of this type of instruction. When used in the classroom, gamification can assist students who are weak in motivation and performing poorly academically. Because of the diversity of challenges and rewards that gaming parts provide, incorporating gaming elements into the classroom may serve as a motivational tool for students to learn. In the study's findings, students who enrolled in educational gamification courses were shown to be more interested and participatory than students enrolled in regular classrooms, on average.

Keywords Gamification · Education · Students · Classrooms · Learning · Games · Academic achievement

M. Ansar (✉)
Research Scholar, Department of Commerce, CHRIST (Deemed to be University), Bangalore, India
e-mail: mifzala.ansar@res.christuniversity.in

G. George
Asst. Professor, Department of Commerce, CHRIST (Deemed to be University), Bangalore, India
e-mail: ginu.george@christuniversity.in

1 Introduction

Gamification has received significant attention and interest in recent years [19]. Although many academicians and students are unaware of it, it is already a part of everyday routine [8]. It is possible to apply game design ideas to non-game situations; according to several definitions found across scientific literature, it has been utilized in various disciplines, including business, employment, health, and the environment. This research aims to look at the use of gamification in educational settings.

It is a term that refers to incorporating game design elements into the classroom to promote student engagement while also assisting them in developing their academic, cognitive, and interpersonal talents. Individuals are engaged, motivated to act, and problem-solving is facilitated via various techniques [1]. When it comes to tasks, the student's attitude displays a sense of empowerment, which increases their appeal and nurtures the good features connected with games [24].

Researchers have shown (2021) show that the rapid increase in the number of publications on gamification in education has been going on for at least seven years. The numbers of countries in which the contributing writers are based and the number of institutions to which they are linked suggests a widespread interest in the subject matter throughout the world. It is evident from the vast number of citations and the large cluster of co-citations in this domain that research communication in this field is successful [25].

Rozhenko et al. indicate that the use of gaming technologies boosts the motivation of students, helps to tilt the scales in favor of active learning, and helps students overcome lethargy, especially while studying mathematical fields that are tough [20].

Manzano-León et al. [17] According to student academic performance, dedication and motivation are all predicted to improve due to the gamification in educational settings. As a result of this study, it is necessary to do more research on the demands and obstacles that students have while using gamified learning approaches [17].

2 Features of Gamification

The goal of gamification is to engage individuals in lucid activities that boost their intrinsic and extrinsic motivation [2] and to do it in a fun and engaging way. While extrinsic motivation is driven by the quest for rewards, intrinsic motivation, on the other hand, motivates people because they like the inherent benefits of their jobs [7]. Finding intrinsic motives for gamification tactics is critical for keeping interest levels high. In the self-determination theory (SDT) [3], three psychological needs are—(1) Autonomy (the extent to which an activity is carried out solely for the sake of one's interests), (2) Competence (the sense of ability and ability to complete an action to a specific level), and (3) Relatedness with others—the definition of capability and ability to complete an activity to a particular grade.

As part of the motivating gamification process used by SDT, players (students) must feel autonomous, in command of their activities, and confident in their ability to complete tasks successfully. On the other hand, gamification must consider the different sorts of players to achieve this purpose.

Relational, Autonomy, Mastery, and Purpose (RAMP) is coined by Kindred and Mohammed [12]. Apart from that, game design tools create gamification confined to points, badge, and leaderboard (PBL). In terms of research, the mechanics, dynamics, and aesthetics (MDA) system developed by Hunicke, LeBlanc, and Zubek [28] has been the most thoroughly explored. Video game designers utilize the MDA model to combine the rules mechanisms, game dynamics, and visually appealing aesthetics systems into their games. It is easiest to categorize the gamification aspects into the following categories.

The game's mechanics are related to actions and control mechanisms available to the player. Players can, for example, draw cards, wager, trade, attack, compete, and collaborate with other players in the game. In addition to the mechanics, the dynamics indicate what the player should be doing at the execution time. Socializing, bluffing, contemplation, status, and attention are just a few of the behaviors that may be seen.

When players interact with a gaming system, "aesthetics" refers to the feelings elicited by the system. Sensation, fantasy, story, challenge, camaraderie, discovery, expressiveness, and enjoyment are all included in this category.

3 Gamification, Education, and Motivation

Education gamification, which includes game aspects and clearly defined goals and rewards, encourages students to engage more actively in class. The researchers at Beemer, Ajibewa, and DellaVecchia [25] discovered that children exposed to gamification in physical education were more than twice as likely as students who received conventional instruction to participate in at least 20 min of daily physical activity. Additionally, this study reveals that extra strategies and maximizing engagement during intermissions between programs are essential. Gamification with instant rewards (points and badges) and a narrative framework may be particularly effective. When courses are designed to be more game-like, it has been demonstrated that student engagement and performance both rise considerably in higher education.

Gamifying schools to encourage healthy behaviors leads to an overall more dynamic and pleasant school atmosphere where children may be more active while having a good time.

Using gamification to engage students in science has increased knowledge retention [26, 6]. According to the findings of the study, it is associated with their perception of being active participants in their education, which is bolstered by the game mechanics' progression and practice, which allow them to practice their academic curriculum in real time and provide them with clues and opportunities for reflection as they encounter new challenges. Commitment and engagement are encouraged in an e-learning university context by implementing a gamified system that includes peer interactions, blogs, challenges, and medals [13], among other features.

The writers can explain it because of a balance between difficulty and students' skills; students can continue through the course with a sense of accomplishment because of this balance between difficulty and students' abilities. Using gamification aspects such as points, badges, and leader boards, for example, students may lose interest in the course and learn over time, and they may become more dissatisfied with their gamified system [17].

4   Problem Statement and Research Objectives

The massive body of empirical evidence on educational gamification, frequently conflicting and refers to various leisure resources, such as instructional video games or game-based learning, are directed. While the implications of educational gamification are still being extensively distributed, more study is needed to define and synthesize present information, ultimately reducing the amount of time necessary to grasp this teaching approach. The use of gamification as a learning technique has been proved in the previous study to generate a significant amount of attention.

The findings of [15] were based on a series of 30 experiments conducted informal educational settings, which revealed that gamification had a more substantial impact on student learning outcomes than the control groups. Student motivation can be increased, can enhance talents, and learning can be maximized using gamification. According to research conducted by Trigueros et al. [27], gamification can encourage students to study, improve their abilities, and maximize their learning. Following these findings, higher education researchers are conducting more studies into the use of game-based learning activities to promote student engagement than was previously thought possible.

4.1   Research Objectives

This study aims to systematically evaluate educational studies that have used gamification in the previous five years, using specified criteria. The following questions are the focus of our investigation. Research Question-1: How can gamification be used in educational environments to benefit students?

Research Question-2: What educational aims do the chosen studies employ gamification? The current study objectives have been selected to get a more profound knowledge of the recent research and application of game-based learning (GBL) in education and discover which educational gamification projects have been most effective and how to reproduce them in future studies.

Table 1 Selection and elimination criteria for articles

Systematic literature review—August–October 2021		
Search area	Web of Science, Google Scholar, Scopus	
Period of study articles reviewed	2010–2021	
Language version	English	
Keywords	"Gamification," "Academic Learning," "School, College, University"	
Elimination criteria		
Total articles sourced		72
Elimination stage-1	Non-specialized websites, Blogs, newspaper articles, Books, Book Chapters, and Theses	(16)
Elimination stage-2	Theoretical and Introspective studies are not considered	(23)
Total articles considered for the study		33

4.2　Research Methods

A systematic review was conducted between August and October 2021 and conducted thorough research to ensure that considered a complete list of relevant studies. The review attempts to collect all available data by previously defined eligibility criteria. This approach results in more reliable findings for concluding and, consequently, decision-making because the findings are more reliable (Table 1).

Used an algorithmic search approach to assemble the systematic review papers [27]. A review of the literature using the databases Web of Science, Scopus, and Google Scholar was asserted. Only English versions of the search keyword utilized were (School or "High School" or "University") and Gamification (Program OR Intervention), and the search period spanned the previous five years (2016–2020).

This systematic review met the following criteria used to narrow the pool of likely candidates: The study materials may be in English only. The study has considered only pieces previously published in peer-reviewed specialized scientific journals for inclusion in the publication. The survey eliminated the following items: non-specialized websites, blogs, digital newspapers, books, book chapters, and PhD theses.

Investigations that were experimental or quasi-experimental were authorized— eliminated articles with a theoretical or introspective bent from consideration. Narrow down this topic to include only research that explicitly uses educational gamification techniques. Instructional video games and escape rooms are excluded from this list of studies since they are classified as "other recreational techniques" by the National Institutes of Health.

Eighty-two studies were identified and considered thirty after meeting the exclusion criteria mentioned above. The results of the extensive review are discussed as follows.

5   Discussion

This study aims to look at gamification systems at various levels of formal schooling to determine their effectiveness. Thorough research of educational gamification programs will benefit both professional instructors who use these approaches and the scientific community that publishes its results. Data suggests that gamification has a favorable impact on student's motivation, engagement, and academic achievement at all educational levels, regardless of their age or educational background. These findings suggest that educational gamification may be a helpful teaching technique in some situations.

Game-based learning has lately gained popularity in the educational setting, and it is expected to continue to do so in the future [8]. Gamification is becoming increasingly popular in educational environments. In education, gamification can take place in several different scenarios. According to the statistics, there is a growing interest in university education, focusing on student success than ever before. The integration of gamification with other instructional methodologies, such as project-based learning [9] and gamification in online learning environments [21], may also be investigated in educational gamification research. Students' and instructors' attitudes toward sustainability are shown to improve by Nurmi et al. [18], who demonstrate that educational gamification enhances student and teacher attitudes toward sustainability. According to the authors, teacher training is essential, ensuring that instructors are familiar with the approach and how to utilize it in the classroom and developing lesson plans that incorporate games and game features appropriate for the classroom environment.

It has been suggested that clever techniques may be directly tied to the emotional component, increasing public social awareness, and fostering emotional connections.

Thereby signaling a shift in pro-environmental behavior [26]. Gamification in the classroom may be a valuable method for long-term learning since it encourages students to actively participate in their education rather than passively listening to lectures about environmental problems. These findings are consistent with the results of one of the publications reviewed [20].

It states that educational gamification has positive repercussions for ethical training since it was fun. It is important to emphasize that while the usage of gamification tactics in these settings is expected to grow, students should have the ability to opt-out of the process if they so want, ensuring that the process is entirely voluntary and driven by their intrinsic desire. Additionally, it is vital to look at other potential uses of gamification in the educational setting.

The most important criteria to consider in this study, according to the findings, are motivation. The magnitude of the research evaluated shows that gamified tactics can increase student engagement and retention. The usage of badges was the sole method used in these trials to increase extrinsic motivation, which is notable. It has been shown that when learners are only awarded medals, their innate drive is suppressed, and they labor purely for the medals [21]. An attractive and engaging approach to education, in general, had been taken [8].

Gamification may be used to educate and strengthen curricular material and competencies. A standard indicator in educational gamification research is an academic accomplishment, which is also commonly employed.

According to the findings of a recent study, educational gamification has been shown to students' academic progress. Students are more motivated to study a subject in a classroom flow that has improved. Teaching methods have evolved to allow students to learn through challenges and significant tasks rather than the traditional one-way transmission of information. Regardless of their academic ability, they are more likely to complete exercises if provided with an entertaining story to read through before beginning them. It should also investigate the impact of gamification on disturbance and absenteeism in the classroom.

Due to the well-balanced design of the many components of gamification, it is thought that increased student interest and engagement will be possible in the classroom mechanics–dynamics–aesthetics (MDA). The most evaluated studies negatively impacted students using the PBL triangle (Points–Badges–Leaderboards) in most studies being assessed.

Though other studies altered their application, which may favor relying solely on extrinsic motivation, which eventually has a detrimental effect on students' intrinsic motivation [28], while PBL is very simple to implement. It is vital to proceed to more intricate gamification designs that include different dynamics, mechanics, and aesthetics that work as reinforcing factors for students' intrinsic drives rather than relying just on the simple mechanics of PBL. "An atmosphere with clear objectives…difficult work and realistic tales in which team spirit is maintained through games, chats, and mechanical arguments" is essential for improving motivation, as is keeping an eye on the surrounding environment and paying attention to design. Our research has provided our research to improve student performance by creating a gamified environment that includes a diverse range of mechanics, dynamics, and aesthetic aspects. The notion of player kinds may have a role in this. Establishing a rapport with the students, discovering what games they enjoy playing, and observing how they engage with one another are essential for building an engaging educational gamification program and delivering a compelling MDA.

Following this investigation, the most significant ramifications of gamification in education are highlighted. The first advantage that gamification provides to students is an enjoyable learning method. In the classroom, gamification can aid pupils who lack motivation and who are failing academically. Due to the variety of difficulties and rewards that gaming aspects bring, including gaming features in the classroom may motivate youngsters. According to the study's findings, students enrolled in educational gamification courses are more engaged and participative than students enrolled in regular classrooms in general.

6 Conclusion

It has been demonstrated that comprehensive literature reviews benefit social science research. One of the most distinguishing qualities of the systematic review approach is its willingness to accept criticism and transparent operations. As with any research procedure, there are inherent limits to both the methodology and the implementation of the study. This review is not without its limitations, though. Included peer-reviewed academic experiments in this analysis; no other types of research, such as those found in gray literature or book excerpts, were considered in this study. As a result, we feel this is realistic given the number of analyzed papers (33 articles reviewed in full text).

Furthermore, only publications published in English were included in the evaluation, removing potentially relevant items from consideration. As a result, there was the risk of publication bias in the systematic review. As a result of these limitations, the study results cannot be expanded upon in-depth. Although this analysis is limited in scope, its significance lies in compelling an extensive body of data demonstrating the usefulness of educational gamification in a formal educational context, emphasizing the specific characteristics of gamification employed in each example.

7 Way Forward

The primary goal of this research was to acquire a better grasp of the concept of educational gamification. According to the findings of a comprehensive research study, gamification can be advantageous at all academic levels, from elementary school to college. Under systematic research, gamified learning can increase students' motivation, engagement their intellectual accomplishment. Furthermore, the results of this poll revealed that points, medals, and rankings are the most often employed gamification features in educational settings. Using simply one or two game-like qualities like points or badges, the effects on student motivation may be minor, if not detrimental, depending on the situation. According to this research, a gamified environment that is diverse and personalized to the player's preferences is more motivating, may match the player's demands based on their player profiles, and is more likely to meet their needs. This project aims to further theoretical and experimental efforts to assess and develop pleasurable learning techniques that will improve the overall quality of education in the classroom setting.

We will examine several mechanics to identify how many most successfully use educational games at various levels of formal education. These mechanics and dynamics will vary depending on students and their training requirements.

References

1. Alsawaier RS (2018) The effect of gamification on motivation and engagement. Int J Inform Learn Technol 35(1):56–79. https://doi.org/10.1108/ijilt-02-2017-0009
2. Buckley P, Doyle E (2014) Gamification and student motivation. Interact Learn Environ 24(6):1162–1175. https://doi.org/10.1080/10494820.2014.964263
3. Black AE, Deci EL (2000) The effects of instructors' autonomy support and students' autonomous motivation on learning organic chemistry: a self-determination theory perspective. Sci Educ 84(6):740–756
4. Campillo-Ferrer JM, Miralles-Martínez P, Sánchez-Ibáñez R (2020) Gamification in higher education: impact on student motivation and the acquisition of social and civic key competencies. Sustainability 12(12):4822. https://doi.org/10.3390/su12124822
5. Dias J (2017) Teaching operations research to undergraduate management students: the role of gamification. Int J Manage Educ 15(1):98–111. https://doi.org/10.1016/j.ijme.2017.01.002
6. Díez Rioja JC, Bañeres Besora D, Serra Vizern M (2017) Experiencia de gamificación en Secundaria en el Aprendizaje de Sistemas Digitales. *Education in the Knowledge Society (EKS)*, *18*(2), 85–105. https://doi.org/10.14201/eks201718285105
7. Fischer C, Malycha CP, Schafmann E (2019) The Influence of intrinsic motivation and synergistic extrinsic motivators on creativity and innovation. Frontiers Psychol 10. https://doi.org/10.3389/fpsyg.2019.00137
8. Frick LT, de Souza DB (2014). Projetos bem-sucedidos em educação em valores: relatos de escolas públicas brasileiras. *Nuances: Estudos Sobre Educação*, *24*(3), 240–245. https://doi.org/10.14572/nuances.v24i3.2709
9. Floryan MR, Ritterband LM, Chow PI (2019) Principles of gamification for Internet interventions. Transl Behav Med 9(6):1131–1138. https://doi.org/10.1093/tbm/ibz041
10. Gatti L, Ulrich M, Seele P (2019) Education for sustainable development through business simulation games: an exploratory study of sustainability gamification and its effects on students' learning outcomes. J Clean Prod 207:667–678. https://doi.org/10.1016/j.jclepro.2018.09.130
11. Huang R, Ritzhaupt AD, Sommer M, Zhu J, Stephen A, Valle N, Hampton J, Li J (2020) A meta-analysis -the impact of gamification in educational settings on student learning outcomes: a meta-analysis. Educ Tech Res Dev 68(4):1875–1901. https://doi.org/10.1007/s11423-020-09807-z
12. Kindred J, Mohammed SN (2006) He will crush you like an Academic Ninja!: exploring teacher ratings on Ratemyprofessors.com. J Comput-Mediated Commun 10(3):00. https://doi.org/10.1111/j.1083-6101.2005.tb00257.x
13. Kyewski E, Krämer NC (2018) To gamify or not to gamify? An experimental field study of the influence of badges on motivation, activity, and performance in an online learning course. Comput Educ 118:25–37. https://doi.org/10.1016/j.compedu.2017.11.006
14. Mahmud SND, Husnin H, Tuan Soh TM (2020) I am teaching presence in online gamified education for sustainability learning. Sustainability 12(9):3801. https://doi.org/10.3390/su12093801
15. Macedo Reis H, de Sousa Borges S, Isotani S (2014). Análise de Usabilidade de Sistemas de Geometria Interativa para Tablets. *RENOTE*, *12*(1). https://doi.org/10.22456/1679-1916.50356
16. Mobasher M (2021) Earthquake in the city: using real-life gamification model for teaching professional commitment in high school students. J Med Ethics History Med. Published. https://doi.org/10.18502/jmehm.v14i1.5316
17. Manzano-León A, Camacho-Lazarraga P, Guerrero MA, Guerrero-Puerta L, Aguilar-Parra JM, Trigueros R, Alias A (2021) Between level up and game over: a systematic literature review of gamification in education. Sustainability 13(4):2247
18. Nurmi J, Knittle K, Ginchev T, Khattak F, Helf C, Zwickl P, Castellano-Tejedor C, Lusilla-Palacios P, Costa-Requena J, Ravaja N, Haukkala A (2020) Engaging users in the behavior change process with digitalized motivational interviewing and gamification: development and feasibility testing of the precious app. JMIR Mhealth Uhealth 8(1):e12884. https://doi.org/10.2196/12884

19. Ouariachi T, Li CY, Elving WJL (2020) Gamification approaches for education and engagement on pro-environmental behaviors: searching for best practices. Sustainability 12(11):4565. https://doi.org/10.3390/su12114565
20. Rozhenko OD, Darzhaniya AD, Bondar VV, Mirzoian MV (2021) Gamification of education as an addition to traditional educational technologies at the university. In: CEUR workshop proceedings, vol 2914, pp 457–464
21. Sailer M, Homner L (2019) The gamification of learning: a meta-analysis. Educ Psychol Rev 32(1):77–112. https://doi.org/10.1007/s10648-019-09498-w
22. Sipone S, Abella-García V, Barreda R, Rojo M (2019) Learning about sustainable mobility in primary schools from a playful perspective: a focus group approach. Sustainability 11(8):2387. https://doi.org/10.3390/su11082387
23. Santos-Villalba MJ, Leiva Olivencia JJ, Navas-Parejo MR, Benítez-Márquez MD (2020) Higher education students' assessments towards gamification and sustainability: a case study. Sustainability 12(20):8513. https://doi.org/10.3390/su12208513
24. Sezgin S, Yüzer TV (2020) Analyzing adaptive gamification design principles for online courses. Behav Inform Technol 1–17. https://doi.org/10.1080/0144929x.2020.1817559
25. Strahringer S (2015) Rezension "Gamification in education and business." HMD Praxis Der Wirtschaftsinformatik 52(6):928–930. https://doi.org/10.1365/s40702-015-0183-8
26. Swacha J (2021) State of research on gamification in education: a bibliometric survey. Educ Sci 11(2):69
27. Trigueros R, Aguilar-Parra JM, Lopez-Liria R, Cangas AJ, González JJ, ÁLvarez JF (2020) The role of perception of support in the classroom on the students' motivation and emotions: the impact on metacognition strategies and academic performance in math and English classes. Frontiers Psychol 10. https://doi.org/10.3389/fpsyg.2019.02794
28. Tsai C, Lin H, Liu S (2019) The effect of the pedagogical GAME model on students' PISA scientific competencies. J Comput Assist Learn 36(3):359–369. https://doi.org/10.1111/jcal.12406
29. Thoemmes FJ, Kim ES (2011) A Systematic review of propensity score methods in the social sciences. Multivar Behav Res 46(1):90–118. https://doi.org/10.1080/00273171.2011.540475
30. Turan Z, Avinc Z, Kara K, Goktas Y (2016) Gamification and education: achievements, cognitive loads, and views of students. Int J Emerg Technol Learn (IJET) 11(07):64. https://doi.org/10.3991/ijet.v11i07.5455
31. Wu B (2019) Hierarchical macro strategy model for MOBA game AI. In: Proceedings of the AAAI conference on artificial intelligence, vol 33, pp 1206–1213. https://doi.org/10.1609/aaai.v33i01.33011206

Suspicious Human Behaviour Detection Focusing on Campus Sites

Mohammed Mahmood Ali[ID]**, Sara Noorain, Mohammad S. Qaseem, and Ateeq ur Rahman**

Abstract Safety and security are the need of the hour for every person in this century. Many countries have acquired and adopted surveillance systems using high-definition CCTV cameras to secure their environment. Thus, automated CCTV surveillance systems can serve as security providers for identifying the behaviour of suspicion humans or intruders based on their actions from previous stages through the CCTV footages is a challenging task. This can be achieved using suspicious behaviour models that use machine learning approaches to a limited extent. It has been found that many of earlier approaches have made use of deep learning, machine learning, IoT and fuzzy logic techniques. The proposed suspicious behaviour detection model (SHBDM) effectively uses CNN model pre-trained on the ImageNet dataset, known as Inception V3 (VGG-16), for image feature extraction. This system was implemented in Python on an open-source platform. The use of Inception V3 + LSTM resulted in an improved precision accuracy of 88.8% when compared with VGG-16 + LSTM and simple CNN model.

Keywords Security · CCTV surveillance systems · Suspicious behaviour detection models (SHBDM) · CNN model · Inception V3

1 Introduction

Human behaviour is one of the most complex subjects in the world with the upcoming challenges for researchers. Studying human behaviour deals with the fact that human

M. M. Ali (✉) · S. Noorain
MuffakhamJah College of Engineering and Technology, Hyderabad, Telangana, India
e-mail: mahmoodedu@gmail.com

M. S. Qaseem
Nawab Shah Alam Khan College of Engineering and Technology, Hyderabad, Telangana, India

A. ur Rahman
Shadan College of Engineering and Technology, Hyderabad, Telangana, India

© The Author(s), under exclusive license to Springer Nature Singapore Pte Ltd. 2023
M. A. Chaurasia and C.-F. Juang (eds.), *Emerging IT/ICT and AI Technologies Affecting Society*, Lecture Notes in Networks and Systems 478,
https://doi.org/10.1007/978-981-19-2940-3_12

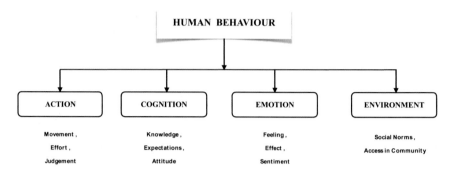

Fig. 1 Basis of human behaviour

test subjects always vary of the fact that they are being observed and can change their conduct in response—whether consciously or unintentionally [1] (Fig. 1).

Many countries have taken up various measures of installation of closed-circuit television (CCTV) surveillance cameras which have the ability to record criminal instances that are saved in storage servers. Strangely, these recordings are of no use unless they are efficiently used for identifying the crimes and tracing the criminals [2]. Now-a-days the surveillance systems are made mandatory everywhere, and further these surveillance systems have also proved to be an essential tool in lot of ways. The biggest example is that the governments in most of the countries use these modern cameras and drones to ensure that the lockdown restrictions imposed are properly followed smoothly and respected by people during the COVID-19 pandemic [3].

Monitoring is an integral part of surveillance systems, and in this process, technology plays an active role. One of the earliest video surveillance technologies to be introduced was closed-circuit television (CCTV). Its major duty is to keep track of all visual activity in a given area, and it continues to do so. It is believed that when criminals look at the CCTV surveillance systems, they are hesitant to intrude because the recordings can help in identifying them. There is a saying 'People Aren't Suspicious, behaviour Is' [8]. Although whenever we say 'a suspicious person', it indicates that their behaviour that is suspicious. Suspicious behaviour might be difficult to characterize, but most individuals agree that they 'know it when they see it'. It is something that is out of place, not quite right, or simply makes you feel 'weird' about it. Suspicious events faced on campuses are usually *theft, vandalism, bullying, and violence.*

In today's world, safety and security is a major concern, wherein everybody faces issues with maintaining their privacy, protecting their close ones and their assets. Crime news covers a major part of our newspapers. Hence, it has become very important to increase the security and safety measures. Even though, the first step towards that is monitoring and spying, but the important objective is to be able to stop such unfavourable event(s) or crime in real time to prevent its drastic consequences. Hence, earlier many security personnel were required for continuous monitoring by either taking rounds or by viewing the CCTV cameras manually in a continuous manner.

Now, with the advancement of technology, real-time monitoring by the system itself is possible. In sensitive areas like Campuses, which usually have open-gates and comprise of large areas wherein almost everyone can get into the campus unnoticed, thereby creating a strong demand for real-time suspicious actions tracking. Abnormal human suspicion activities executed collectively in smart cities in a particular locality is another issue which is addressed by Asma belhadi and et al., using deep learning and data mining techniques by excessively training the system. This system assisted in identifying the trajectory movement in sequence of suspicion activities of those humans in smart cities which helped in improving the discipline in smart cities [5]. The movements of pedestrians on roads are complex issues where the trajectory movements of the disabled persons can be easily traced using computer vision and deep learning strategies. This system can also be efficiently used for monitoring tool for video surveillance of persons moving on roads and suspicion human behaviour which assists to find the exact location [6].

A university should be a safe and secure haven for both students and staff members. The protection of students and faculty must be the top priority for any educational institution administration. Apart from the criminals, sometimes the students also behave in a suspicious manner. It could be due to *distress* or *anxiety*. Hence, monitoring the human behaviour on campuses helps in preventing any bigger undesirable events caused by mob mentality.

The following section explains about the significance and its necessity to detect suspicious events from the behaviour of persons from CCTVs footages that are captured and precautionary steps taken up by raising alarms before the criminal activity takes place. Ultimately, this impact towards maintaining the peace by avoiding crime-related activities in the society. In Sect. 2, we define the suspicious behaviour definition based on previous studies and the problems faced if suspicious events are not detected even after recordings of CCTVs exist. In Sect. 3, a brief literature overview based on existing campus detection models is presented. In Sect. 4, we proposed SHBDM system that may improve the existing suspicious behaviour detection models by embedding the concept of generation of alarms not only through SMS, emails, WhatsApp, but alerting the security personal through automated phone calls and tracing the profile details (such as name, photo, address, location and job) using IoT and enhanced online suspicious detection models. Finally, in Sect. 6, it is observed that every safety and security measures adapted the use of high-end cameras that captures the recording with excessive storage servers, which has become one of the important parts of surveillance systems without which surveillance is not possible. It is also concluded that many of intelligent CCTVs lack the features of alerting mechanism which is an immense requirement and the current need to mitigate crimes especially in campuses.

2 Problem Statement and Previous Knowledge

2.1 Suspicious Behaviour Definition

"Suspicious events could be a consequence of suspicious activities. A suspicious behaviour or activity may be any action that is out of place and does not fit into the same old day-to-day activity that might indicate an individual may be involved in or about to commit a criminal offense" [2].

2.2 Suspicious Behaviour

2.2.1 Person Running

A person running generally indicates that he is in a hurry or urgency and tensed about it. It is more suspicious if the person was looking here and there nervously as if he was being watched.

2.2.2 Unidentified Person Carrying Property

A person with suspicious behaviour can be carrying noticeable items or carrying something during unusual hours or locations. The major question is "Could the person be a suspicious person involved in theft?"

2.2.3 Unusual Behaviour

- A person seen doing anything in an environment where he/she does not belong or may seem suspicious. It could be an inappropriate attire, or the individual is being unable to hold on to a meaningful conversation. A suspicious person is someone who exhibits suspicious conduct or who is in an unusually suspicious environment and circumstance.
- Other unusual behaviour includes nervousness. Actions such as nervous glancing or/and sweating followed by repeated entering and exiting from a facility.
- Attempts by an individual to conceal one's face by turning away when another person approaches.
- 'Hiding' behind objects in an attempt to keep himself/herself from being identified.

2.3 Problem Statement

Suspicious behaviour needs to be verified at an early stage before it turns into a serious event, by prompting an appropriate safety measure in real-time scenarios at run time. It is observed that most of the criminal cases are unsolved because the surveillance system records without generating alarms or alerts to near-by police department or neighbourhood security personnel. The technique developed for tracking a suspicious person in a campus is implemented using VGG-16 + LSTM model [7]. Similarly, many approaches to detect and track suspicious events and persons were proposed, and most of them make use of machine learning (i.e., classification technique) and deep learning techniques.

The strategy is developed for detecting suspicious behaviour of people by comparing daily CCTVs surveillance footage from a campus, in which the captured images are trained using CNN model. Once the detection model is learned, it identifies a suspicious behaviour with the help of LSTM [8]. Then, the detection model alerts the concerned authorities which can help in preventing subsequent bigger problems or crimes.

Places of learning like institutions and educational campuses are usually open to all and connected to large areas with various departments and sections due to which there is an imminent need for a dynamic automated suspicious behavioural tracking through advanced intelligent CCTVs.

In this section, we thoroughly brief the problems that are faced if automated CCTVs are not efficiently deployed and utilized by embedding advanced technologies of deep learning models that have the capability to extract the behavioural features from captured images using high-definition cameras, which assist in identifying the suspicious behaviour resulting in avoiding of criminal activity.

3 Literature Overview on Campus-Based Behaviour Detection Models

Similar works suggest various methodologies for detecting human behaviour in campuses from video footages. The main aim of this work is to understand any anomalous or suspicious behaviour from those videos.

The ICSS [2] surveillance system recognizes objects and individuals in the surveillance area with pinpoint accuracy. If an suspicion is detected, the system checks to see if any valuable items are moved from their usual location. In addition, the suspicion activities in the specific location are measured. Subsequently, the user is alerted for any unexpected behaviour. The system's real-time alerts via WhatsApp microblogs and audio conversations are a standout feature. Both day and night modes are supported by the system.

STAM-CCF [9] improves human identification and avoids errors caused by information loss in cases of object occlusion and overlapping while tracking cases by using the geographical information of cameras and the YOLO object recognition framework. To better identify such challenges of identification, many cameras were installed that initializes the use of camera correlation system along with the two-stage gait identification technique.

Radar sensors are also being used to see if radar systems can be used for security surveillance. An end-to-end neural architecture that can take real-time radar data inputs and distinguish between human and nonhuman targets, as well as classify diverse human behavioural gestures, is presented [10]. The model is monitored in real time.

An AI-based desktop application is built and developed to initiate recording when a person or a human face is detected. This technique increases the system efficiency by reducing not only the amount of storage required for recording storage but also the processing and searching time in the recordings. The proposed system [11] is based on Linux and incorporates deep learning techniques and OpenCV libraries. Human body detection and human-face prediction are the two detection mechanisms. The appropriate detection algorithms will be triggered based on the mode selected. Various organizations that use surveillance systems are assessing this system.

The deep spatio-temporal approaches are applied to crowd behaviour, and the videos were divided into three distinct categories: pedestrian future path prediction, holistic crowd behaviour, and destination estimation. A CNN layer was used to extract spatial information from video frames. The LSTM approach was utilized to learn the temporal motion sequences. CUHK, UCY, ETH and PWPD datasets were employed in the proposed system. By adopting deeper architectures, the system's accuracy can be increased [12].

The optical flow in each zone was evaluated using the Lucas–Kanade method [13] after abnormal occurrences from a university were separated into zones. The magnitude histogram of optical flow vectors was then produced. The content of a video is analysed using software algorithms to classify events as normal or abnormal.

In an academic setting, a deep learning approach [7] is utilized to detect suspicious or normal behaviour which sends an alarm message to the concerned authorities if suspicious activity is predicted. Monitoring is frequently carried out by extracting successive frames from a video footage. The framework is split into two sections. In the first phase, video frames are used to compute the characteristics, and in the second phase, the classifier uses the features to predict whether the class is suspicious or normal.

An accuracy of 87.15% was achieved for KTH and CAVIAR dataset along with some YouTube videos and real-time videos.

For recognizing human behaviour from videos, most of the research work discussed above uses computer vision, various algorithms or/and neural networks. Backdrop subtraction, for example, is based on a static background hypothesis that is frequently inapplicable in real-time circumstances. In the real world, many problems arise in crowded places. When it comes to dealing with crowds, the solutions mentioned above are not effective. Thus, based on the literature review, a deep architecture for suspicious behaviour detection in campuses is modelled using Inception V3 [14] and LSTM, allowing the system's accuracy to be enhanced.

4 Proposed Model

The suspicious behaviour is detected through the proposed SHBDM. Since this is a real-time suspicious event detection system, it is a continuous process carried out until stopped. The SHBDM takes the video surveillance footages and then extracts images. Then, if any kind of motion is detected in the frames, it is classified into suspicious or normal. Further, in case of suspicious behaviour, an automated alarm is raised to security personnel for prompt action (Fig. 2).

4.1 System Architecture

For detecting the suspicious human behavior in campuses few notations are used such as 'L' is used to represent real-time video footages. To address this problem, we define 'y' as dataset extracted from pre-existing real-time surveillance footage, 'vF' as the frames extracted, 'M' for the trained model and then propose the model to detect suspicious behaviour in successive frames. SHBD model has four modules that help in detecting suspicious behaviour. The detailed architecture of the proposed model is presented with the help of Fig. 3.

(i)   *Interface Module*: It utilizes real-time surveillance footage that is denoted by 'yr'. This real-time footage is passed to the classification module for prediction of behaviour and gains the labelled video 'vL'.

(ii) *Visual Module*: It utilizes the pre-existing real-time video surveillance footage 'y' to create a behavioural database 'vF'. This behavioural database will contain preprocessed and normalized frames of each behavioural category that can be passed to network training module for training the model.

(iii) *Network Training Module*: This module utilizes 'vF' for creating and training a model M that can be further utilized by the classification module.

(iv) *Classification Module*: This module classifies 'yr' into different behavioural features $f_1, f_2, \ldots, f_6$ and provides a labelled video 'vL' as outcome.

Fig. 2 Workflow of proposed model

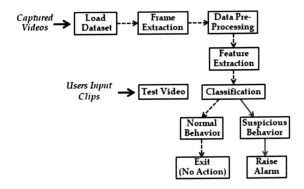

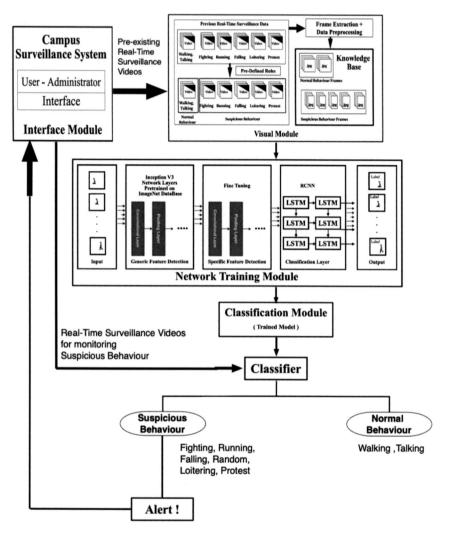

Fig. 3 System architecture

The first step is loading the footage captured with the help of surveillance cameras in real time and checking whether the footage is blank. The processing in implementation is carried out by using the frames of the video footages, and hence, the videos are converted to the frames. The implementation can be done on live videos as well as by saving those videos and loading them later. The frames can then be preprocessed according to the requirements of the network training model. The most important step is when they are passed as input to the model for classification. Classified label of each frame is achieved if there is any suspicious behaviour detected in successive frames; then, the authorized person is notified or alerted with the video clip. This

Table 1 Feature categorization

S. No	Category	Features
1	Normal behaviour	Walking, talking to other people and talking on phone
2	Suspicious behaviour	Running, loitering, fighting, falling, protest and throwing papers

whole process is repeated continuously in real time for automatic surveillance until the system is shut down.

4.2 Camera Placement and Video Surveillance

The CCTV cameras that are placed around the whole campus record the behaviour and activity of students in the campus in real time. The footage gained from this is continuously passed to the proposed model, and it keeps detecting regular behaviour or abnormalities.

4.3 Feature Categorization and Dataset Description

Suspicious behaviour detection is a long process and has many challenges. To overcome these challenges, each step or module in architecture must be defined clearly. The foremost thing is to categorize features or attributes by defining what is suspicious and what is normal. The list of features mentioned in Table 1 will help in categorizing human behaviour.

Various videos are captured from different cameras, covering the outer area of like corridor and parking of the campus site. The implementation is carried out by converting the videos to frames. There were many standard and benchmark datasets available. A custom dataset which has collection of sequences representing six actions and has got 100 sequences was created. The whole dataset is manually labelled like UCF data set format. The model is trained on this dataset for normal behaviour (walking) or suspicious behaviour (running, falling, fighting and loitering). Table 2 represents the different actions extracted from their respective standard datasets.

4.4 Implementation

Frames are extracted from captured videos as part of preprocessing. Two labelled folders for storing behaviour-based frames and an output folder are created. The complete 100 videos are converted to frames, which are then saved as jpg images.

Table 2 Standard datasets used in making of custom human behaviour dataset

S. No	Name of the dataset	Videos extracted
1	Avenue dataset for abnormal event detection [15]	Normal behaviour in a campus and suspicious behaviour related to throwing items like paper in air
2	The VIRAT video dataset [16]	Normal behaviour in a campus parking and outdoors
3	IIT-B corridor dataset [17]	Normal behaviour in a campus and suspicious behaviour like running, loitering, fighting, and random instances like chasing or protesting
4	UR fall detection dataset [18]	Suspicious detection—Falling

Each frame is then stored after being scaled to the desired size for CNN architecture. In addition, the 12 testing videos are also converted to frames. In Google Colaboratory, the proposed system was constructed in Python. In order to extract image features, a CNN model pre-trained on the ImageNet dataset known as Inception V3 is used. Convolution layers, SoftMax layer activation function, max pooling layers, and fully connected dense layers are all included, and thus, the model can be fine-tuned. Therefore, the model's last layer is removed. After that, the model is trained using the LSTM architecture. In sequence prediction problems, LSTM networks are a type of RNN that can learn order dependency. The number of neurons in the last layer is equal to the number of classes we have, and thus, there are six neurons (Fig. 4).

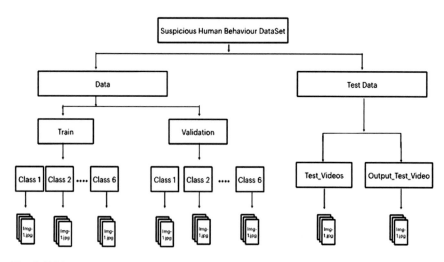

Fig. 4 Folder structure

5 Result Analysis

The proposed SHBD model is compared with [7] which uses the Python OpenCV package for video pre-processing. A pre-trained CNN model known as VGG-16 was trained on the ImageNet dataset for image feature extraction, and the last layer of this model was removed. The model was then trained using an LSTM architecture.

The entire system consisted of collecting the video sequences from CCTV footage, extracting frames from video footages, preprocessing the images, and preparing the images for training and testing. When the system detects suspicious activity, it sends an SMS to the concerned authority. The system has been developed in Python on an open-source platform. Creating an account with Twilio by installing the Twilio library in Python was used to send SMS.

In comparison with VGGNet, we have proposed Inception networks to be used for the suspicious human behaviour detection because they have proven to be more exceptionally efficient, in terms of the features generated by this network with low cost, less memory usage and minimizes resource utilization.

In the Inception V3 model, several ways for enhancing the network have been presented to loosen the constraints and make model adaption easier. Factorized convolutions, regularization, dimension reduction and parallelized calculations are among the approaches used.

To gauge the SHBDM, we have considered 12 scenarios or video clips which contain either single or multiple behavioural anomalies. Based on these scenarios, we have calculated precision, recall and F1-score. The proposed model gained higher F1-score percentage when compared to VGG-16 + LSTM model (Tables 3 and 4).

Table 3 Comparison of results using different approaches

	Simple CNN model (%)	VGG-16 + LSTM (%)	Inception V3 + LSTM (%)
Accuracy	66.66	72.2	88.8
F1-score	72.76	79.9	91.61

Table 4 Comparison of time efficiency based on different approaches

Models	Accuracy (%)	Time taken for prediction of single event (average) (Colab GPU time) (s)
Simple CNN model	66.66	15.58
VGG-16 + LSTM	72.2	17.11
Inception V3 + LSTM	88.8	16.62

6 Conclusion

In today's world, practically everyone understands the importance of CCTV footage, yet in most circumstances, video footages are utilized to investigate after the crime or incident has occurred. The advantage of a surveillance system is preventing a crime from occurring by tracking and analysing in real time from suspicion behaviour [3]. The analysis obtained and the conclusion is a directive to the concerned authorities to take action.

The important factor in identifying the success of surveillance systems is how well they detect suspicious human behaviour which could lead to a security breach [19]. Regrettably, present surveillance techniques are primarily reliant on human observers. As a result, these technologies' ability to become front-line crime-fighting tools is limited. The present systems use a variety of strategies to get started.

To demonstrate the usefulness of the proposed model for identifying suspicious human behaviour, a comparative experiment was conducted between the proposed systems and existing systems discussed in Sect. 5. The proposed system was able to perform more accurate detections as a result of this research since it employs Inception V3 network.

The experiment carried out demonstrates that the proposed system performs better when it has previously acquired knowledge of the current situation. The addition of LSTM to Inception V3 also increased the model's accuracy. The system was proved to be capable of detecting unexpected events in the final experiment by accurately identifying suspicious human behaviour in campuses. Experimental results show that the proposed model can improve the detection performance when compared to [7] and achieved of precision and of accuracy.

References

1. (Online). Human Behaviour. https://www.nu.edu/resources/ask-an-expert-can-human-behavior-be-studied-scientifically/
2. Khodadin F, Pudaruth S (2020) An intelligent camera surveillance system with effective notification features. UOB J (6)
3. Ali MM, Qaseem MS, Rahman A (2022) Strategies and tools for effective suspicious event detection from video: a survey perspective (COVID-19). In: Chaurasia MA, Mozar S (eds) Contactless healthcare facilitation and commodity delivery management during COVID 19 pandemic. Advanced technologies and societal change. Springer, Singapore
4. (Online). Suspicious Event Types. https://dpss.umich.edu/content/services/report-a-crime/suspicious-behavior/
5. Belhadia A, Djenouri Y et al (2021) Deep learning for pedestrian collective behavior analysis in smart cities: a model of group trajectory outlier detection. J. Inform Fusion 65:13–20. Elsevier
6. Brunetti A, Buongiorno D et al (2018) Computer vision and deep learning techniques for pedestrian detection and tracking: a survey. J Neurocomput 300:17–33. Elsevier
7. Amrutha C, Jyotsna C, Amudha J (2020) Deep learning approach for suspicious activity detection from surveillance video. In: International conference on innovative mechanisms for industry applications (ICIMIA)

8. (Online). LSTM. https://en.wikipedia.org/wiki/Long_short-term_memory
9. Sheu R-K, Pardeshi M, Chen L-C, YuanKe S-M, Kim HD-S (2019) STAM-CCF: suspicious tracking across multiple camera based on correlation filters. Sens (Basel) J 19(13)
10. Kaushik P (2019) Radar as a security measure real time neural model based human detection and behaviour classification. In: IEEE global 2019 conference on signal and information processing (GlobalSIP)
11. Alajrami E, Tabash H, Singer Y, Astal M-E (2019) On using AI based human identification in improving surveillance system efficiency. In: International conference 2019 on promising electronic technologies (ICPET)
12. Li Y (2019) A deep spatiotemporal perspective for understanding crowd behavior. IEEE Trans Multimedia 20(12)
13. Kain Z, Youness A, El Sayad I, Abdul-Nabi S, Kassem H (2018) Detecting abnormal events in university areas. In: International conference on computer and application
14. Zahid Y, Tahir MA, Durrani MN (2020) Ensemble learning using bagging and Inception-V3 for anomaly detection in surveillance videos. In: IEEE international conference on image processing (ICIP)
15. (Online). Avenue Dataset. http://www.cse.cuhk.edu.hk/leojia/projects/detectabnormal/dataset.html
16. Oh S, Hoogs A, Perera A, Cuntoor N, Chen C-C, Lee JT, Desai M (2011) A large scale benchmark dataset for event recognition in surveillance video. CVPR
17. Rodrigues R, Bhargava N, Velmurugan R, Chaudhuri S (2020) Multi-timescale trajectory prediction for abnormal human activity detection. In: IEEE 2020 winter conference on applications of computer vision (WACV)
18. (Online). Ur Fall Detection Dataset. http://fenix.univ.rzeszow.pl/~mkepski/ds/uf.html
19. Ali MM et al (2020) ESMD: enhanced suspicious message detection framework in instant messaging applications. In: Fourth Fourth international conference on inventive systems and control (ICISC). IEEE, pp 777–784

The Security in Networks Through Short Normalized Attack Graph Modeling

Gouri R. Patil

Abstract In this chapter, a proposal is made to design a new attack graph model based on shortest path normalization. The main aim of the proposed method is to provide the security for the dynamic network conditions. As the conventional attack, graph models are limited to only static networks, and they cannot ensure a secure data communication in the network. The proposed attack graph model tackles this problem by the accomplishment of a dynamic stochastic modeling on the network. To ensure the robustness, the proposed model is employed in two different cases; they are halt condition and varying condition. Simulation is conducted through several attributes, and the performance is measured through computational time and link probability.

Keywords Attack graph · Network security · Multiple prerequisite · Dynamic network · Computational time · Link probability

1 Introduction

With the new technologies developing day by day, new mode of communications is upcoming. Data are communicated via wireless mode from a source to long-distance receiver through the network. In these entire network layouts, nodes are used to generate, receive or exchange data from one node to other. During the exchange of data packets, various communication protocols are used. With the advancement of such protocol for data exchange, these packets are getting more vulnerable to attackers. An attacker tries to modify or corrupt the originality of the content, giving wrong information's to the nodes or generate defective operations at the core level, and this results in issues with network security. The network security is measured with the robustness to resist the external attacks made by the attacker to preserve data integrity. It is a prime requirement in current network communication to develop robust approaches to achieve high security in data communication.

G. R. Patil (✉)
Muffakham Jah College of Engineering and Technology, Hyderabad, India
e-mail: gourirpatil@gmail.com; gouripatil@mjcollege.ac.in

© The Author(s), under exclusive license to Springer Nature Singapore Pte Ltd. 2023
M. A. Chaurasia and C.-F. Juang (eds.), *Emerging IT/ICT and AI Technologies Affecting Society*, Lecture Notes in Networks and Systems 478,
https://doi.org/10.1007/978-981-19-2940-3_13

Toward provisioning of network security in current communication architectures, security measures are developed at packet forwarding level or network level. In the coding level, various coding approaches such as security header insertion, key level encoding, packet alignment, encryption, and decryption. Among different topology-driven security approaches, attack graph approach [1] is a proposed approach of security provisioning in networking. An attack graph is defined as a pre-defined routing structure, which defines the level of vulnerability observed in a link and route. The graph is formulated as a group of terminal nodes, which are interlinked with each other to exchange data from a host to sink. Wherein attack graph is defined for a given topology, the network is monitored and the security measures are recorded for a period of time, which intern gives a security weight factor to each link node to define the level of vulnerability expected per node per packet transmission.

Various network modeling scenarios are done by aiming at this concept [2–5]. But the performance of the network is getting degraded when the data passing through the overall nodes in a network is suffered from attacks by various attackers at various instants. Attack graphs are able to specify attack scenarios, in [3] a new type of attack graph model to detect an intrusion is proposed called as multiple prerequisite (MP) attack graphs. But the problem associated with this model, it is able to evaluate the network security under static conditions only; i.e., there is no any information about the varying conditions of network structure and network content. To address this problem, new proposal is made, wherein an attack graph model is built considering the volatile parameters [1]. The proposed approach is called as short normalized attack graph model, and it solves the problem with tradition MP attack graph model.

2 Literature Survey

The previous developments' outcome for security measures following attack graphs is having the limitation of network variability, online node updation, quality measures, and multiple attack resilience in the network. In developing security approach using attack graph, different methods of attack graphs modeling and evaluation of network security are presented in past [4]. Different methods for graph-based attack modeling techniques such as graph-based intrusion detection approach (GrIDS) [5], attack trees [6], network diffusion tests or red teams [7], attack graphs [8], collaborative attack modeling [8], multistage attacks [10], attack plan tracking [9], multi-host arrangement vulnerability checker [10], and quantitative assessment of operational security [11] are related to the graph-based methods. The attack tree method has been applied in resistance test such as complicated diffusion test [12] and organized red team approach. It has been shown that graph-based modeling techniques are applied for studying simulated attack scenarios, for analysis of network vulnerabilities, and building a defensive mechanism around networks. Recently, a new kind of attack graph model to detect an intrusion is stated called as multiple prerequisite (MP) attack graphs [3]. But the issue associated with this model, it is able to evaluate the network security under static conditions only, the varying conditions of network formation and network content were discussed [1].

3 Proposed System: Short Normalized Attack Graph Modeling

This section explores the details of proposed new attack graph model. Initially, we discuss the details of MP graph. Next, we discuss stochastic nearest path model and then the details of dynamic stochastic modeling. Further, we explore different case studies.

3.1 MP Graph

The network representation of a MP graph contains several stationary hosts, all of which define one or additional interface with observing addresses. Four kinds of rights on a host are considered in the model: "*root,*" "*user,*" "*dos,*" and "*other,*" where "*dos*" states an importance of "denial of service" and "*other*" shows a confidentiality or a integrity loss. The large size of graph is controlled by the optimization of graph modeling developed as illustrated.

The graph model is generated as a full graph, where all possible paths from a source to host are explored. A generation of graph model for a given network is shown in Fig. 1. A full graph model for the network model is depicted in Fig. 2.

Such networks become large with the increase in node density. To optimize attack graph, non-redundant links are removed from the full AG to obtain a reduced graph called "predictive graph" as shown in Fig. 3.

Further optimized graph of predictive graph is MP graph as shown in Fig. 4. MP graph is a simple network which has edges and three types of nodes, namely state, prerequisite, and vulnerability nodes. State nodes represent an attacker's level. Prerequisite nodes represent the reachability of an attacker. Vulnerability instance nodes represent a particular vulnerability on a specific port. However, the graph optimization is developed over a topology in assumption of node been static or fixed, with exploits been fixed. However, the node positioning or exploit in the node dynamically changes with the time and the graph optimization w.r.t. to exploit variability,

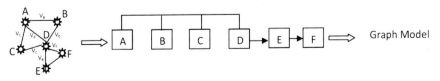

Fig. 1 Generation of a graph model

Fig. 2 Full graph
representation with its
parameters

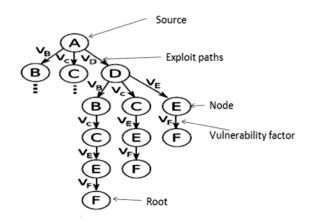

Fig. 3 Predictive graph
plots for the full graph
representation

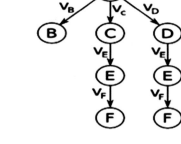

Fig. 4 MP graph
representation of the
predictive graph

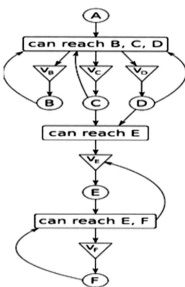

and node position needs to be developed. In developing variability monitoring, a cost optimization in terms of variability and exploit factor is proposed.

Algorithm 1: Probabilistic graph modeling

Step 1: Compute the link probability with range constraint.
Step 2: Derive all probable path states from source to host.
Step 3: Compute cost for each path given by $C_i(u)$ as,

$$c_i(u) = \frac{1}{T_S} \sum_{j=1}^{k} p_{ij}(u) \gamma_k a_{ij}(u) \tag{1}$$

Step 4: Record the variability of node in the network, $(x \pm 1, y \pm 1)$.
Step 5: Compute cost for Halt of variable condition.
Step 6: Converge the path selection by satisfying min (C_{Pi}) constraint.
Step 7: Derive the new optimize attack graph model.

Algorithm 2: Stochastic nearest path model

Step 1: Define the primary policies $\{\mu_0, \mu_0, \mu_0 \ldots\}$ Let $k = 0$.
Step 2: Calculate cost vector $x(\mu_{k_r})$ for μ_{k_r}.

If the obtained cost vector value is not proper value, then
Step 3: Get the new policies $\{\mu_{k+1}, \mu_{k+1}, \ldots\}$ by satisfying,

$$\mu_{k+1,}(i) = c_i(u) + \sum_{j=1}^{n} p_{ij}(u) x_j(u_k) \quad \text{for } i = 1, 2, 3 \ldots n \tag{2}$$

Step 4: Calculate again the cost vectors and finally select an optimal attack path having less cost.

The algorithm above-mentioned gives the complete illustration about the network security evaluation in the view of cost. But the major drawback with this MP graph is it does not specify any dynamic conditions of the network, i.e., various volatile conditions of network, various structures of network, and nobilities of network nodes. The method above-mentioned did not specify any network security evaluation parameters in regard of dynamicity. Thus, to overcome this issue a novel attack graph which also considers the varying conditions of the network is illustrated in next section.

3.2 Dynamic Stochastic Modeling

The MP attack graph proposed in the above section is failed to evaluate the network security under the varying conditions [3]. This section gives the complete illustration about the varying cost factors under various time conditions. In this attack graph model, considering n nodes having n states, and k intermediate move-able nodes having the dynamicity of γ_k[15]. Nodes are first destination node when it stops at the DP. In this work, the dynamicity factor γ_k is going to play a vital role. It is going to vary with network structure, network conditions, and with number of nodes. Variability in network is defined by the change in node condition, where the location in the node changes. For observing of variability parameter in the network a monitoring parameter γ_k is introduced.

γ_k is defined as a parameter of variability defined by, $0 \leq \gamma_k < 1$.

The value of γ_k defines two conditions of a network, varying or halt.

Under varying node condition the cost function is updated as,

$$c_i(u) = \sum_{j=1}^{k} p_{ij}(u)\gamma_k a_{ij}(u) \tag{3}$$

If the time period for traveling state is considered as T_s the cost function described in Eq. (3) can be modified as

$$c_i(u) = \frac{1}{T_S} \sum_{j=1}^{k} p_{ij}(u)\gamma_k a_{ij}(u) \tag{4}$$

where

T_s defines the time period for which the node position changes.

γ_k defines the rate at which the node changes its position.

The cost function is defined w.r.t. rate of variability of node for a time of Ts time period at γ_k rate. Each change in position results in change in link probability p_{ij} and its associated action model a_{ij}. The change in probability and action model generates a new cost value. This variation tends to generate new possible paths to reach to host from a source with varying exploits. The optimization of probability is hence bound for convergence to a minimal cost function given as;

$$\text{Path}_{\text{opt}} = \min(C_i(u)) \Rightarrow \min(P_{i,j}, a_{i,j})_{\gamma_k=0\text{to}1} \tag{5}$$

The second case arises as halt state where the node has moved with γ_k rate and halt to T_w period. The cost factor during the halt state can be written as follows.

$$c_i(u) = \frac{1}{T_w} \sum_{j=1}^{k} p_{ij}(u)\gamma_k a_{ij}(u) \tag{6}$$

As the problem is determining, the "nearest" path is beneficial to the attacker. But this cost factor is a varying one due to extra added factor called dynamicity. Because of this varying nature, the attacker is not able to attack in the fixed manner. There is a probability to get the attack on some other node even if it is aimed on a particular node.

3.3   Illustration: Case 1: Halt Condition

A network model under halt condition and its derived graph model are shown in Fig. 5. The full graph for source A reaching to F is shown in Fig. 6.

There exist nine probable paths to reach host F. For each of the link (L) to move from one state to other, the exploit is defined by a vulnerability score V_L and vulnerability score is defined from the CVE vulnerability scoring [97] In this case, the vulnerability scoring given is listed in Table 1.

Illustrating for three outlined paths, (P_1, P_2, P_3) the cost computed is,

$$C_{P1} = \frac{1}{T_W} \left(\sum_{j=1}^{k} p_{i,j} \gamma_k a_{i,j} \right) \tag{7}$$

Taking halt period T_W for 10 ms and γ_k as 0.1
Probability $p_{A,D} = 1$ and $a_{i,j} = V_D = 1 \Rightarrow p_{i,j} \gamma_k a_{i,j} = 1 \times 0.1 \times 1 = 0.1$

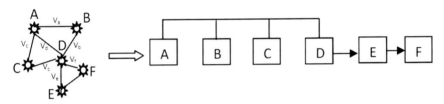

Fig. 5   Graph model under halt condition

Fig. 6   Probable path of flow

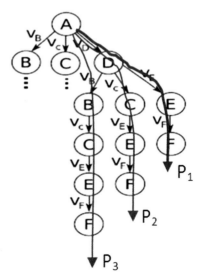

Table 1 Vulnerability scoring for exploits types

Exploit type	Vulnerability index	Score
Tcp	V_A	2
Html	V_B	4
Web	V_C	5
Host	V_D	1
http	V_E	7
ftp	V_F	3

Iterating the process for each of the link in the path (P_1), the cost value is obtained as,

$$C_{P1} = \frac{1}{10 \times 10^{-3}}(0.1 + 0.5 + 0.3) = 900$$

Similarly, iterating for P_2 and P_3 the cost value obtained is,

$$C_{P2} = 600 \text{ and } C_{P3} = 1200$$

The convergence is defined as min(C_{Pi}) result for the selection of C_{P1} as optimal state flow. The optimized MP graph for this full graph given by the existing approach is shown in Fig. 7.

Where the states are merged based on the transition, without consideration of their exploit level. The optimized graph by the proposed approach is shown in Fig. 8.

Fig. 7 MP graph representation

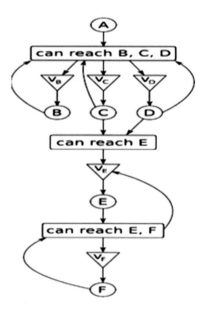

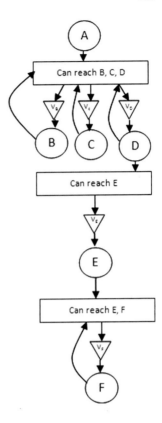

Fig. 8 Optimized MP graph representation

In this, the probability of transition from state *B*, *C*, to *E* is eliminated, as the links are more vulnerable. The reduced graph has a probability of three possible paths to reach *F* which has offer lower vulnerability factor.

3.4 Case 2: Variability Case

This case has node changing its location for Ts = 10 ms time period. The graph in this case is developed as shown in Fig. 9.

Here the node *C* moves outwards. As the node is displaced outward and comes out of reachability, the link probability changes. The full graph model in this case evolves as shown in Fig. 10.

Here a total of 4 possible paths exist to reach *F* from *A*. Applying the cost optimization as presented in the above illustration, 2 optimal paths are derived satisfying as min(C_{Pi}) as shown in Fig. 11.

The optimized MP graph is obtained is shown in Fig. 12. For a node (*B*) moving inward to the network, the network obtained is shown in Fig. 13.

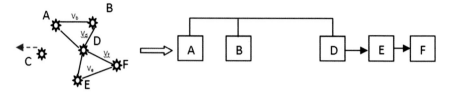

Fig. 9 Graph modeling under variable condition

Fig. 10 Full graph
representations for the graph
model

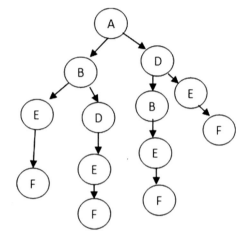

Fig. 11 Probable path of
flow

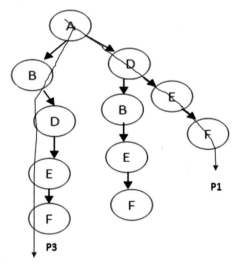

Fig. 12 MP graph
representation of the path
flow

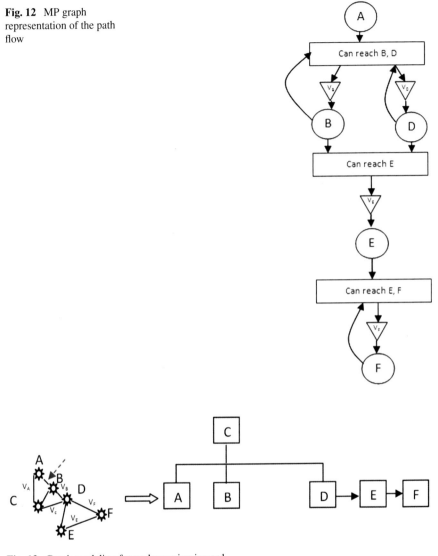

Fig. 13 Graph modeling for node moving inward

The full graph model obtained is shown in Fig. 14. The number of path state transition is decreased to 4. Applying the cost optimization approach over the attack graph, the path selected is shown in Fig. 15.

The optimized MP graph obtained is shown in Fig. 16. The proposed approach has a dual advantage of providing lesser state transition and more secure probable paths to reach host. The proposed approach of graph optimization is summarized in the given algorithm. To optimize the path selection for data exchange over the

Fig. 14 Full graph modeling
for the graph model of node
moving inward

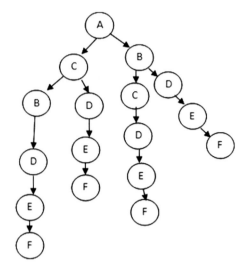

Fig. 15 Probable paths of
flow

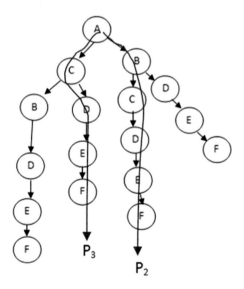

attack graph, an online updation of path selection based on risk factor monitoring is
proposed.

Algorithm 4: Summary of proposed approach

Step 1: Compute the link probability with range constraint
Step 2: Derive all probable path states from source to host
Step 3: Compute cost for each path given by $C_i(u)$ as,

Fig. 16 Optimal MP graph
plot for the path flow

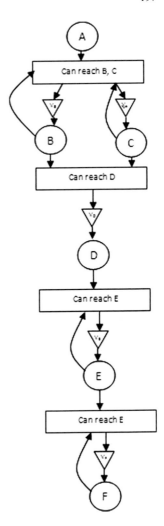

$$c_i(u) = \frac{1}{T_s} \sum_{j=1}^{k} p_{ij}(u)\gamma_k a_{ij}(u)$$

Step 4: Record the variability of node in the network, $(x \pm 1, y \pm 1)$,
Step 5: Compute cost for halt of variable condition.

$$c_i(u) = \frac{1}{T_s} \sum_{j=1}^{k} p_{ij}(u)\gamma_k a_{ij}(u)$$

Step 6: Converge the path selection by satisfying $\min(C_{Pi})$ constraint.
Step 7: Derive the new optimize attack graph model.

4 Simulation Results

The results had drawn give the illustration about the performance evaluation of the proposed with the previous work having static nodes [3]. Two networks were compared, where movable network is a proposed work and static network is a previous work. The computation time for the network simulated under static and movable conditions is presented in Table 2. An average difference of 20 ms is observed for the proposed approach under movable and static conditions.

The proposed attack graph model includes computing all sources and all targets graph up to 100 nodes within 8 min 20 s. From the above observations, it can be concluded that the performance evaluation of the proposed attack graph model is able construct an attack graph within less amount of time for any network, under any network conditions as shown in Table 3. The obtained risk factor and cost value of

Table 2 Observation of computation time (ms) over varying number of nodes in the network at $\gamma = 0.01$

No. of nodes	Computational time (ms)	
	Movable	Static
10	70	70
20	120	120
30	180	180
40	240	240
50	290	300
60	350	360
70	410	420
80	450	470
90	510	520
100	600	580

Table 3 Observation of computation time over varying number of nodes in the network at $\gamma = 0.1$

No. of nodes	Computational time (ms)	
	Movable	Static
10	50	50
20	100	100
30	150	150
40	200	200
50	240	250
60	230	300
70	320	350
80	360	400
90	410	450
100	450	500

Table 4 Observations for the risk factor and cost value for different node variation conditions

γ_k	Node variation	Computation time (ms)	Risk factor	Cost value
0.01	Slow	23	115	0.22
0.1	Medium	26	131	0.24
0.2	Medium	24	154	0.21
0.5	Faster	29	127	0.27
0.6	Faster	26	113	0.22
0.9	Faster	28	142	0.25

Table 5 Observations for link probability for different node variation conditions

γ_k	Node variation	Link probability ($P_{i,j}$)
0.01	Slow	6
0.1	Medium	8
0.2	Medium	12
0.5	Faster	23
0.6	Faster	26
0.9	Faster	25

the simulated network under varying node condition are presented in Table 4. Table 5 presents the link probability for different node varying conditions.

5   Conclusion

In this effort, a stochastic model is projected to approximate network security. This reproduction deals with security evaluation scenarios at various network conditions. Based on node, dynamicity stochastic approach of graph prediction and selection in the network security estimation using on the observed attack paths is presented. The projected method allows for the unstable parameters, i.e., addition or fall in the number of nodes, increase or decrement in distance between the nodes and dynamicity of the nodes which are able to move from one end to other end to transfer the data.

References

1. Patil GR, Damodaram A (2014) A short-normalized attack graph based approach for network attack analysis. In: IOSR J Comput Eng (IOSR-JCE) 16(3). p-ISSN:2278-8727
2. Vigna G, Kemmerer R (1999) Netstat: a network-based intrusion detection system. J Comput Secur 7

3. Noel S, Jajodia S, O'Berry B, Jacobs M (2003) Efficient minimum-cost network hardening via exploit dependency graphs. ACSAC IEEE Comput Soc 86–95
4. Sheyner O, Haines JW, Jha S, Lippmann R, Wing JM (2002) Automated generation and analysis of attack graphs. IEEE Sympos Secur Privacy 273–284
5. Jajodia S, Noel S, O'Berry B (2005) Topological analysis of network Attack vulnerability. Massive Comput 5:247–266
6. Zhang Y, Fan X, Xue Z, Xu H, Two stochastic models for security evaluation based on attack graph. In: The 9th international conference for young computer scientists
7. Noel S, Jajodia S, Understanding: complex network attack graphs through cluster adjacency matrix. In: Proceedings of the 21st annual computer security applications conference (ACSAC), pp 72–84
8. Wang L, Yao C, Singhal A, Jajodia S (2006) Interactive analysis of attack graphs using relational queries. In: Proceedings of the 20th IFIP WG 11.3 working conference on data and applications security, pp 119–132
9. Ingols K, Lippmann R, Piowarski K (2006) Practical attack graph generation for network defense. In: Proceedings of the 22nd annual computer security applications conference (ACSAC), pp 121–130
10. Noel S, Jajodia S (2004) Managing attack graph complexity through visual hierarchical aggregation. In: Proc. ACM workshop visualization and data mining for computer security, pp 109–118
11. Wang L, Singhal A, Jajodia S (2007) Toward measuring network security using attack graph. In: Karjoth G, Stølen K eds ACM, pp 49–54
12. Ammann P, Wijesekera D, Kaushik S (2002) Scalable, graph-based network vulnerability analysis. In:CCS '02: proceedings of the 9th ACM conference on computer and communications security. ACM, NewYork, NY, USA, pp 217–224
13. Salfer M, Eckert C (2018) Attack graph-based assessment of exploitability risks in automotive on-board networks. In: Proceedings of the 13th international conference on availability, reliability and security, pp 1–10
14. Somak Bhattacharya SK (2008) Ghosh: an attack graph based risk management approach of an enterprise LAN. Dyn Publish J Inf Assur Secur 2:119–127
15. G, Hong, Zong-Yuan M (2002) Immune algorithm. World Congress Intell Control Automation 3:1784–1788
16. Patil GR, Damodaram A (2015) Quality restrain coding in network security using optimal attack graph modeling. IJAER 10. ISSN 0973-4562
17. Kai Z, Xiangru M, Zhiqiang M (2008) Research on intrusion detection technology based on immune algorithm. In: International symposium on knowledge acquisition and modeling, pp 759–762

A Conception of Blockchain Platform for Milk and Dairy Products Supply Chain in an Indian Context

Dharun Vincent, M. Karthika, Julie George, and Justin Joy ⓘ

Abstract The potential for adulteration in the Indian dairy supply chain process is immense. The possibility of incorrect information recorded by middlemen cannot be ruled out and found to be rampant. The reality is that the data required to assess the safety and quality of milk produced is inadequate in the existing setup. The current set of checks and balances to fight adulteration of milk and dairy products in India is studied and articulated. An elaborate and daunting set of procedures marks these checks and is still significantly found wanting. To increase the product's safety and traceability of the product an alternate pathway to deploy Blockchain technology in the milk and dairy product supply chain has been proposed. Despite the proposal requiring drastic changes in the milk and dairy industry, the authors believe the benefits of implementing a Blockchain platform far outweigh the challenges involved.

Keywords Blockchain · Blockchain platform · Milk products · Dairy supply chain

1 Introduction

Indian dairy industry is a 2.5-billion-dollar industry in India with a 6.4% growth every year [21]. India's 150 million small dairy farmers, local co-operatives, and networks of small-scale vendors have made India the world's largest producer of milk because of the growing market both national and international players are entering in this market [8]. The Indian government implements many rules and regulations for the safety of consumers. But still, milk-borne diseases are affecting the consumers in the market. The reason mainly is because of less traceability and control in the supply chain of dairy products. "*68% milk & milk products in India not as per FSSAI standard: Official*" [16]. Milk in India was found to be a highly-adulterated food

D. Vincent · M. Karthika · J. Joy (✉)
Christ University, Kengeri Campus, Bangalore, India
e-mail: justin.joy@christuniversity.in

J. George
Dairy Consultant, Kochi, Kerala, India

© The Author(s), under exclusive license to Springer Nature Singapore Pte Ltd. 2023
M. A. Chaurasia and C.-F. Juang (eds.), *Emerging IT/ICT and AI Technologies Affecting Society*, Lecture Notes in Networks and Systems 478,
https://doi.org/10.1007/978-981-19-2940-3_14

product despite FSSAI assurances of safety: the northern part of India has more milk adulteration than the other parts of India [15]. Most of the adulteration in milk happens in the supply chain part.

2   Organization of This Article

This article discusses the current procedures in place to check the safety of milk and dairy products and the challenges in meaningfully enforcing these. The prospect of implementing Blockchain Technology seems to be providing a promising solution to these challenges.

The article begins with an introduction of the current problem and is followed by the literature review on the prospect of Blockchain Technology and on Blockchain technology specific to the milk and dairy industry. The objectives of this article are framed in the process.

This is followed by the methodology pursued, where a thorough analysis of the process was carried out, and immense insights were gained from the interviews with experts.

The study led to propose a Blockchain platform specific to the milk and dairy supply chain in India complemented by a module structure of the logistics platform.

3   Context and Setting

The research setting includes the various elements of the milk and dairy supply chain of several famous milk and dairy brands in South India. Stakeholders of these brands involved in the various processes of its supply chain were the sources of information that formed the base of this article. In the process, several secretaries of Co-operative milk collection centers spread across 3 states in India were interviewed along with officials in charge of receiving and dispatching products at the various points of the milk and dairy supply chain.

4   The Prospect of Blockchain Technology

Makarov et al. [13] examined the international and domestic use of Blockchain technology. The study examined the conceptual approaches to its application in the management of dairy supply chains. It inferred that the current quality control mechanisms in place at the various technological stages of feed production, final products at the manufacturing facility, and incoming inspection in retail organizations are no longer adequate. The world practice has a positive experience of employing Blockchain technology in some fields of activity to control the adherence of defined standards, based on the transparency of the complete technological process of making products with many phases. The benefits of implementing were found to be data security, protection of the repository of documents from hacking, elimination of the possibility of making changes to information on the progress of transportation, reduced delivery delays, reduced chance of unauthorized use of banned or unwanted ingredients and allow retailers and even consumers to browse the history of the product creation: to determine the origin dairy farm, the dairy plant, and the date the product was created.

Cohen [6] acknowledges that Blockchain technology, though it currently does not seem to have a significant impact on society, the term as such is familiar with a majority of the community as inferred from the several interviews with stakeholders. These stakeholders were from multiple industries, including transportation, food, medical and health, finance, and more. The standard expectation noted from these interactions was the appreciation of the benefits realizable in integrating Blockchain in supply chains. There are advantages to implementing Blockchain in the supply chain. It reduces paperwork and administrative costs, eliminates fraud and counterfeit products, facilitate origin tracking and recall of the product in a time-efficient way [2].

Some companies have started integrating Blockchain with IoT and using IoT as a bridge between the virtual and real world. An agro-food supply chain company uses RFID with Blockchain for better traceability systems. With trusted information in the entire agro-food supply chain, which would effectively guarantee food safety by gathering, transferring, and sharing the authentic data of agro-food in production, processing, warehousing, distribution, and selling links [20]. One research study tells that organizations should not blindly believe Blockchain; in each and every step of the supply chain, there should be a traditional quality check and auditing process to create a legit transactional record. This will thereby help to trust the whole transaction. Suppose a party obtains or owns more than 50% of the network hashing power. In that case, it could present a threat to the consensus protocol, thereby allowing the miners themselves to enter illegitimate transactions into the Blockchain. Significant advance by any falsification of the material source has to be done prospectively in real-time, which is a much more complex challenge, than at any time retrospectively falsify a physical transactional record, where hard copy documents can be simply substituted with new versions containing different facts [1].

Blockchain technology has been recommended as a solution for networking issues, trust concerns, and traceability in the supply chain. Blockchain remedies allow supply chain parties and stakeholders to track bottlenecks in the flow of products. The system can discover if the products are kept in one place for a long time or placed at the wrong location, which is particularly significant for refrigerated goods [5]. Moreover, Blockchain provides the right and precise information about potential suppliers and customer's liquidity, and current financial status. Blockchain remedies, with their features capable of traceability and transparency enhancement, are considered an approach for the supply chain monitoring concerns [5]. So potential challenges such as supply chain sustainability can be tackled through applying Blockchain solutions. In this way, all company transactions are recorded in the ledger, which makes it possible to prove responsibility and company dishonesty in sustainability-related issues.

Lin [12] The implementation of Blockchain technology is encouraged by this study since it offers a high level of data consistency and validity. It also discusses the technology's early adopters and key users. In 2017, IBM announced a partnership with many large food manufacturers and retailers, including Dole, Nestlé, Tyson Foods, Kroger, Unilever, and Walmart, to use disruptive technologies like Blockchain to improve quality control, food safety, management, and traceability. Despite the benefits and opportunities of Blockchain, it also discusses implementation challenges such as technical capacity and infrastructure gaps, scalability and implementation costs, global standardization politics, cybersecurity and data protection, and Blockchain's technologically inherent limits. According to this article, such regulatory issues may necessitate a rethinking of the forms and contents of traditional food law and policy, as well as data protection, antitrust, and trade law. In light of this, this article advocates for a more technologically informed policy-making approach before jumping into the frenzy of Blockchainizing food legislation.

This paper studies and examines the prospect of creating a Blockchain platform for the supply chain of milk and milk products to help reduce milk-borne diseases caused by adulteration of milk. It also explores the current challenges that make the proposal all the more relevant and the need for the industry to be increase traceability, clarity, and control over the supply chain.

5 Objectives of the Study

The objectives of the study are stated as follows:

- To analyze the supply chain of milk and milk products and its challenges in an Indian context.
- To propose a pathway to deploy Blockchain technology for the Indian dairy ecosystem to increase safety and reduce waste.
- To propose a Blockchain logistics platform to help the traceability and transparency of the supply chain of milk and dairy products targeting the suppliers, manufacturers, and customers.

6 Literature Review on Blockchain in Dairy Industry

Subramanian [19] develops a Blockchain-enabled food supply chain framework, including future prospects and current constraints, based on a comprehensive literature analysis and semi-structured case interviews from the context of emerging economies, using the people, process, and technology (PPT) model. The report also shows that combining Blockchain technology with other artificial intelligence technologies can help supply chains deal with trust, traceability, and collaboration challenges. Several sectors have implemented comparable technologies. This report describes a few Blockchain pilot projects carried out by a few significant organizations and start-ups. The development of Blockchain applications also allows the entire food-safety chain to be more responsive in the event of a food-safety incident.

The purpose of this study [7] is to determine if Blockchain technology can assist farmers, food suppliers, and investors share information in the food industry. The primary purpose of this study is to investigate the use of Blockchain technology in the food industry from the standpoint of businesses. The author also advises businesses to use such technologies in conjunction with effective performance management tools, noting that the Blockchain cloud solution is neither time-consuming nor expensive. In order to construct a Blockchain solution that is compatible with them, organizations can try to foresee how legislation or standards will be formed. As a result, a more effective decision-making process is possible, with goods information traceable without danger.

Kay Behnke [3] infers that consumers expect quality and transparency in the process, in addition to product quality, in today's competitive market. Blockchain has been promoted as a way to improve traceability by establishing trust. Because the adoption of such technology necessitates some basic infrastructure to connect supply chain participants, the purpose of this article is to identify boundary requirements for sharing assurance information in order to increase traceability. Furthermore, the study's findings show that before Blockchain can be successfully deployed, supply chain systems must be adjusted, and organizational actions must be done to meet the boundary criteria.

Kasten [10] study examines the potential for adulteration and incorrect information to be recorded by middlemen in the supply chain process in the dairy sector. However, the United States government and the Food and Drug Administration (FDA) have issued recommendations for the dairy industry's safety and standard criteria. The goal originates from the reality that the data required to assess the safety and quality of dairy producer's milk is frequently held by the same companies that process the milk into various products. The paper also details the cost, performance, security, and storage requirements for Blockchain technology implementation. The research also makes a case for cost efficiency by claiming that lower inspection costs would offset any cost increases on the state's behalf. Therefore, implementing such a design will improve the visibility of the process without compromising the organizational costs.

The level of adulteration in milk in Varanasi is discussed in this study [11]. According to the results of a survey conducted with 50 samples in one of the milk mandis in Varanasi district, 20% of the samples tested positive for starch content, and 80% of the samples tested positive for acidity/alkalinity. Adulteration with neutralizers, sodium chloride, and urea was found in 28, 80 and 60% of these samples, respectively. In the third group, formalin was detected in 30% of milk samples, while hydrogen peroxide was detected in 36%. Similarly, detergents were found in 44% of milk samples. Formalin and peroxide: These compounds were either added to extend the shelf life of the milk or to profit from it in an unethical manner. Detergents may have been discovered due to poor tank maintenance. According to the findings of this inquiry, milk marketed for public consumption is produced and handled in an unsanitary manner. The loosening of milk control rules may allow greedy retailers and producers to adulterate manufactured milk to maximize their profits.

Holmberg and Aquist [9] infers that as the dairy sector moves toward globalization, it becomes increasingly challenging to maintain control over and trust information. As a result of the food crises, customer awareness is growing, as are measures such as certificates to address the issues due to the knowledge asymmetry. This research looks into the obstacles and opportunities of implementing a traceability system based on Blockchain technology. The study concludes that Blockchain technology is still in its infancy in the context of food supply chains, with some of the most significant challenges being the development of a culture that encourages collaborations, information sharing, and standardizations, as well as understanding the true value it provides to all stakeholders before implementing it in the dairy industry.

Shingh [18] study focuses on using Blockchain technology to improve the dairy supply chain system by increasing supply chain transparency and traceability. This research was carried out in response to widespread public concern about human health, environmental sustainability, and welfare issues. This research emphasizes the importance of dairy traceability in the supply chain. Dairy traceability is defined as the capacity to track the movement of milk and milk products through various routes from production to final consumption. RFID technology is also being recommended for the deployment of Blockchain. The system can only be efficiently implemented if each member updates the relevant information in the network, which can also be automated using IoTs. Certification agencies and regulatory authorities can also be integrated into the supply chain system to simplify regulation and save money. This paper explains how this technology can be employed in the dairy supply chain system and the possible benefits to various stakeholders and the entire dairy industry.

Rambim and Awour [17], Biscotti et al. [4]: These studies claim a proposed system may reasonably assure food safety during food production, handling, preparation, and storage. However, because of the difficulties in implementing such technology in an automated form, the authors recommend using IoT and Blockchain technology instead. The Internet of Things (IoT) solution collects temperature data automatically and makes it available as soon as the pasteurization process is completed. Local operators would no longer be burdened with the error-prone task of collecting measurements on a regular basis. The Blockchain is used to assure the non-reliability of recorded data. As a result, basic technologies can be used to maintain high traceability, transparency, and standardization. Furthermore, the quality of milk (i.e., the water and fat content, as well as cases of milk adulteration, such as when a farmer adds chemicals to his milk delivery to enhance amount, etc.) is automatically measured and recorded using quality sensors. Due to the farmers' lack of education, an additional accessibility design is proposed, in which farmers can access a farmer's milk delivery record through SMS, and simple biometric technologies are used for farmer authentication. The solution uses Blockchain to enable the development of permanent and non-repudiable records for all stakeholders.

To create a Blockchain platform for the milk and dairy industry in India, the authors had to study the industry and understand the processes, regulations, and standards executed.

7 Methodology

No one Governmental agency in India centrally maps all the procedures, certificates, and standards required for the companies that are involved in dairy product processing and dairy supply chain. To understand all the procedures, a methodology that not only refers to secondary sources but validates by collecting primary information too was called for. The methodology used is interviews with experts. Some of those interviewed work as secretaries in the co-op milk collection centers in South India. Some of the documentation and certification requirements in the milk products supply chain are gathered by interviewing industry experts, retailers, and FSSAI documents. The set of questions and the answers collated are summarized below.

7.1 Inference from the Interviews

Question 1: How does the supply chain of milk from farmers till the packaging work?

When a local farmer brings his milk from a milched cow to the co-op milk collection center, it will be checked for quantity and fat percentage in it. The fat percentage is the critical key to deciding the milk price. For example, if the milk contains more than 13% of fat, it will get the highest price. Then this milk will be stored in the milk container, which will be transported to the processing plant. From this plant, the raw milk is processed and made into different products, packed, and sent to other places for the customers.

Question 2: Is there any system or cloud data to store and track which farmer gave a particular batch of milk?

No, we don't have any technology to store the data which contains who and where the particular batch of milk is produced. But there will be identification cards for the farmers, and the cows will have a tag number that will be enrolled in the local government veterinary hospital because the farmers who rear cows should have a record on whether the cow is treated with proper medicines and flu injections. These data will be readily available in the local hospital. Every farmer will be having identification with him because the government is giving incentives to them for agricultural purposes.

Question 3: How will you trace the milk from origin till the end? Do you have a system to track them from end to end?

No, we cannot trace if the milk is mixed with other batches because all the milk from one collection point will be filled and transported in a milk container. Once *it reaches the processing plant, all the milk from different collection points will be mixed and processed to desired products. So, we cannot track which product is from which batch of milk. There is no system that can follow from end to end.*

Question 4: If there is any contamination in milk or if a cow from a particular region is diseased, will you reject all the products?

Yes, if there is any contamination, we will be rejecting all the products because there is no way we can trace and point that this particular batch only had that issue.

Question 5: If there is a platform that can trace and contain all the data of milk such as origin, which cow's milk and who is the donor of the milk, at what time and date it is collected, etc., will that be useful for you?

Yes, because in the present situation, we cannot trace or can see the origin of the milk. And there are many problems. So, if there is a platform that can show or record all the data to make it more visible, it will be helpful to us. This will increase more transparency in the system, so there will be no adulteration in the milk and will be safer for the customers.

The Blockchain platform was also reviewed by industry experts and offered appreciation for ideation of the same, but admitted the limitation in the detailed know-how of Blockchain technology.

7.2 Process Analysis

The processes involved in maintaining the set standards are pretty elaborate, and the interviews and the secondary records revealed that the certificates and documentation required for adhering to the safety norms of FSSAI are several. These are shown in Fig. 1.

Documents and Certificates in a Milk & Dairy Supply Chain are listed as below:

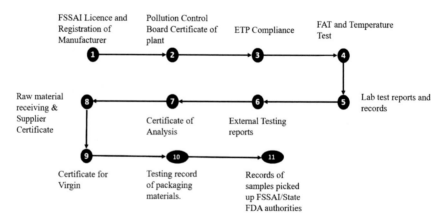

Fig. 1 Documentation involved in milk and dairy supply chain (Framed by authors from process analysis)

1. Legal

   (a) FSSAI License and Registration of Manufacturer/Supplier/Dealers/Retailers
   (b) Pollution Control Board Certificate of plant/manufacturing unit
   (c) Record of Discharge Effluent and its Compliance with statutory requirements—ETP Compliance

2. Procurement/Quality

   (a) Raw material receiving and traceability records (including records for milk received from Milk Collection Centers, BMCs, Chilling Centers).
   (b) Receiving records for raw materials and additives (other than milk)
   (c) Quality Control/Lab test reports records/Compositional analysis/Microbial test records—raw milk, processed milk, and milk products.
   (d) External testing reports—Microbiological/chemical test reports about milk and milk products, water, other food ingredients, additives, etc.
   (e) Certificates of Analysis/COA
   (f) Internal and external audit records/ Corrective action.
   (g) Records for receipt of packaging materials and Supplier certification.
   (h) Certificate for Virgin/food grade Packing material
   (i) Certificate of Ink approved for use for milk and milk products packet.
   (j) Testing record of Packaging materials.
   (k) Records of samples picked up FSSAI/State FDA authorities.

Despite these numerous documentation and certificates, cases of adulteration continue in their occurrence. The authors have proposed a Blockchain platform to tackle the same by significantly improving traceability and transparency.

8 Proposed Blockchain Platform

8.1 Blockchain Platform for Milk and Dairy Supply Chain: The Model Explained

Currently, there is no existing platform for the milk & dairy industry. Figure 2 given below is a model the authors propose for the Milk & Dairy Blockchain Platform. This platform consists of all the blocks which can trace a batch of milk from start to end. Using this platform, the traceability of milk products will be easier and faster in comparison with traditional methods.

The process is explained as follows:

Milking Cow:

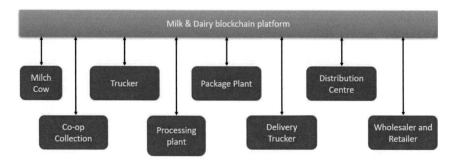

Fig. 2 Blockchain platform for milk and dairy products (Model proposed by authors)

For milk, the process starts from milking cows. This process will happen on a big farm or by an individual cow. Every cow has a certificate and every farm has a license from the state and central authorities. When the milk is milked from a cow, the milk quantity is measured. The farmer should upload all the certificates of the cow, such as medical and other certificates, in the platform, which will be used for tracking.

Cooperative Collection:

Each and every state in India have a co-operative collection center to collect milk. Almost all the villages and towns have this collection area where farmers bring their milch milk and test it for standards and sell it. Every farmer has their identity card and has a unique number on it. The co-operative collection officer will register his entry and start testing the milk. In this testing, the temperature, Fat percentage, and quantity of the milk are calculated. This will be entered into the platform. This method of collecting data will be secure and can be used and referred for future processes. By storing certificates and test results in the Blockchain platform, a checking officer can see and verify the milk product for authenticity.

Trucker:

In this stage, the co-operative collection will send the collected milk from the farmers to the processing plant for producing dairy products from it. The co-operative center has assigned a trucking company that is capable of collecting and delivering milk to the processing plant at a controlled temperature. In this process, the truckers will come to the co-op collection point and load all the milk containers into their truck and deliver it to the processing plant. During this process, the trucker has to have many documents and license to transport milk in the truck. Even if there is one document missing, the state authorities will seize the truck and take custody of the same, which leads to contamination of milk that is inside the truck.

The truck will be equipped with temperature sensors, security cameras, RFID sensors, etc. These sensors are connected to the Blockchain platform, monitored directly by the persons who are tracking the trucks. This process and technology will help supply chain stakeholders and customers to access real-time data that is constantly being collected. Through this technology, a processing plant can monitor whether the milk that is transported in a truck has been maintained at the specified temperature. This will help companies analyze the data generated and monitor milk throughout every stage of transportation. By doing this, companies can recommend trucks on the scope to improve quality and eliminate errors detected in the process.

Processing Plant:

Milk is one of the most consumed items that are devoured by individuals from over the world. There are various milk handling plants that have been set up to take into account the necessities of the shoppers. Improvements are occurring in the mechanization of dairy preparing plants in the nation. These have functioned admirably to complement the quality and the increase in the amount of milk generation in India.

The plant will record all the processes which are done during the production of milk and milk-based products, such as ingredients, temperature, etc., and will be stored in the Blockchain platform. This process of storing will happen in batches so that the product batches can be differentiated and the replacement cost of products reduced if they get contaminated. After this, the processing plant will upload the documents and certificates which show proof that the milk is processed by the right standards.

The certificates consist of all the documents such as the equipment and machinery used for the process, such as their manufactured date, their service date, and their temperature and substances used during the process. This will enable the companies to track the manufacturing process entirely without latency and false data. Blockchain entry in these processing stages can capture the details of temperature changes batch-wise and keep it transparent for the downstream processes.

Packaging Plant:

The milk goes to the filling and bundling process when the item fabricating procedure is done. Milk can be stuffed into various kinds of bundles: containers, glass, pockets, PET jugs, and so forth. Disinfected milk that requires a long timeframe of realistic usability ought to be filled and pressed utilizing aseptic advancements. For this situation, past cleansing of the bundle ought to likewise be finished. Blockchain entry at these stages of milk reception, blending, filtering, filling and bundling, etc., helps trace and track the origin of issues should they get reported. All these procedures for the various sorts of milk produced can be documented.

Delivery Truck:

Completed items must be appropriately secured during delivery. The kind of vehicle or holders required relies upon the item and the conditions under which it must be moved. Except if successful control measures are taken, completed items may get defiled during delivery or may not arrive at their goal in a reasonable condition for utilization. This can happen in any event when legitimate cleanliness control measures were taken when you made the item.

Transportation control measures ought to include:

- Shielding item from potential wellsprings of sullying.
- Isolating items from non-perfect items on a similar burden.
- Shielding items from harm that could make them unsatisfactory for utilization.

- Keeping the item at the correct temperature to forestall the development of deterioration smaller scale living beings that may abbreviate the timeframe of realistic usability of the item.

The vehicles and compartments used to ship nourishment should be kept clean and in good condition. If a similar vehicle or compartment for moving various nourishments or for shipping non-nourishment items is utilized, it must be successfully cleaned and sterilized between loads. These cleanliness prerequisites also require a capture in Blockchain.

Criteria for receiving:

At the point of fixings or bundling materials, the vehicle is to be outwardly assessed to affirm that it is appropriate for shipping nourishment. It ought to be fabricated, kept up in great condition, spotless, and encased to shield the heap from the residue, exhaust, and climate. The Blockchain entry could record whether the following are in place. A sample checklist that could be adapted and replicated in other processes too is provided below:

- It is spotless
- The administrator can demonstrate when it was last cleaned
- It doesn't have any smells
- There are no indications of rat movement
- On the off chance that different things are carried on a similar vehicle, they are:
- Not things that might taint your fixings or bundling materials
- Appropriately isolated from your fixings or bundling materials (for instance, by utilizing physical dividers or plastic overwrapping)
- On the off chance that the fixings or completed item must be kept at a particular temperature:
- The vehicle has kept them at that temperature
- The temperature was observed and recorded all through the excursion
- Make a record of the state of the vehicle for every conveyance got.
- Sign shipment solicitations and keep them on the document as a record of receipt.
- Criteria for shipping:
- Before items are stacked onto the vehicle, outwardly investigate the vehicle to ensure:
- It is at the correct temperature for the item being delivered
- It can keep the item at the correct temperature all through the excursion
- The temperature can be checked during the excursion
- The vehicle is in acceptable condition, spotless, liberated from scents, and liberated from any indications of rat or bug exercises
- The item is adequately shielded from defilement, including residue and vapor
- Cleaning records are accessible
- No different things being dispatched in the vehicle could sully your item
- Your item will be adequately isolated from some other kinds of nourishments or non-nourishment things in the vehicle during the excursion
- Record the after-effects of your review.

- Ensure the completed item being stacked is in sound condition with no open, spilling, or harmed compartments. Burden the vehicle in a manner that keeps your item from getting damaged or sullied.
- For items requiring refrigeration, pre-cooled the trailer, turn off the reefer preceding stacking to forestall ice develop, load rapidly, and afterward walk out on the following stacking.

Although these elaborate steps are in place in the documentation, a validation of these being adhered to can be required in the Blockchain entry during the receiving and shipping stages.

Distribution Centre:

As distribution is the last stage in the promoting of milk and milk items, the offices for dispersion ought to guarantee that the nature of the items is kept up, alongside its auspicious inventory. The report finds that presently around 80% of the whole milk delivered is conveyed through the profoundly divided chaotic division, which incorporates nearby milk merchants, wholesalers, retailers, and the makers themselves. Then again, the composed dairy industry appropriates 20% of the whole milk produced. The report has likewise distinguished a few difficulties in conveying milk items in India. As per the report, perhaps the dairy business's most significant test is the lopsided circulation of cold chain offices all through the store network. This frequently brings about the perishability of the item because of more extended occasions taken in arriving at the office and the end shoppers. Different difficulties in disseminating milk items incorporate—low infiltration into the rustic market, absence of appropriate transportation foundation, lack of mindfulness on great conveyance rehearses, and so forth. As per an expert the authors interviewed, *"despite the fact that the sorted out dairy area in India has a few retailers and merchants, there is as yet the requirement for a composed system of wholesalers to help decrease the circulation costs by institutionalizing the appropriation procedure"*.

Aside from the dissemination framework, the article likewise gives an understanding of different parts of the dairy business in India. The article, which depends on interviews on work area-based insights, provides a beneficial awareness for any who intends to wander into the dairy business in any structure and especially to understand the entry points of a Blockchain platform.

Wholesaler and Retailer:

Wholesalers and retailers will procure their stocks from a distribution center. After buying, they stock the milk, and milk products will be sent to the stores located in different parts of the region. These stores will be found in the high-demand areas such as in the heart of the city or where market demand is high. Some retailers will have a delivery van or persons who can deliver the products directly to the customer's house. A module structure of a logistics Blockchain platform is also proposed by the authors below.

8.2 Module Structure of Logistics Platform

The logistics module structure is divided into three layers. The top layer is the transaction layer which is used to complete the operation of forward and reverse logistics services throughout the process, including transportation, tracking, customs declaration, bill payment, after-sales service, and so on.

The middle layer is the collaboration layer. According to the information obtained in the transaction layer, forwarders take part in a particular task through cooperation with other logistics providers. They consider the uncertainty and risk of logistics jobs, design an optimal plan, and send it to the transaction layer users.

The bottom layer is the management layer. The central control system manages the integrated task and stores the data from the above two layers to the database. At the same time, it acquires the latest management orders going through the Decision Support System analysis.

The Blockchain platforms proposed by the authors can conceptually improve the safety and healthiness of milk products while traversing through the supply chain.

9 Findings and Challenges

As per the objectives of the study, the supply chain of milk and milk products and its challenges in an Indian context was studied. The challenges specific to this industry include the elaborate set of Governmental checks, certifications, and documentation that give a perception of false safety but does not prevent adulteration of milk and milk products in highlighted. A pathway to deploy Blockchain technology for the Indian dairy ecosystem to increase safety and reduce waste is proposed with the various entry point requirements elaborated in Fig. 2. A Blockchain logistics platform to help the traceability and transparency of the supply chain of milk and dairy products targeting the suppliers, manufacturers, and customers is also proposed in Fig. 3.

As any new technology will go through a hype stage, Blockchain technology is no different and is in danger of being currently portrayed as a flawless solution. In reality, it takes a lot of time to overcome all the challenges involved. Implementing Blockchain at one stretch in the milk products supply chain would be a challenge as the milk supply chain consists of a huge network spread all across the country and mostly in the rural parts. Most of the producers of milk would be from the village, and the data entry part feeding the database with the cow, place, and farmer details will take a substantial amount of time. Moreover, the importance of proper procedures in precisely entering data during the collection of milk in co-operative milk collection centers will need adequate training and monitoring of practices so that the process is efficient and effective. The user-friendliness of the data entry interface will also require constant revisits to check for accuracy and make it foolproof.

Another challenge in this domain is that there is no specific regulation currently in place, and no one set of rules is specified for all to follow. This is characteristic of

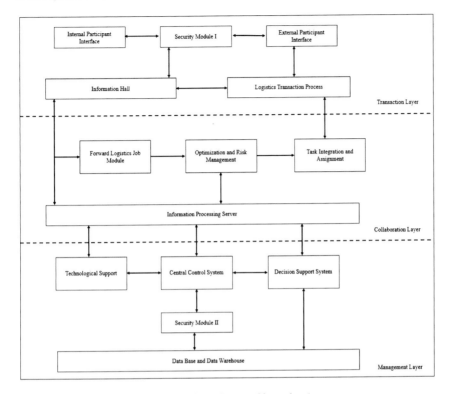

Fig. 3 Module structure of logistics platform (Proposed by authors)

any new technology, but governments and extremely controlled sectors may need to create regulations for Blockchain soon.

Another one of the challenges of implementing Blockchain is scalability. In reality, Blockchain works fine for a small number of users. But when mass integration takes place and as user numbers increase on the network, the transitions take longer to process, as seen in mining the most popular Blockchain application—Bitcoin. As a result, the transactions cost higher than usual, thus restricting more users on the network. Hence the prospect of Blockchain is enhanced if this challenge is dealt with sooner.

However, when weighing the balance of concerns regarding the safety of milk products and the health of the citizens in India, the benefits of implementing a Blockchain platform in the dairy supply chain far outweigh the challenges involved. New developments in this technology and headways being made in this domain do make this solution very implementable.

10 Conclusion

Compared to the traditional supply chain with the newly proposed Blockchain platform for the milk and dairy supply chain, the proposed Blockchain platform might have significant advantages. The proposed Blockchain-based supply chain can improve the degree of automation and transparency; the study shows that there are many chances that a manufacturer can have control over the supply chain. They can use all the information available in the Blockchain platform for the safety of the consumers. This can increase the transparency in the supply chain and reduce waste and increase the security of the milk and milk products. Blockchain technology is an emergent and potentially game-changing technology for supply chains. Blockchain technology is available in the market, but many barriers exist in adapting it to milk and dairy supply chain management. Many companies are still trying to adapt this technology by using it to its full capacity and potential. In this article, the authors have shown the model and different methods to adapt and implement Blockchain in the milk and dairy supply chain. The authors had our study reviewed by expert panel members who have worked in the milk industry for many years. The authors couldn't make a trial run for the proposed model because of time and human resource constraints. Hence, the experimental part of this model can be taken up for future study and research.

References

1. Apte S, Petrovsky N (2016) Will Blockchain technology revolutionize excipient supply chain management?
2. Azzi R, Chamoun RK, Sokhn M (2019) The power of a blockchain based supply chain. Comput Ind Eng 135:582–592. https://doi.org/10.1016/j.cie.2019.06.042
3. Behnke K, Janssen MF (2020) Boundary conditions for traceability in food supply chains using blockchain technology. Int J Inform Managem 0268–4012
4. Biscotti A, Giannelli C, Keyi CF, Lazzarini R, Sardone A, Stefanelli C, Virgilli G (2020) Internet of things and blockchain technologies for food safety systems. In: IEEE international conference on smart computing (SMARTCOMP), pp 440–445
5. Casey MJ, Wong P (2017, 3 13) Harvard Business Review. Retrieved from Global Supply Chains Are About to Get Better, Thanks to Blockchain: https://hbr.org/2017/03/global-supply-chains-are-about-to-get-better-thanks-to-blockchain
6. Cohen M (2019) The use of blockchain technology to solve common challenges in the supply chain. digitalcommons.usm.maine.edu. Retrieved from digitalcommons.usm.maine.edu
7. Dehghani M, Popova A, Gheitanchi S (2021) Factors impacting digital transformations of the food industry by adoption of blockchain technology. J Business & Indust Market 0885–8624
8. Grain (2019, 6 19) Indian dairy under threat from new trade deals. Retrieved from Grain.org: https://grain.org/en/article/6257-indian-dairy-under-threat-from-new-trade-deals
9. Holmberg A, Åquist R (2018) Blockchain technology in food supply chains: a case study of the possibilities and challenges with an implementation of a blockchain technology supported framework for traceability. Digitala Vetenskapliga Arkivet 79
10. Kasten J (2019) Blockchain application: the dairy. J Supply Chain Managem Syst pp 45–54
11. Kumar A, Goyal SK, Pradhan RC, Goya RK (2015) A study on status of milk adulterants using in milk of district VaranasI. South Asian J Food Technol Environ 2454–6445

12. Lin C-F (2019) Blockchainizing food law: promises and perils of incorporating distributed ledger technologies to food safety, traceability, and sustainability governance. Food and Drug Law J 586–612
13. Makarov EI, Polyansky KK, Makarov ME, Nikolaeva YR, Shubina EA (2019) Conceptual approaches to the quality system of dairy products based on the blockchain technology. Russia, Springer, Cham
14. Mehta DP (2004) Action research in policy making: a case in the dairy industry in Gujarat. Springer link, pp 344–363
15. Neo P (2019, 6 14) Milk in India found to be most highly-adulterated food product despite FSSAI assurances of safety. Retrieved from https://www.foodnavig ator-asia.com/: https://www.foodnavigator-asia.com/Article/2019/06/14/Milk-in-India-found-to-be-most-highly-adulterated-food-product-despite-FSSAI-assurances-of-safety
16. PTI (2018, 9 5) 68% milk & milk products in India not as per FSSAI standard: Official. Retrieved from The Economic Times: https://economictimes.indiatimes.com/industry/cons-products/food/68-milk-milk-products-in-india-not-as-per-fssai-standard-official/articleshow/65689621.cms
17. Rambim D, Awour FM (2020) Blockchain based milk delivery platform for stallholder dairy farmers in Kenya: enforcing transparency and fair payment. In: 2020—"IST-Africa Conference Proceedings, pp 1–6
18. Shingh S, Kamalvanshi V, Ghimire S, Basyal S (2020) Dairy supply chain system based on blockchain technology. Asian J Econom Business Accounting 13–19
19. Subramanian N, Bhatia MS, Dora M (2021) Food supply chain in the era of Industry 4.0: blockchain technology implementation opportunities and impediments from the perspective of people, process, performance, and technology. Prod Planning & Control 1–21
20. Tian F (2016) An agri-food supply chain traceability system for China based on RFID & blockchain technology. In: 13th international conference on service systems and service management (ICSSSM). https://doi.org/10.1109/icsssm.2016.7538424
21. Wood L (2019, 3 22) BusinessWire.com. Retrieved from The Dairy & Milk Processing Market in India, 2018–2019 & 2023 - ResearchAndMarkets.com: https://www.businesswire.com/news/home/20190322005336/en/The-Dairy-Milk-Processing-Market-in-India-2018-2019-2023---ResearchAndMarkets.com

Blockchain Technology in Financial Sector and Its Legal Implications

K. S. Divyashree and Achyutananda Mishra

Abstract The blockchain technology has reached the tipping point. Blockchain is a zero-level technology on different applications such as crypto, smart contract, tokens, and DAO function. The applications of blockchain technology have generated new paradigms posing significant challenges to the world. Although blockchain technology is evolving and is in its nascent stage, the financial world has started relying upon it and is heavily investing in blockchain applications. In such a scenario, there is every possibility of its being misused and used for illicit purposes like money laundering, hacking, privacy breach, Ponzi investment scheme, terrorist financing, etc. Keeping this in view, this paper aims to explore how blockchain technology is used in the financial sector and analyse the legal implications thereof. For this, the chapter will explain what blockchain technology is, how it is being applied in the financial sector, need for regulation, and lastly, draw conclusions and suggest a few corrective measures.

Keywords Blockchain · Technology · Financial sector · Regulation

1 Introduction

Disruptive technology in the form of 'blockchain' has come to the financial world throwing open many speculations, issues, and apprehension. 'Blockchain technology is a decentralized distributed ledger that records transactions between two parties. It moves transactions from a centralized server-based system to a transparent cryptographic network. The technology uses peer-to-peer consensus to record and verify transactions, removing the need for manual verification' [1]. Blockchain is

K. S. Divyashree (✉) · A. Mishra
School of Law, Christ University, Hosur Road, Bangalore, Karnataka 560029, India
e-mail: divyashree.ks@res.christuniversity.in

A. Mishra
e-mail: achyuta.mishra@christuniversity.in

a digital ledger system used to securely record transactions. The blockchain eliminates the need for a third party, instead relying on technology to conduct business [2]. Blockchain implies reconstruction of credit, a system that allows individuals to trust without social relations. It is used to reduce costs and helps in bookkeeping and settlement work. It can control risks more effectively by tracking and monitoring loan use. It looks for creative profit-making methods. Highlighting its significance WTO said 'The world as we know it has been shaped by technological innovations. A new technology, b—lockchain—a distributed ledger technology—has been greeted by many with enthusiasm and excitement as the next big game-changer' [3].

Blockchain has found its way into the financial technology ecosystem and is making its presence felt in profound ways. According to CNBC, 1 in every 10 people are investing in crypto. 10% of global GDP is stored on blockchain technology. The Fourth Industrial Revolution will have a profound impact on the nature of state relationships and international security. According to a 2020 report released by PriceWaterhouse Coopers, 'Blockchain is expected to reach a tipping point by 2025 hitting the mark of $422 billon GDP value in blockchain industry' [4]. Blockchain application in the financial sector is growing abundantly in trade, payment, digital identity verification, cross-border transactions, asset management and capital investment, etc. Regardless of its growing significance, there is no clear acceptance of it by the nations across the globe, and hence, it eludes regulation in most of the countries.

The chapter aims at providing an understanding of blockchain technology and its application in the financial sector. Further, it seeks to examine the challenges of such application and regulation thereof and offers few suggestions. Moreover, it will serve the cause of bringing awareness among the investors in the financial market through blockchain technology. For the chapter, scholarly articles, blogs, conference reports, and newspaper articles have been relied upon.

This chapter is about the application of blockchain technology in the financial sector and its implications on individuals and society. It starts with an explanation of what blockchain technology is and then illustrates how blockchain technology is being applied in the financial sector. In continuation of this, it explains the areas to be regulated, i.e. Taxation, Payment Regulation (KYC and AML), IP Laws on Mining, Laws on Tokenization, Laws on ICO (Initial coin offering) IDO (Initial Dex Offering) IEO (Initial Exchange Offering) and STO (Security Token Offering), Laws on Smart Contract, Laws Related to Privacy, Laws on Defi Composability, Consumer Protection Law, Regulation of DAO, and Laws on Crypto Currency. Then, it examines the impending challenges of the application of blockchain technology in the financial sector and draws a conclusion and offers few suggestions for adoption.

1.1 Blockchain Technology: Meaning and Nature

Stuart Haber and W. Scott Stornetta [5] published a paper called 'How to time stamp digital assets' which started the concept of blockchain even before Satoshi Nakamoto [6] introduced blockchain. Blockchain is literally just a chain of blocks, but not in the traditional sense of those words. The words 'block' and 'chain' in this context, we are actually talking about digital information (the 'block') stored in a public database (the 'chain'). Like each individual has a unique fingerprint every block in the blockchain has a unique hash function. Blocks on the blockchain store data about monetary transactions or record valuable data. Blockchain works on consensus mechanism or p2p data security which is unique and makes it immutable. But it turns out that blockchain is actually a pretty reliable way of storing any type of data transactions.

Blockchain is immutable because it works on a hash function which is unique for each block. Hashing is quite complex, and it is impossible to alter or reverse it. 'NONCE' which is an abbreviation for 'number only used once' and refers to a number that is added to a hashed or encrypted block in a blockchain that increases the difficulty level or restrictions when rehashed. In order to receive cryptocurrency, blockchain miners must solve for a nonce. There is no chance to alter or hack the data because it is impossible to corrupt every data stored on every node in the network. There is no central authority in blockchain; therefore, no one can alter its characteristics. Using encryption gives another layer of security. The blockchain is decentralized and provides transparency, and every change in data can be viewed publicly. Traditional banking systems are slow, and the blockchain facilitates fast transactions and settlements. Blockchain works on consensus mechanisms. Blockchain therefore is the future where people put their trust on technology instead of institutions. It is called the age of deinstitutionalization [7].

The blockchain technology application can be studied by understanding the layers of technology that transactions are performed on. There are three main layers; at the base there is a computer hardware network. Through the network nonce, the distributed ledger technology is created. The second level comes from such distributed ledger technology, also known as blockchain technology, whereby the side chains are created. With this the first layer of blockchain gets limited and continues through second layer side chains. The final level is the scaling level which is ever growing. Here, the technology is used for various applications in different sectors like gaming, health care, etc (Fig. 1).

1.2   Opportunities with Blockchain Technology

Blockchain technology benefits not only cryptocurrencies, but also many other industries that must store and manipulate massive amounts of data. It not only supports the financial system, but also public and social services such as land record management, asset management, educational services, energy conservation, citizen registration systems, patient management systems, taxation systems, security and privacy enhancement of mobile devices, and associated services. Banking, cybersecurity, academia, marketing and advertising, supply chain management, ecommerce, voting, supply chain networks, finance, asset management, health care, real estate, Internet of things, government record keeping, and health industries, among others, use blockchain technology to improve their operations [8].

How does a transaction get into the blockchain?

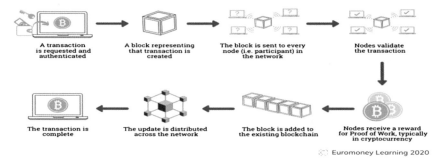

Fig. 1 Blockchain transaction. *Source* Team, D., 2022. *What is cryptocurrency? And why you should care—Pipe & Piper*. [online] Pipe & Piper. Available at: https://pipeandpiper.co.uk/2021/08/16/what-is-cryptocurrency/ [Accessed 6 January 2022]

In general, blockchain technology can be used for the following scenarios:

1. Asset digitization in order to provide data-driven business models.
2. Digitization of business-partner processes and transactions.
3. Immutable records of transactions and assets are provided.
4. Data storage that is decentralized, permanent, and secure.

2 Blockchain Technology in Financial Sector

Blockchain technology is radically transforming the economy. Moreover, in a complex financial market setup blockchain technology enables greater personal control over our finances replacing financial institutions. The crypto market runs parallel to the stock market, therefore affecting the global economy. Economists have been exploring the behaviour of the crypto market and new technology blockchain that has fundamentally changed how we exchange value. Blockchain can revolutionize the financial market; therefore, it is called disruptive technology. For the purpose of this study, the application of blockchain can be studied by dividing various aspects of the finance market like capital/crypto market, asset management, payments and remittance, and banking and lending. Blockchain technology has the potential to improve a wide range of financial services such as quick payment process, remove intermediaries through DAO, automate the execution of agreements through smart contracts, generate capital with ICOs, and store financial data. Lenders will be able to fund loans faster, vendors will receive payments sooner, and stock exchanges can settle investments and sales almost instantly. The WTO report entitled 'World Trade Future 2018: How Digital Technologies Transform Global Trade' covers various topics related to international trade and how blockchain technology can be utilized to develop it [9]. The financial services industry is a highly regulated one, and the pace of blockchain adoption will depend on how the changes are supported by regulatory bodies across the globe. Some regulations connected to blockchain are already in force [10].

2.1   Benefits of Blockchain Technology in Financial Sector

1. Risk Control: Due to the distributed nature of blockchain technology, the technology is generally resilient. Settlements become risk free with blockchain, which saves both parties a substantial amount of time and money. There is no delay in transactions with the help of blockchain.
2. Reduced Cost: Blockchain technology eliminates a large portion of the risk that the intermediary party will be unable to satisfy which reduces significant cost.
3. Transparency: Blockchain ensures more transparency among financial firms, which results in improved regulatory reporting and monitoring systems in the financial sector.
4. Secure and Auditable: Data recorded on the blockchain is immutable and secures from attacks. Any data stored on a blockchain may be traced in real time, providing a comprehensive audit trail.
5. Innovative Financial Alternatives: Crypto or digital currencies or tokens have increased the number of financial solutions available in times of crisis [11] (Fig. 2).

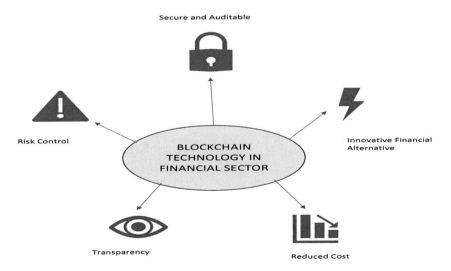

Fig. 2 Benefits of blockchain technology in financial sector. *Source* Leeway Hertz—Software Development Company. 2022. *Blockchain in Finance: Blockchain Use Cases.* [online] Available at: https://www.leewayhertz.com/10-use-cases-of-blockchain-in-finance/. [Accessed 6 January 2022]

3 Areas to be Regulated

World over blockchain technology is being used in the financial sector but without any regulation except in a few countries. The following areas have been identified for regulation which will be beneficial for individuals as well as countries.

3.1 Taxation

Blockchain applications such as crypto shall be taxed in three ways 1. Mining process of crypto 2. Buying or selling crypto 3. Crypto as legal tender [12]. The process of producing blockchain or cryptocurrency is called mining. Mining is not an easy process as it requires a huge amount of electricity and Internet access [13]. There are still people who have made a huge amount of profit from mining. The miners are paid a small amount of commission that the people pay for transaction fees on blockchain. In recent days, there is a pool of miners who share the mining commission. There are no clear tax regulations yet. Crypto can be taxed under gift or capital gains. Depending on the type of transaction [14], regulators must fix the threshold limit to acquire crypto when it is used as legal tender.

3.2 Payment Regulation KYC (Know Your Customer) and AML (Anti Money Laundering) Laws

Taxation on crypto is not clear. Should minors be taxed for huge amounts of profits? Most crypto exchanges fall under taxation preview [15]. ILLEGITIMATE use of crypto is very evident, e.g. for buying drugs just like money can be used for legal or illegal purposes the same way cryptocurrency can be used. Crypto can be used by KYC and AML regulation on crypto so it can be controlled. Crypto works on blockchain technology, and there are crypto attacks such as Ruyk Ransomeware attack or exchange attack that leads to investor losses. There is a need for proper regulation for the use of crypto [16]. There is an increased rate of scams in the crypto market and trading. 95% of crypto coins are fake [17]. The KYC regulations will not suffice to prevent the exploitation of digital currency for unlawful purposes such as drug purchases on the dark web, money laundering, terrorist financing or illegitimate betting, among other malicious activities. There is a need for strict laws to check the illegal activities and illegal exploitation of blockchain technology [18].

3.3 IP (Intellectual Property) Laws on Mining

Bitcoin, the most well-known cryptocurrency, is not protected by a patent. Experts believe that the lack of IP protection is what allows cryptocurrencies to grow at such a rapid rate. Crypto coins, decentralized applications (Dapps), and other blockchain software products hold intellectual property rights. There are no defined laws to bring crypto trademark, patent, and design under intellectual property regime. Can software be patented or given copyright? How does AI violation of copyright be treated? etc. There are a number of issues before World Intellectual Property Organisation (WIPO) related to this new technology invention and its governing IP Laws [19].

3.4 Laws on Tokenization (NFT—Non-fungible Token)

An NFT is a type of crypto asset wherein each item, or token, is unique. It is not easily exchangeable for another value or item. NFT is a digital certificate of ownership of any designated digital asset [20]. Creators and users of NFTs need to understand the rights granted to token holders. When the token holders have the right to profit-sharing, what they have is a security token. As such, it is subject to financial regulations [21]. NFTs do not grant such rights, although they grant access to future content or give rights to royalties. The dilemma here is that ownership of NFT does not translate into ownership of an original work. In other words, buying an NFT does not mean that one is buying the underlying IP rights in a given content. There are no clear laws set for NFT. There is no regulation to check international transfers on tokens [22].

3.5 Laws on ICO (Initial Coin Offering) IDO (Initial Dex Offering) IEO (Initial Exchange Offering), and STO (Security Token Offering)

An ICO is also known as a token generation event (TGE), which refers to a revolutionary technique of raising funds through the issuance of crypto tokens [23]. Companies that issue tokens or coins through an ICO, IDO, IEO, STO typically organize the offering so that the tokens or coins are tied to a smart contract on the blockchain or a comparable platform [24]. An initial coin offering (ICO) is an unregulated method of raising funds as there is no proper regulation. There is a critical need for regulation of ICO and other crypto-related fundraising schemes to protect consumers against risky or fraudulent investments [25]. The current boom in ICOs has resulted in an increase in pump and dump scams in which the tokens have no inherent worth. It is challenging to audit an ICO to determine the authenticity and value of a token [11]. When there is token swap, it creates liquidity. It creates arbitrage and hedging of funds. Regulators must check the fundraising methods done through blockchain platforms to protect investors.

3.6 Laws on Smart Contract

Smart contracts execute automatically after meeting or completing predefined criteria. It essentially means that you cannot go to the next transaction until the previous one on the ledger has been completed [26]. Smart contracts employ cryptography to code into the ledger-based system. Smart contracts also make use of digital signatures for authentication and limited access. Blockchain technology generates digital signatures on its own. This implies that smart contracts can be used for any purpose where a document, information, or form require authentication through the affixation of signatures, but they are not certified under the governing legislations. The effect is considered ineffective [27]. Smart contracts are used by various industries such as insurance, health care, and supply chain contracts, and there is a need to bring awareness among the public regarding these types of contracts to reap benefit of blockchain technology.

3.7 Laws Related Privacy

Use of blockchain technology is fraught with issues of privacy and information security. Further, the decentralized nature of blockchain technology not only affects the application of various jurisdictions' laws, but it also causes problems with those that prohibit cross-border data transfers. Moreover, rights of data correction and data erasure, also known as the right to be forgotten, bring apparent conflict with blockchain technology's immutability characteristics. Therefore, to protect privacy and information security, one must ensure compliance with three critical components, i.e. confidentiality, integrity, and availability, which are related data in blockchain [28].

3.8 Laws on Defi (Decentralized Finance) Composability

This is when you mix two different features of different blockchain applications and come up with new features. Decentralized finance is basically finance as we know it, like banks, pooled entities, negotiable instruments and entire space which includes finance companies. When the entire finance entities decentralized without centralized authority. Defi is a technological phenomenon backed by a blockchain. It is the first moment; later, it might hit other sectors as well. Removing intermediaries like banks and NFT, etc., releases value for consumers [29], e.g. streaming directly instead of cable operators. There is a problem if we remove centralized authority when people in power are removed, e.g. removing the CEO of the company. The government sees it as a threat, and we must find a middle ground of indirect governing of crypto. If all banks and financial companies enter Defi, the growth potential is huge. There are different stages of decentralized finance from 0 to 10. We are still in the kindergarten stage in Defi [30].

3.9 Consumer Protection Law

Consumer protection is another area which needs to be focused on. Variety of scams are done under blockchain applications to fool people into investing. Capital fixation or the value of crypto keeps varying and cannot be fixed like that of gold, and the government needs to work on this capital fixation on crypto assets. The exchanges can be registered and controlled by the government, and crypto finances can be traced to be taxed. Does stable coin serve any purpose? Stable coin has its own blockchain and is limited in supply. The real question arises is their innovation and benefit involved in the coin. People start accepting shit coin/joke coins without proper knowledge about it. The meme coin which imitates the properties of famous coins which may or may not have similar properties of original coin fools buyers into risking their investments to it, e.g. Dogecoin and Elon coin [31].

3.10 Regulation of Decentralized Autonomous Organization (DAO)

DAOs are informal groups of participants relying on smart contracts to implement collective choices without human mistakes or manipulation. DAOs are often used for decentralized financial governance, fundraising, exchanges and property loans and all generating billions of dollars in non-intermediaries' transactions. The DAO is a decentralized venture capital fund that serves as a hub for investors interested in funding blockchain-based projects. Court finds that the investors are joined for the single enterprise of investing in the crowdfunded venture for a limited duration, and it is possible that the court will take a conservative approach to the agreement and define the DAO as a joint venture. Due to a lack of a centralized individual and/or group managing the business, DAOs disrupt the economy in contrast to a company which is established and governed by regulations [32].

3.11   Laws on Cryptocurrency

Cryptocurrency is anti-currency. Some people consider this better than currency, and some people say it fails as currency. The evaluation of crypto as currency to buy goods and services is still under debate. Crypto appreciates faster than that of other currency as it is limited, e.g. Buying 1 kg rice in one year uses 10 crypto, but next year 5 kg rice should be exchanged for 10 crypto [33]. Government is nervous about making it an official payment system. El Salvado is the first country to officially accept cryptocurrency as official currency [34]. The biggest risk in blockchain and crypto is not just the pricing volatility and hacking but the regulation and government's control over the economy. Some nations who do not circulate their currency outside the territory can be bypassed through cryptocurrency anonymously but can be traced through wallet addresses so it can be used for illegal purposes. Crypto has entered mainstream finance. Governments around the world are working on how to regulate, tax, trace, and record it.

4   Impending Challenges and Need for Regulation of Blockchain Technology

Blockchain technology is still a new concept for regulators. Regulatory frameworks are often built-in stages and are intended for centralized and intermediated sectors. The blockchain sector must explain how and why the technology is unique so that policymakers understand the benefits and risks of widespread use. The fact that blockchain technology is a disruptive technology that allows organizations to implement new business models that are more efficient and safer in the process of digital transformation is an obvious truth and a future bet. Through Digital Banking and Fintech, block technology has been adapted to the financial sector to manage the emerging ecosystem of cryptocurrencies, and other blockchain transactions are difficult. IoT continues to have an impact on all facets of daily life. Data is the oxygen that keeps the Fourth Industrial Revolution going. Blockchain, which allows for decentralized and secure data storage and movement, has already shown to be an effective tracking and transaction tool [35]. Blockchain will bring significant change in data storage, digital identity, certification, supply chain integration, smart contracts, and currency and monetary systems. We are currently looking at a new wave of computing in society [36]. Consumers will benefit from cheaper costs and better services because of blockchain technology.

Regulation of technological innovations such as blockchain is difficult, and it gets even more difficult when crossing borders, as each country sets its own standards [37]. Measures should be taken to protect personal data. Many investors lose money due to high market volatility of the crypto market and Ponzi investment schemes. Taxation and intellectual property laws are still unclear. The technology risk exists where it is difficult to debug smart contracts and prevent exchange hacking. Since blockchain technology is transparent and immutable, it is difficult to bring privacy measures. The DAO and Defi ecosystems must be indirectly controlled by the governments. The mining activity and creation of NFT need to be recorded to control illegal activities, the applications of blockchain are growing through Dapps, stable coins, etc., and there is a need for strict rules to legalize and exploit the technology for the benefit of people.

5 Conclusion

Blockchain technology has the potential to alter organizations by allowing them to rethink procedures and business models and explore fundamentally disruptive use cases. Traditional legal frameworks are ill equipped due to the virtual and decentralized nature of blockchain technology. Furthermore, due to the lack of a defined legal framework, regulating blockchain technology and its application in the financial sector will be difficult, and it can be misused. In such a grey scenario, it is highly essential to bring awareness among the people regarding this technology and its impact.

The Fourth Industrial Revolution, characterized by emerging technology such as blockchain, will have an enormous influence on the financial sector. To address the issues highlighted above, the following few points are suggested to be taken into account by governments and international institutions:

a. To frame adequate laws for regulation of blockchain technology application.
b. To create independent institutions to govern blockchain technology applications.
c. To conduct due diligence before investing in blockchain applications and consider the level of risk that will be incurred.
d. Blockchain is a global technology; it is important to have global guidelines for its use.
e. To create digital security exchanges in line with security exchanges.

References

1. Manjunath P, Shah P (2019) Exploratory analysis of blockchain security vulnerabilities. Aust J Wirel Technol Mob Sec 5–10
2. Werbach K (2018) Trust, but verify: why the blockchain needs the law. Berkeley Technol Law J 33:487–550
3. Chang V, Baudier P, Zhang H, Xu Q, Zhang J, Arami M (2020) How blockchain can impact financial services—the overview, challenges and recommendations from expert interviewees. Technol Forecast Soc Chang 158:120166
4. 2022. [online] Available at Financial Services Technology 2020 and Beyond: Embracing disruption. https://www.pwc.com/gx/en/financial-services/assets/pdf/technology2020-and-beyond.pdf. Accessed 6 January 2022
5. Haber S, Stornetta W (1991) How to time-stamp a digital document. J Cryptol 3(2):99–111

6. Nakamoto N (2017) Centralised bitcoin: a secure and high performance electronic cash system. SSRN Electron J
7. Torres de Oliveira R, Indulska M, Zalan T (2020) Guest editorial: blockchain and the multi-national enterprise: progress, challenges and future research avenues. Rev Int Bus Strategy 30(2):145–161
8. Sial MFK (2019) Blockchain technology—prospects, challenges and opportunities. IEEE Blockchain Technical Briefs
9. Maltseva V, Maltsev A (2019) Blockchain and the future of global trade (Review of the WTO report "Can Blockchain revolutionize international trade?"). Int Organ Res J 14(4):191–198
10. Kapadia J (2021) Blockchain technology: application in the financial industry. Scholedge Int J Manage Dev 7(8):130. ISSN 2394-3378
11. Syzdykova A, Abubakirova A (2020) Application and prospects of blockchain technology in the financial sector. Econ Ser Bull L.N. Gumilyov ENU 130. https://doi.org/10.32523/2079-620x-2020-1-149-157
12. Mazur O (2021) Can blockchain revolutionize tax administration? SSRN Electron J
13. Laurie L, Susan S, Jerry S (1997) How to make a mint: the cryptography of anonymous electronic cash. American University Law Review, January 1997
14. Gupta A (2021) Investing in cryptocurrencies? Know the tax implications
15. Abubakar Y, Ogunbado A, Saidi M (2018) Bitcoin and its legality from Shariah point of view. Seisens J Manage 1(4):13–21
16. Steblianko A, Riepin D (2021) Cryptocurrency as a modern phenomenon: advantages, disadvantages, problems of legal regulation. Legal Horiz 26:97–101
17. Bovaird C (2019) 95% of reported bitcoin trading volume is fake, Says Bitwise
18. Zwitter A, Hazenberg J (2020) Decentralized network governance: blockchain technology and the future of regulation. SSRN Electron J
19. Singh UP, Khurana and Khurana (2021) India: Indian cryptocurrency Saga: regulation and IP protection
20. Cornelius K (2021) Betraying blockchain: accountability, transparency and document standards for non-fungible tokens (NFTs). Information 12(9):358
21. Aksoy PÇ, Üner ZÖ (2021) NFTs and copyright: challenges and opportunities. J Intellect Property Law Prac
22. Kugler L (2021) Non-fungible tokens and the future of art. Commun ACM 64(9):19–20
23. Martino P, Bellavitis C, DaSilva C (2019) Blockchain and initial coin offerings (ICOs): a new way of crowdfunding. SSRN Electron J
24. Lee J, Li T, Shin D (2021) The wisdom of crowds in FinTech: evidence from initial coin offerings. The Review of Corporate Finance Studies
25. Vaidyanathan N (2018) Acca 34, ICOs: real deal or token gesture? Exploring initial coin offerings. Blemus S, Guegan D (2019) initial crypto-asset offerings (ICOs), tokenization and corporate governance. SSRN Electron J
26. Governatori G, Idelberger F, Milosevic et al (2018) On legal contracts, imperative and declarative smart contracts, and blockchain systems. Artif Intell Law 26(4):377–409
27. Ene C (2020) Smart contracts—the new form of legal agreements. In: Proceedings of the international conference on business excellence, vol 14(1), pp 1206–1210
28. de Haro-Olmo F, Varela-Vaca Á, Álvarez-Bermejo J (2020) Blockchain from the perspective of privacy and anonymisation: a systematic literature review. Sensors 20(24):7171
29. Cumming DJ, Johan S, Pant A (24 July 2019) Regulation of the crypto-economy: managing risks, challenges, and regulatory uncertainty
30. Metjahic L (2018) Cardozo law review deconstructing the Dao: the need for legal recognition and the application of securities laws to decentralized organizations
31. Bhattacharjee S (2018) Consumer's dilemma: vulnerability of banks in India versus crypto currency adaptability. Indian J Comput Sci 3(4):37
32. Hassan S, De Filippi P (2021) Decentralized autonomous organization. Internet Policy Rev 10(2)

33. Nelson B (2019) Commentary: digital currencies and payments: financial stability and monetary policy implications. J Invest 28(3):70–72
34. Taylor L (2021) El Salvador's adoption of bitcoin hits further problems. New Sci 252(3358):16
35. The advantages and disadvantages of blockchain technology. In: 2018 IEEE 6th workshop on advances in information, electronic and electrical engineering (AIEEE), November 2018
36. Haro-Olmo F, Valencia-Parra Á, Varela-Vaca Á, Álvarez-Bermejo J (2021) Data curation in the Internet of Things: a decision model approach. Comput Math Methods
37. The international conference on financial innovation and economic development (ICFIED 2020) application analysis on blockchain technology in cross-border payment

Artificial Intelligence in Education

Venkata Rajasekhar Moturu and Srinivas Dinakar Nethi

Abstract Artificial intelligence-deployed applications for learning are in use and made way into the field of education for quite some time. Therefore, its requirement forces teachers, learners, and stakeholders more than ever in the past. The aim of the current study is to provide focus on the matter that implementation of AI is not a choice but a need. As education technology evolves as a new standard, all the stakeholders involved in education must deploy AI to obtain the basic education goals, i.e., it must be individualized, effective, transformative, output based, integrative and long lasting. The current research is to portray the transformation in methods of education and put forward the current directions in incorporating artificial intelligence. The investigators contemplate on this issue and attempt to address how advancement of AI is contemporaneous with advancement in education. The pursuit to examine this circumstantial believes to add productive surface for a fruitful consideration on probing the power of artificial intelligence-based e-learning applications.

Keywords Education · Artificial intelligence · Online learning · Education technology

1 Introduction

In recent years, the educational environment has undergone a significant transition. Changing cutting-edge technologies has transformed teaching and learning across the globe. Every sector is benefiting from artificial intelligence, so is education sector. Significant advancements have developed in the education field, and with the combination of artificial intelligence approaches in teaching styles, educational institutions

V. R. Moturu (✉) · S. D. Nethi
Assistant (Academics and Research), Indian Institute of Management Visakhapatnam,
Visakhapatnam, Andhra Pradesh, India
e-mail: rajasekhar.m@iimv.ac.in

S. D. Nethi
e-mail: srinivas.dn@iimv.ac.in

have modified their methods of offering education from the past times. Artificial intelligence is one of the revolutionary approaches that allows varied learning groups, professors, and instructors to have their experiences tailored to their needs. Machines with artificial intelligence can analyze and make accurate judgments much like humans. Applications of AI include robotic technology, deep learning, and natural learning process. AI is omnipresent, from voice assistants in electronic devices, i.e., TV, fans, lift, chatbots, which we use in daily life to robots in industries.

Learners, educators, and all the stakeholders have no choice but to adapt themselves in these newly advanced modalities of virtual learning. The instructing means applied for virtual learning are basically through assignments, pre-captured lectures, videos, online case studies, multiple answer questions, etc. Of late, the most accessed education platforms are Udemy, Skillshare, Teachable, Coursecraft, EdX, Byju's, Vedantu, etc. A few other ways of the EdTech platforms being used are Zoom, Microsoft Teams, Skype, Google Classroom, and Google Meet. Several learning enthusiasts across the globe are making the most of these EdTech platforms accolades to the information technology (IT) aids which have proliferated into the classical classroom to make learning and participation more lively, effective, engaging, individualistic, and inclusive. This redefining practice of combining educational practices and information technology (IT) for active and engaged learning is how 'education technology,' 'EdTech,' is described.

Until 2019, EdTech was one of the least supported industries, but the advent of remote learning in the COVID ridden 2020 has given it new wings. The world EdTech market was estimated at USD 85 billion in 2021 and is poised to reach USD 218 billion in 2027, increasing at CAGR of roughly 17% [1], while the Indian EdTech business was valued at US$ 750 million in 2020 and is predicted to grow at a compound annual growth rate (CAGR) of 39.77 percent to reach US$ 4 billion by 2025 [2]. Despite the fact that everything seems to be going well, the EdTech business is confronted with substantial hurdles. Cybersecurity is one problem that could slow down the growth of the EdTech market, while some writers say that the user interface and user experience could be problems for the company.

The chapter is organized as follows. Background of artificial intelligence in education history and objectives is provided in Sect. 2. Artificial Intelligence (AI) development in education. Section 3 presents tools and technologies in education sector. Evolution and impact of AI are presented in Sect. 5. Practical implications and conclusion are provided in Sects. 6 and 7, respectively.

2 Background of Artificial Intelligence in Education History

Artificial intelligence in education, AIED, questions the normalcy and is reviewed suspiciously by traditional advocates in the education field on results of potent student or learner categorization and emotional engagement and active involvement with students or learners. This gives way to plenty of chances to combine not only competence to excite facilitators or teachers and allow student or learner categorization but also advances emotional engagement [3, 4]. Nonetheless, an aspect on the premise of which the advocates and skeptics meet is the fact of the matter which the range and pace of advancement are swift, and change is inevitable [5, 6]. Although traditional universities are not yet been replaced by e-learning universities, the stakeholders involved in the learning system have begun to focus and contemplate what the current normalcy will be.

A holistic study of two concerns, viz. MOOCs and Proximate Education, would be of great help to differentiate the norms for AIED in the future. We witnessed a digital revolution toward the final decade of the twentieth century. The invention of the Internet encouraged mechanization and integration of education globally [7]. This type of education or learning was instantly followed in several universities with the intention of substituting traditional lecture halls, course materials, video lectures, and digital learning material with regard to administer affirmative instructions for participants across the globe. Yet, it missed smart aspects like human-to-human interaction, physical instruction and touch, and live feedback and analysis. Severe critique followed and then [8], it was challenged that proximate education was advancing to sale of education, hindering the learning ability of professionals, and lastly online platforms were just 'diploma disposing outlets.'

In the early 2010, MOOCs, Massive Online Open Courses, were started which targeted at bulk amount of lecture material or content [9]. MOOCs targeted at streaming the learning and educational technology toward skill-focused digital learning or education distinct from the grade-based education in live environments. Universities and institutions moved at greener meadows and began catering to learning needs of participants or students looking to learn courses and earn points of credit depending on their interests. The huge rate of participants leaving the MOOCs was disturbing. The confirmed participants for MOOCs were natives of developed markets, and most of them were well educated. The credits for the courses so earned were observed as simple extensions and not as replacement of deeply practiced modes of education.

Presenting artificial intelligence in education is observed as a transition where the students will be equipped with inconceivable more data and information which would have been likely given to them by any individual instructor [10]. It is to be noted that an intelligent tutoring system, based on AI, will begin to change the dimensions of education [11]. Few schools of thought opine that human interface might not appreciate and recognize the competence and potency of brilliances, but adaptive systems which rely on AI can definitely create wonders [12]. We can visualize the quantum of influence and impact of AI on education, by deploying approaches of technological prediction that considers both the qualitative and quantitative elements, wherein we will have realistic goals and advisory narrative at both the ends, respectively.

The current inquiry is a pursuit to understand, acknowledge, and identify the purview of how AI has proliferated into education and the kind of transformation it is about to create. The study comprises five sections.

The initial section is introduction and discussion on the background. The next portion accords with a comprehensive review of accessible literature on adopting AI for digital learning. The next part in the sequence explains proliferation of AI on various standards. The penultimate aim affirms on the results, and lastly the final objective mentions the feasible suggestions and conclusion.

2.1 Objectives of the Study

The aims were to:

- Analyze the different means by which digital education is utilizing AI
- Examine the outreach and effectiveness of AI-oriented digital learning applications to various users
- Provide space for predicting the future of digital education.

2.2 Rationale for Education Technology

It is vital to initially investigate into the information why education technology has grown in prominence:

Person-centric learning experience: The means by which we know, learn, and engage with colleagues or classmates and instructors, and our comprehensive enthusiasm for the erstwhile courses or subjects is distinctive. In such a case, there is vast possibility for the upgradation of person-centric learning. The education technology applications make it smooth for instructors to design person-centric course outlines and learning engagement adventures that propel a feeling of inclusiveness and enhance the learning competencies of students or participants, irrespective of their learning competencies or age. In particular, the experiences created are not confined to live environment between the instructors and pupil; such an experience is served with every individual participant in the process of learning as per their own flexibility and convenience.

Enhanced communication, participation, and comprehensive value of education: enhanced expertise in learning: AI intensifies similar learning interests in shorter span of duration [13]. The intention of EdTech is to transfigure the sphere of learning and the processes of learning. EdTech aims on unfolding skills and abilities in participants or students that ought to be useful in the future.

Education frameworks in several nations are not transformed for many years. Education technology, EdTech, assists in collaborating technology and education for the reason that if participants or students are hidden from technology and its applications since schooling, by collaborating technology with learning, the gap between ones with online skill sets and those who were hidden to such digital education is going to be immense.

3 Artificial Education (AI) Development in Education

Debate about advancements of artificial intelligence in education field drives us rationally to a conversation on changing dimensions of education. This portion contemplates on the equal lines and makes an attempt to discuss how advancement of education is equivocal with advancement in artificial intelligence.

3.1 Results of Learning

The objective of new age education is to convert students or learners into active on the job specialists and flexible experts but not necessarily prepare a mechanistic workforce for a specific task. Institutions need to design syllabus which is aimed on application of knowledge, integration, and self-operative learning abilities [14]. An investigator states, knowledge is what we do, a verb, rather than what we have, a noun [15]. It is essential to mention that assessments or examinations will also revise after the learning results change. Therefore, assessments or examinations will face the challenge to record the learning processes and positions instead of quantifying the knowledge phase of the learner or participant. Subsequently, examinations or assessments have transformed from being an aggregate quantification of achievement to a continuous developmental measure which gives scores or grades instantly [16]. For example, the AI-supported ExamSoft platform offers a good cooperation amid the two notions of examination or evaluation, by initially evaluating learners on the necessary knowledge for graded exams, and then by person-centric assistance as required.

3.2 Lecture Room Practices

A distinct set of changing dimensions of education is the main reason for the varying aesthetics of the lecture room methods. There is a paradigm shift toward technological applications and usage of information systems [17]. Another dynamic change is from person-centric to collaborated teams' interactions [18], and the new normal will be individual-oriented learning graph as learners or participants represent various backgrounds, experiences, and cultures [16]. Even in this case, artificial intelligence will be an enabler, an impetus, and an application for different partners in this environment. As instructors, it is vital to appreciate and recognize that although digital education is a very essential mode of delivering experience and skill; however, it cannot accommodate the social and interactive environment among learners and the instructors needed to build a learner's psychology, abilities, and personality to facilitate social communication.

3.3 Environment

When we refer education, previously it was described within the model of a place, duration, and structure. As a result of fast-moving dynamics happening in all walks of life, education by description involves career-long and career-wide engagement and learning. This transformation in prospect has provided ground to the increasing flow of open online courses to which hundreds of thousands of learning enthusiasts enroll annually [3]. This has transformed the view of courses convenience and learner populace [13]. Indeed, it is known source to obtain certificates of open online course providers such as Edx, Swayam, Great Learning, Coursera, SkillShare, and UpGrad. Therefore, the instructors are not anticipated to have and deliver knowledge and skills to participants or learners, and they are donning the hat of creative intellectuals for the learners in knowing, finding, obtaining, and collaborating knowledge [19].

4 Technology Tools in Education and its Applications

Education technology bots are making it effortless for learners to be involved and engaged via fun and exciting ways of learning. Internet of Things (IoT) appliances are being praised for their effort to generate online lecture rooms for participants, if they are present in an institution, on the transport system or at their private place. Blockchain tools and machine learning are helping instructors with evaluating assessments and holding participants liable for take home assignments. For learning vocabulary, **Knowji** is a novel audio and visual vocabulary application. A well-known designed digital collaboration tool is **Padlet** which makes it easier for creative collaboration using a wide range of varied communication sources. For providing good understanding to learners about real-world mathematical concepts, **QuizNext** is used. It creates concept-based questions, instead of offering basic practice assignments, makes learning mathematics fun, exciting, and immersive. The methods of teaching an experimental-oriented or numerical-based course and a theory-oriented course on a digital platform must be distinct. A course on problem-solving approach, for example mathematical science, is clearly complex and a bit uneasy to understand and hence needs real-time-based examples, assessments, projects, and experience-based learning with the help of various frameworks. Digital instruction for a theory-oriented course demands more interactive live sessions. Thus, there is a need for the synergy of hands-on importance and pedagogical discussion in digital learning. The limit of educational technology platforms offering digital learning to enthusiastic participants is infinite. Due to the advancements of information technology and various modes of digital instruction tools, the traditional ways of education may at certain point in the future face crisis.

5 Evolution of Artificial Intelligence's Impact on Education

Artificial intelligence has been concentrating, to a greater extent, on resolving the two-way dilemma by designing approaches which are competent as one-on-one instruction; [20] as it is noted, a key role of artificial intelligence is to deliver a personalized engaging learning experience. Impressively, the learner or participant is not only personalized, [21] but also competent, which means exhibiting equivalent learning outcomes in a short span of interval. As we exceed the expansion of learning results from field-level data to design thinking, critical analysis, and associations, these transformations in education field can be viewed as a chance: Present educational frameworks vouch for personalization and agency [13]. Artificial intelligence has made significant influence on education via system evaluation and description, various categories of learning behaviors, if exploratory or step based, [22] associated designs that could be autonomous or with contemporaries, technology application formats in terms of software applied and institutional settings which could be legal.

As mentioned previously, education is moving from concentrating on product or service to processes, growing above field knowledge to combine collaboration, self-sustenance, and intrinsic motivation with a viewpoint of learning outcomes. On one side, big percentage of investigation focusses on field-level learning, and considerably a less percentage of research papers investigate into self-sustenance learning, [23] intrinsic motivation [24], and learners' satisfaction from an artificial intelligence (AI)-assisted learning atmosphere. Self-effectiveness is another component which is required to be incorporated into AI-supported learning experience [21].

5.1 Limitations to Artificial Intelligence's Application in Education

Few of the specifications that are not considered in artificial intelligence initiatives involve considering sufficient capital or funding, instructor training and their continuous professional advancement, applying additional coursework support and teaching methods, involving parent community as a key participant in the steps of learning, etc. [25]. A structure which has evolved as an answer to jump over the limitations is the design-oriented application research. This method considers on multi-participant collaborative design which propels sustainable competence [26]. The practitioners who are a part of the process infuse conceptual, territorial, and demographical components which are essential for fruitful design and application.

Another limitation is information technology symmetry. To incorporate artificial intelligence in education field, it is needed which information technologies advanced must capture the appropriateness to social teams so as to become fair as well known and certain. This requires a non-symmetrical association among technological abilities and the implementation of the novel AI-oriented educational technological systems [27]. For competitive advantage of any artificial intelligence technology, innovation and transformation by learners and instructors will be mutual ground. This truth rather than astonishing the designers must in fact inspire and motivate them. An application designed for one use may be applied for another, and the demanding essence of technology will be interrupted [28]. Technological symmetry is also a limitation when viewed in space of social events. Information technology primarily classifies the fundamental criteria of education—course outline, pedagogy, learning goals, regional culture only to quote a few—and later strengthens and retains them [29]. If we can counter and diffuse these limitations, non-symmetry of technology will be exposed.

The difficult limitation in applying technology in education field is the instructor–learner interaction. Traditionally, range of assessing quality of education was performed by placing greater significance to material over structure, and the authentic framework was taken away in the form of scales of measurement. Even educational advocates, designers, and assessors have stressed that online learning platforms are no way substitutes to the lecture halls we are habituated to learn [30]. While few investigators debate although elementary ability can be obtained by digital education [31], genuine competence can only be obtained only when human interaction is involved. Participants experience better learning when they have a chance to present, discuss, debate ideas, experience and analyze phaseouts, and acknowledge networking dynamics. This truth has stated that massive open online courses offer a limited participant interaction [32]. Learning skills such as building propositions, networking in private capacity, feeling intrinsically motivated, and gaining expertise need the support of a human instructor; Web is a minimal proximate of the said skills. There is no second thought that actions are being shown by few instructional strategies which enhance social and motivational interaction like alerts, notifications, emails, reminders, lectures, quizzes, forum discussions, etc., by using big data and analytics, tracking participant's clicks, tracing involvement to video lectures, and enhancing MOOC design and delivery [33]. Distributed Open Collaborative Courses, DOOCs, will be the substitute for MOOCs wherein importance will be on decentralized instead of central subject expertise, collaboration and involvement instead of seclusion, and inclusivity of opinions instead of unnormal opinions.

The last limitation to be overcome is personalization. Although this is a trademark of a physical instructor, it is evident by the lack of presence with MOOCs. This incorporation in an artificial intelligence-oriented framework depends hugely on the coursework chosen by the student. These applications require to be measure oriented and competent of collective-emotional involvement.

5.2 Ethical Concerns in Educational Technology Platforms

There is an entire transformation from conventional ways of teaching to transferring of knowledge with the help of online platforms in todays' world. Though the mechanisms and efforts of driving education in times of COVID-19 are laudable at the exact time, educational technology platforms built for delivering learning have been charged of being complaint with terms and conditions which are not in tune with the privacy and digital laws as prescribed by a specific country. For example, the most vital concept of a digital or online class room is video conferencing as it acts as a platform between the participants and instructors and enables better communication. Nevertheless, they carry a chord of privacy and security threats associated with them. These platforms, at times, may not guarantee that participant information would not be stolen, shared, or sold.

Raise in the usage of proximate-learning tools must bring consciousness among universities and educational schools regarding privacy of data and security. There are numerous forms through which education technology firms can address issues related to these privacy and security. It is essential to verify the privacy concerns regarding the third-party operators or vendors with whom they are associated with. This will also involve the application of top range of encryption for all of their applications, databases, and secure cloud supported tools. It is compulsory to be compliant with the data privacy terms and conditions of that specific nation. Students, parents, and all the others involved should have prior knowledge that their information is being collected, stored, and used and also must be aware of the logic for the data compilation. Users must have the privilege to obtain, access, and edit their data whenever needed. All the data should be carefully secured via access control systems, encrypted, and safely stored. Strict compliance and vigilance need to be carried out so in order to reprimand misuse and manipulation of data, for instance, complete closure of an educational technology firm in a specific nation or prohibiting its financial operations to continue further transactions. Experts in cybersecurity can highlight the ambiguities from the origin of leakage and manipulation of confidential information experts by using tool kits such as vulnerability scans and tests of penetration.

Firms, such as FlaskStack and G Suite (for education), are main players which assist in coping with the variety infrastructure problems of data or information collection with the help of advanced interoperability. Google's G Suite issues a privacy and security note to help parent community in appreciating the ways in which compiled data could be applied to use.

The contemporary scenario over the globe specifically defines the ambiguity of place and time when participants would be coming back to the traditional school. To strike a balance among safety and secureness of the instructor and the participants and progress of academics, security and technology experts should maintain the flag post of information security for cloud-based tools, applications, and platforms, applied during distance learning soaring high.

Therefore, we can conclude that the artificial intelligence (AI) in the education encompasses a series of actions which creates efficient, effective, individualized, and interactive learning experience.

6 Practical Implications

Educational technology platforms are incrementally increasing since last decade. The COVID-19 pandemic has forced conventional education methods to be replaced with digital education conducing everything remotely. Even though pandemic effects have subsided, it is still uncertain reopening educational institutions with their full capacity in the near future. In this context, we can safely state that artificial intelligence will be central hub aiding and shaping the educational system. The main advantage in digital education is its flexible nature of timings, place making it adaptable by teachers as well as students according to their needs. AI offers the necessary transformation required for the digital platforms to function more efficiently in terms of integration, collaboration, and decentralized learning ecosystems. In a study conducted in 2018 noted that 65% of school going children will be graduating to a new era of jobs which are non-existing at present. A study conducted in the UK in the year 2017 states that recruitment and retainment of talented populace are the challenges to be succeeded. Pioneers in the industry speculate that AI will be the cutting-edge revolution that can take digital education to new heights, but in reality, only minimal preparations have been done to accommodate it. The skeptical notion of AI bots replacing humans can be changed, only when students become adept in AI functionality and counter further complications arising in the predictable future. The synchronization of developments in the areas of artificial intelligence and education is obscure due to the impact on the primary education level. According to a MIT study on technology in 2020, nearly one hundred and ninety-nine percent of educators are in agreement that the implementation of AI is the unmatched solution to face the competition given by EdTech companies, and nearly ninety percent of educators have already begun to execute the necessary steps of integrating AI into educational realm.

7 Conclusion

Due to COVID-19 pandemic, the whole world has faced an impasse. This gave an opportunity for the technological educational institutions to flourish by catering to the educational needs through digital learning platforms, filling the deficiency created by lack of offline operational institutions which are closed indefinitely. This is a big breakthrough for the digital learning methodology. A large consumer base consisting of distinct groups of students, faculty, curriculums, and steadfastly advancing technologies will be outreached to many more people after the relaxation of imposed lockdown. Flexible and adaptable learning is preferential to all categories of users. The career-oriented conceptual digital education is in high demand and is being progressively looked up by different organizations to meet such demands and deliver them to digital platforms. This gives an exponential surge in the online education enrollment giving rise new companies in the view of unpredictable and uncontrollable circumstances like pandemics, etc. This change in mode of learning will be preferred even the after the pandemic gets under control. The need of soft skills cannot be replaced and will be complacent with new method of learning. The practice of being able to learn new concepts and relearn the updated skills accordingly new necessary requirement for all kinds of students. Artificial intelligence has transformed the education realm and will aid in the betterment of all stakeholders involved.

References

1. (Online) BusinessWire: Global Ed tech Market Report 2021: Industry Trends, Market Size, and Forecasts 2019–2027—ResearchAndMarkets.com Available at: "https://www.businessw ire.com/news/home/20211118005651/en/Global-EdTech-Market-Report-2021-Industry-Ana lysis-Trends-Market-Size-and-Forecasts-2019-2027---ResearchAndMarkets.com
2. (Online) IBEF (2022) Available at: "https://www.ibef.org/blogs/india-to-become-the-edtech-capital-of-the-world
3. Evans HK, Cordova V (2015) Lecture videos in online courses: a follow-up. J Political Sci Educ 11:472–482
4. Scagnoli NI, Choo J, Tian J (2019) Students' insights on the use of video lectures in online classes. British J Educ Technol 50(1):399–414
5. Agre PE (1999) Information technology in higher education: the global academic village and intellectual standardization. On the Horizon 7(5):8–11
6. Hawkins BL (1999) Distributed learning and institutional restructuring. Educom Rev 34(4):12–15, 42–44
7. Feenberg A (2002) In: Transforming technology: a critical theory revisited. Oxford University Press
8. Noble D (1998) F: Digital diploma mills: the automation of higher education. Science as Culture 7(3):355–368
9. Kenneth KL, Dedrick J, Sharma P (2009) One laptop per child: vision versus reality. Communications of the ACM 52(6):66–73
10. Arroyo I, Royer JM, Woolf BP (2011) Using an intelligent tutor and math fluency training to improve math performance. Int J Artif Intell Educ 21(2):135–152
11. Woolf BP, Lane HC, Chaudhri VK, Kolodner J (2013) L: AI grand challenges for education. AI Mag 34(4):66
12. Hao K (2019) China has started a grand experiment in AI education. it could reshape how the world learns. MIT Technology Review
13. Liyanagunawardena TR, Williams S, Adams AA (2014) The impact and reach of MOOCs: a developing countries' perspective. eLearning Papers 38–46
14. Tomlinson CA (2000) Reconcilable differences? standards based teaching and differentiation. Educational Leadership: J Department of Supervision and Curriculum Development, N.E.A. 581:6–13
15. Van Lehn K (2011) The relative effectiveness of human tutoring, intelligent tutoring systems, and other tutoring systems. Educational Psychol 46(4):197–221
16. Collins A, Halverson R (2010) The second educational revolution: rethinking education in the age of technology. J Comput Assisted Learn 26(1):18–27
17. Heffernan NT, Heffernan C (2014) L: The assistments ecosystem: building a platform that brings scientists and teachers together for minimally invasive research on human learning and teaching. Int J Artif Intell Educ 24(4):470–497
18. Dillenbourg P (2013) Design for classroom orchestration. Comput Educ 69(1):485–492
19. Morrison CD (2014) From 'sage on the stage' to 'guide on the side': a good start. Int J Sch Teach Learn 8(1)

20. Toner P (2011) Workforce skills and innovation: An overview of major themes in the literature. OECD Education Working Paper, OECD Publications No. 55
21. Banfield J, Wilkerson B (2014) Increasing student intrinsic motivation and self-efficacy through gamification pedagogy. Contemporary Issues in Educ Res 7(4):291
22. Koedinger KR, Corbett AT (2006) Cognitive tutors: technology bringing learning science to the classroom. In: Sawyer K (ed) The Cambridge handbook of the learning sciences. Cambridge University Press, pp 61–78
23. Baker RSJD, Corbett AT, Koedinger KR, Evenson SE, Roll I, Wagner AZ, Naim M, Raspat J, Baker DJ, Beck J (2006) Adapting to when students game an intelli- gent tutoring system. In: Proceedings of the 8th international conference on intelligent tutoring systems
24. Roll I, Wylie R (2016) Evolution and revolution in artificial intelligence in education. Int J Artif Intell Educ 26(2):582–599
25. Bingimlas K (2009) Barriers to the successful integration of ICT in teaching and learning environments: a review of the literature. Eurasia J Mathem Sci Technol Educ 5(3):235–245
26. Fishman B, Penuel WR, Allen A, Cheng BH, Sabelli N (2013) Design-based implementation research: an emerging model for transforming the relationship of research and practice. In Fishman AJ, Penuel WR, Allen A-R, Cheng BH (eds) Design-based implementation research: theories, methods, and exemplars. National Society for the Study of Education Yearbook, vol 112(2), pp 136–156
27. Rosenberger R (2017) The ICT educator's fallacy. Found Sci 22:395–399
28. Feenberg A (2017) The online education controversy. Found Sci 22(2):363–371
29. Winner L (1980) Do artifacts have politics? In: Kaplan DM (ed), Readings in the philosophy of technology (pp 289–203). Albany: Rowman & Littlefield
30. Dreyfus HL (2002) Anonynmity versus commitment: the dangers of education on the internet. Educ Philos Theory 34(4):369–378
31. Boulay B (2011) Towards a motivationally intelligent pedagogy: How should an intelligent tutor respond to the unmotivated or the demotivated? In: Clavo R, D'Mello S (eds) New perspectives on affect and learning technologies. Explorations in the learning sciences, instructional systems and performance technologies, vol 3
32. Christensen G, Steinmetz A, Alcorn B, Bennett A, Woods D, Emanuel E (2013) The MOOC phenomenon: who takes massive open online courses and why?. Available at SSRN 2350964
33. Simonite T (2013) Search under way for gold in online education data trove. MIT Technology Review
34. Christensen G, Steinmetz A, Alcorn B, Bennett A, Woods D, Emanuel EJ (2013) The MOOC phenomenon: who takes massive open online courses and why?
35. Francesco I (2019) "Do Artifacts Have Politics?" by Langdon Winner. A reflection" Available at: "https://francescoimola.medium.com/do-artifacts-have-politics-by-langdon-winner-a-reflection-bde891f9f546" online (2019)
36. Bingimlas K (2010) Evaluating the quality of science teachers' practices in ICT-supported learning and teaching environments in Saudi primary schools (Doctoral dissertation, RMIT University)

An Impact of COVID-19 a Well-Being Perspective for a New World

Thanveer Jahan

Abstract One of the data science issue is Covid-19, which had created a major hit massively is on public health. It resulted on a major health issues and thereby resulted to deaths. The structure of the society and community is affected to concentrate on important issues such as affordability of a health care, availability of medicines, rights for worker's and freedom to move. Some parts of the population in the world were exposed to the complications of anxious, depression and symptoms of post-traumatic as these are related to stress. The situation had become crucial for a data scientist, as there were many questions started to rise and trust the data, curves that were plotted by social media Web sites. The situation made them to scare or think worst that could even happen in society. It made society more panic and empowered to handle this situation. This survey paper concentrates on the issues and problems on society, children, students, teenagers such as physical fitness and psychological and social evaluation effects of pandemic. It also focuses on a new perspective on the usage of digital devices that effected mental health.

Keywords Data science · Psychological · COVID-19 · Mental health

1 Introduction

Traditional health surveillance systems are well known for major time lags. The current situation clearly indicates that the systems are critically needed locally and are robust [1]. In this situation, the collection of data is very difficult for such an infectious disease. In real-time data, analysis of such a high resolution data has become a difficult task for a data scientist. They work in domains such as public health and also learn from various domains. Coronavirus is a disease which is contagious, where mild infection is treated in home quarantine or thereby rely on hospitals or a practitioner to estimate the spread that can mislead the early stages of disease progression. The people in the society are lesser, who have actually made their

T. Jahan (✉)
Vaagdevi College of Engineering, Warangal, Telangana, India
e-mail: tanveer_j@vaagdevi.edu.in

© The Author(s), under exclusive license to Springer Nature Singapore Pte Ltd. 2023   245
M. A. Chaurasia and C.-F. Juang (eds.), *Emerging IT/ICT and AI Technologies Affecting Society*, Lecture Notes in Networks and Systems 478,
https://doi.org/10.1007/978-981-19-2940-3_17

presence at health facilities to test or care. The report of it can lead to focus on morbidity and mortality. The fact is that many countries do not show actual count of people having virus. The increase in the number of test will increase the count. Countries like Iceland have done systematic sampling including the people having asymptomatic symptoms [2]. The prevalence of the infection is indicated along with the containment areas. Keeping apart the conspiracy theory of the government is that the test for coronavirus is expensive. The count collected from a country is dependent on the widespread of the virus and the financial status of a local healthcare facility for testing. The problem for data sampling has become concern for a data scientist in many cases. The concern of pandemic had also affected the society wherein they are separated from their loved ones, less freedom as well as uncertainty of the spread of the disease. The concern in the general public is increased working in healthcare centers as to understand spread of number of cases. The families in the society were panic in storing long shelf of food items. It became more fear for them to shop in supermarkets or public places. Lockdown was a stressful period that made a society living style have changed completely. The affect of COVID-19 also affected children in the society with obesity problems [3]. The lockdown in many countries has also imposed children by stopping the physical activity. Children who stay in urban or in small houses or apartments were having limited space for any physical activities. These were one of the waves of COVID-19 that affected children very badly [4].

Data collection had become essential part in this situation, which it rely on accuracy and limitations.

1.1 Data Modeling and Prediction

The WHO gave an information from the country China where the symptoms of the virus may vary from first day to 14 days after the exposure [5]. It is been issued from the center of disease control and prevention that the highest alerts from Italy.

The countries that restricted to travel such as Iran, China and South Korea. The virus outbreaks are shown in Fig. 1. The impact of virus had made data scientist to model the data and predict its impact on the society. The coronavirus disease made conditions worst with an outbreak of sense of fear, stress, anxiety and mental disorders. The overwhelming of the disease has caused to develop more emotions among adults and children. Overcoming with stress led the society, people made stronger [6].

To protect the mental health of the people in the society, WHO updated these measures.

1. The sense of fear is created by the digital media, reading and watching news.
2. To protect their dear ones by seeking relevant information.
3. The trigger of fear and anxiety id from the sources such as social media.

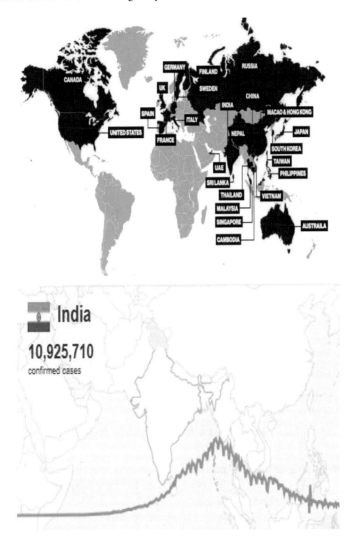

Fig. 1 COVID-19 affected countries and number of cases in India

2   COVID-19 Impact on Mental Health

The most attractive online platforms for a young generation were used even before pandemic [7]. It has before a cup of coffee after pandemic. The schools were closed during pandemic. The communication among young children increased with this platform by playing online games and access social media platforms also. The virtual learning is been made compulsion on many school children. The physical activity was decreased as there prolonged sedentary periods, as they were using screens for long time. The screen time is been increased day by day while using these platforms [8,

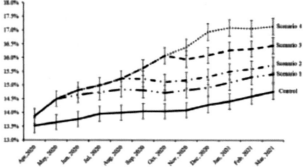

Fig. 2 Four alternative scenario of children obesity during COVID-19

9]. The increased weights in the children have been increased day by day. This in turn had led to sedentary habits which have further increased in the risk for complications such as fear, anxiety and depression. It was then predicted by many data scientists that the prolonged closure of schools will proportional rise the obesity rate in children. The data scientists have also predicted and recorded that in December 2020, there will be an rapid increase on new obesity cases. The statistics are inferred by data scientists as shown in Fig. 2.

3   Impact of COVID-19 on Unchangeable Environmental

The lifestyle during COVID-19 pandemic should have a family environment, which can change the behavior in child [10]. The fetal environment plays an important role in life course of children. As many women had a problem of obesity which is linked with childhood that can even lead to diabetes and cardiovascular diseases [11]. Pregnant women were also made to lower hospital visits as they are more vulnerable for the spread of COVID-19 [12, 13]. The containment zones were avoided more by taking measures. The government made a mandate to stay at home. The routine checkups are canceled temporarily. This had in turn led a pregnant woman to extra pressure and stress-related issues. There was a remote antenatal care available in many remote areas [14, 15]. The stop of maternity checkups and unavailability of resources during pregnancy, due to prioritizing the disease covid-19 that made data scientists to predict increase in the risk of death, maternal morbidity and mortality [16, 17].

4 An Impending Recommendation for Post COVID-19

The number of new challenges in the upcoming year 2021 is been faced many countries in the world. The year 2020 has ended up with a new scholastic year approach. The social economic conditions and emotional conditions such as stress had fallen down abruptly. The Child education is been hanged with a termination of physical learning. Children are re-introduced to schools and can facilitate the physical activity also [11]. This will not provide benefit to child behavioral health but also solves the problems related to childhood obesity. The threshold and safety measures indicators are crucial to avoid the spread of corona transmission within the school, colleges and educational institutions [18].

The wide spread of the curbing disease covid-19 and to protecting the health have become a highest priority till a good effective vaccine for Covid-19 is available. Various adhesive problems also exist in the society such as stress, mental and obesity issues [19, 20]. The above problems are uncontrolled if there is a long-term extreme health and economic results [21]. There must be more support and manage system that can be dealt in the problems in obesity in children. The issues related to availability and choosing food on low budget can be handled wisely by educating parents. A necessary need of physical activity and maintaining physical distance is a need of every teenager, child and parents in the society. A very special attention should be given in the form of counseling to the pregnant ladies, who need lot of care [22, 23]. They should be educated with the problems of obesity and the preventive measures taken to avoid obesity before the child is unborn. The issues should be considered as priority based by an every individual in the society, community during the pandemic [24, 25].

The pandemic can be ended if there is a large share in the world that needs to get immune to the disease COVID-19. Vaccines are the only one technology that is dependent in the past to lesser the death rate. The challenging task is to make these vaccine available for all people in the countries [26]. In this connection, data scientists are making their efforts in constructing the datasets for an international vaccination.

References

1. Bansal et al., J Infect Dis 214:S375–S379
2. https://www.government.is/news/article/2020/03/15/Large-scale-testing-of-general-population-in-Iceland-underway/
3. World Obesity Federation (2019) Global Atlas on Childhood Obesity [Internet] London. Available from: https://www.worldoty.org/nlsegmentation/global-atlas-on-childhoodobesity. Accessed 3 Sept 2020
4. González-Muniesa P, Mártinez-González M-A, Hu FB, Després JP, Matsuzawa Y, Loos RJF et al (2017) Obesity. Nat Rev Dis Prim 3:17034 [Internet]. Nature Publishing Group. Cited 27 May 2020. Available from: http://www.nature.com/articles/nrdp201734. Accessed 3 Sept 2020

5. World Health Organization (WHO) (2020) WHO Director-General's opening remarks at the media briefing on COVID-19—11 March 2020 [Internet]. Cited 15 Apr 2020. Available from: https://www.who.int/dg/speeches/detail/who-director-general-sopening-remarks-at-the-media-briefing-on-covid-19%2D%2D-11-march-2020. Accessed 3 Sept 2020

6. Cuschieri S (2020) COVID-19 panic, solidarity and equity—the Malta exemplary experience. J Public Health (Bangkok) 1–6 [Internet]. Springer. Cited 3 Jun 2020. Available from: https://link.springer.com/article/10.1007/s10389-020-01308-w. Accessed 3 Sept 2020

7. Rundle AG, Park Y, Herbstman JB, Kinsey EW, Wang Y (2020) COVID-19—Related school closings and risk of weight gain among children. Obesity 28 (Silver Spring) [Internet]. Cited 28 May 2020). Available from: https://pubmed.ncbi.nlm.nih.gov/32227671/. Accessed 3 Sept 2020

8. Pietrobelli A, Pecoraro L, Ferruzzi A, Heo M, Faith M, Zoller T, et al (2020) Effects of COVID-19 lockdown on lifestyle behaviors in children with obesity living in Verona, Italy: a longitudinal study. Obesity [Internet]. Wiley. Cited 28 May 2020. Available from: https://onlinelibrary.wiley.com/doi/full/10.1002/oby.22861. Accessed 3 Sept 2020

9. Ribeiro KD da S, Garcia LRS, Dametto JF dos S, Assunção DGF, Maciel BLL (2020) COVID-19 and nutrition: the need for initiatives to promote healthy eating and prevent obesity in childhood. Child Obes 16:235–237 [Internet]. Mary Ann Liebert, Inc., Publishers, New Rochelle, NY USA. Cited 19 Sep 2020. Available from: https://www.liebertpub.com/doi/10.1089/chi.2020.0121. Accessed 3 Sept 2020

10. Asigbee FM, Davis JN, Markowitz AK, Landry MJ, Vandyousefi S, Ghaddar R et al (2020) The association between child cooking J diabetes Metab Disord involvement in food preparation and fruit and vegetable intake in a Hispanic youth population. Curr Dev Nutr 4:nzaa028 [Internet]. Cited 19 Sep 2020. Available from: http://www.ncbi.nlm.nih.gov/pubmed/32258989. Accessed 3 Sept 2020

11. Leddy MA, Power ML, Schulkin J (2008) The impact of maternal obesity on maternal and fetal health. Rev Obstet Gynecol 1:170–8 [Internet]. Med Reviews, LLC. Cited 28 May 2020. Available from: http://www.ncbi.nlm.nih.gov/pubmed/19173021. Accessed 3 Sept 2020

12. Guan H, Okely AD, Aguilar-Farias N, Del Pozo Cruz B, Draper CE, El Hamdouchi A et al (2020) Promoting healthy movement behaviours among children during the COVID-19 pandemic. Lancet Child Adolesc Heal 4:416–418 [Internet]. Elsevier. Cited 28 May 2020. Available from: http://www.ncbi.nlm.nih.gov/pubmed/32458805. Accessed 3 Sept 2020

13. Sport New Zealand IHI Aotearoa (2020) Guidance for physical activity at COVID-19 alert level 3 | Sport New Zealand—IHI Aotearoa [Internet]. Sport New Zeal. IHI Aotearoa. Cited 22 Sep 2020. Available from: https://sportnz.org.nz/about/news-and-media/media-centre/guidance-for-physical-activity-at-covid-19-alertlevel-3/. Accessed 3 Sept 2020

14. Tripathi M, Mishra SK (2020) Screen time and adiposity among children and adolescents: a systematic review. J Public Health (Bangkok) 28:227–44 [Internet]. Springer. Cited 28 May 2020. Available from: http://link.springer.com/10.1007/s10389-019-01043-x. Accessed 3 Sept 2020

15. Marsh S, Ni Mhurchu C, Maddison R (2013) The non-advertising effects of screen-based sedentary activities on acute eating behaviours in children, adoles- cents, and young adults. A systematic review. Appetite 71:259–273. Appetite [Internet]. Cited 2020 May 28. Available from: http://www.ncbi.nlm.nih.gov/pubmed/24001394. Accessed 3 Sept 2020

16. Inchley J, Currie D, Budisavljevic S, Torsheim T, Jåstad A, Cosma A et al (2020) Spotlight on adolescent health and well-being. Findings from the 2017/2018 Health Behaviour in School-aged Children (HBSC) survey in Europe and Canada, International report, vol 1. Key findings, Copenhagen

17. Nagata JM, Abdel Magid HS, Pettee Gabriel K (2020) Screen time for children and adolescents during the coronavirus disease 2019 pandemic. Obesity 28:1582–1583 [Internet]. Wiley. Cited 19 Sep 2020. Available from: https://onlinelibrary.wiley.com/doi/abs/10.1002/oby.22917. Accessed 3 Sept 2020

18. Centers for Disease Control and Prevention (2020) COVID-19—School reopening: indicators to inform decision making I CDC [Internet]. Centers Dis Control Prev. Cited 19 Sep 2020. Available from: https://www.cdc.gov/coronavirus/2019-ncov/community/schools-childcare/indica tors.html. Accessed 3 Sept 2020

19. An R (2020) Projecting the impact of COVID-19 pandemic on childhood obesity in the U.S.: a microsimulation model. J Sport Heal Sci [Internet]. Elsevier. Cited 28 May 2020. Available from: https://www.sciencedirect.com/science/article/pii/S209525462030065X. Accessed 3 Sept 2020

20. Jogdand SS, Naik J (2014) Study of family factors in association with behavior problems amongst children of 6–18 years age group. Int J Appl basic Med Res 4:86–89 [Internet]. Wolters Kluwer—Medknow Publications. Cited 22 Sep 2020. Available from: http://www.ncbi.nlm.nih.gov/pubmed/25143882. Accessed 3 Sept 2020

21. Nicola M, Alsafi Z, Sohrabi C, Kerwan A, Al-Jabir A, Iosifidis C et al (2020) The socio-economic implications of the coronavirus pandemic (COVID-19): a review. Int J Surg 78:185–93 [Internet]. Elsevier. Cited 28 May 2020. Available from: http://www.ncbi.nlm.nih.gov/pub med/32305533. Accessed 3 Sept 2020

22. Franckle R, Adler R, Davison K (2014) Accelerated weight gain among children during summer versus school year and related racial/ethnic disparities: a systematic review. Prev Chronic Dis 11 [Internet]. Cited 28 May 2020. Available from: https://pub-med.ncbi.nlm.nih.gov/24921899/. Accessed 3 Sept 2020

23. von Hippel PT, Workman J (2016) From kindergarten through second grade, U.S. children's obesity prevalence grows only during summer vacations. Obesity 24:2296–2300. (Silver Spring) [Internet]. Cited 28 May 2020. Available from: http://www.ncbi.nlm.nih.gov/pubmed/27804271. Accessed 3 Sept 2020

24. Centers for Disease Control and Prevention (2020) Others at risk for COVID-19 I CDC [Internet]. Centers Dis Control Prev. Cited 22 Sep 2020. Available from: https://www.cdc.gov/coronavirus/2019-ncov/need-extra-precautions/other-at-risk-populations.html. Accessed 3 Sept 2020

25. Esegbona-Adeigbe S (2020) Impact of COVID-19 on antenatal care provision. Eur J Midwifery [Internet]. E.U. European Publishing. Cited 22 Sep 2020. Available from: http://www.journalss ystem.com/ejm/Impact-of-COVID-19-on-antenatal-careprovision,121096,0,2.html. Accessed 3 Sept 2020

26. UNICEF (2020) Framework for reopening schools [Internet]. Available from: https://www.unicef.org/sites/default/files/2020-06/Framework-for-reopening-schools-2020.pdf. Accessed 3 Sept 2020

Machine Vision Systems for Smart Cities: Applications and Challenges

Shamik Tiwari and Anurag Jain

Abstract A smart city employs information and communication technology (ICT) to boost efficiency and productivity, share data with the public, and promote government service and citizen satisfaction. Smart cities use a combination of low-power sensors, cameras, and AI algorithms to observe the city's operation. Machine vision has advanced in terms of recognition and tracking thanks to machine learning. It provides efficient capture, image processing, and object recognition for vision applications. Governments benefit greatly from the use of machine vision and other smart applications. This technology allows city administrators to easily integrate and utilize resources. As the "eyes" of the city, computer vision plays an important role in smart city management. The chapter begins with a brief review of machine vision, smart cities, and real-world machine vision applications in smart cities. Lastly, we highlight several smart city difficulties and prospects discovered through a comprehensive literature review.

Keywords Smart city · Machine vision · Deep learning · Smart city application

1 Machine Vision

Machine vision systems interpret images captured by cameras to produce image feature data that assists robotic and automation devices in comprehending the physical environment shown in the image. Computer vision is generally used to automate image processing; however, machine vision is the area of applied computer vision in real-world interfaces, such as a factory line. Artificial intelligence (AI) is a technology that enables machines to mimic human behaviour [1]. Machine learning is a subset of artificial intelligence that enables a machine to learn from prior data without

S. Tiwari · A. Jain (✉)
Systemics Cluster, School of Computer Science, University of Petroleum & Energy Studies, Dehradun, India
e-mail: anurag.jain@ddn.upes.ac.in

S. Tiwari
e-mail: shamik.tiwari@ddn.upes.ac.in

© The Author(s), under exclusive license to Springer Nature Singapore Pte Ltd. 2023
M. A. Chaurasia and C.-F. Juang (eds.), *Emerging IT/ICT and AI Technologies Affecting Society*, Lecture Notes in Networks and Systems 478,
https://doi.org/10.1007/978-981-19-2940-3_18

Fig. 1 Steps in machine
vision systems

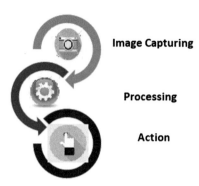

having to design it explicitly. Machine vision and machine learning are combined in
a collection of techniques that provide consumer and business hardware unparalleled
ability to see and analyse their surroundings. A computer's capacity to comprehend
the environment is known as machine vision. With the help of a computer, a machine
vision system uses a sensor in the robot to observe and recognize an object. Material
inspection, object classification, pattern classification, electronic component examination, signature identification, text recognition, and money identification are all
examples of how machine vision is employed in industry [2]. To know how machine
vision works, there are three basic stages to perform as shown in Fig. 1.

Image acquisition is the first step in machine vision systems. The image is captured
using vision sensors, digital cameras, infrared, or thermal cameras. This is a snap-
shot of a single or several occurrences. The image is captured and converted into
digital data by the device. Image processing algorithms can be used to analyse the
digital data generated by the hardware. Image processing is used in machine vision;
however, the two concepts are not interchangeable. Image processing is the practice
of producing images from old ones. Filtering (smoothing and sharpening), segmentation, edge detection, and geometric operations are all activities that image processing
is employed for in the early phases of machine vision. In machine vision systems, not
all image processing methods are used. De-blurring, image stitching, and image and
video compression are some of imaging techniques that are of secondary relevance
to machine vision. The machine is trained to do the relevant action based on the
information extracted in the previous phase [3].

1.1 Components of Machine Vision System

Lighting, lens, image sensor, vision processing, and communications are the main
components of a machine vision system. The part to be inspected is illuminated,
allowing its details to shine out and be viewed clearly by the camera. The image is
captured by the lens and presented to the sensor as light. This light is converted into
a digital image by the sensor in a machine vision camera, which is then delivered

Fig. 2 Components of machine vision system

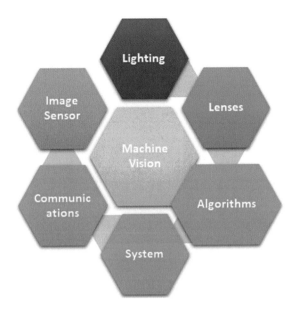

to the processor for processing. Algorithms in vision processing examine an image and extract relevant data, do the necessary inspection, and draw conclusions. Lastly, communication is usually achieved by sending discrete I/O signals or data through a serial connection to a device that is logging or using the information [4]. Figure 2 illustrates the different components of a machine vision system.

- Lighting
- Lenses
- Vision processing
- Image sensor
- Communications

2 Smart City

A smart city is a high tech metropolitan region that gathers data using various electronic technologies, voice activation systems, and sensors. The concept of a smart city has grown to encompass a far broader notion of interaction and communication than simply installing separate solutions for multiple departments. In other terms, smart cities rely on various services cooperating for a single goal, with data serving as the backbone. A smart city is one in which technology is used to deliver services and deal with problems [5]. Excess of water, guaranteed supply of electricity, hygiene, along with solid waste management, effective urban transport and public mobility, low-cost housing, particularly for the poor, reliable IT communication and automation, and effective governance are the some of the core infrastructure components

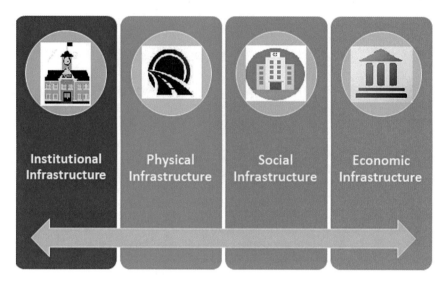

Fig. 3 Four pillars of smart city

in a smart city. A smart city enhances mobility and connectivity, enhances social services, promotes sustainability, and empowers its residents. To acquire data, smart cities employ vision sensors, IoT devices such as embedded devices, lights, and metres. Social facilities, physical infrastructural facilities, administrative infrastructure, and economic infrastructural facilities are the four foundations of a smart city. Machine vision applications play a critical part in each of the smart city's pillars [6].

Figure 3 shows the all four pillars of smart city. Machine vision systems support in each pillar of smart city. Smart Cities must adopt agile strategies that allow breakdown of traditional barriers, facilitate cooperation and trust, and integrate with existing infrastructure and systems to deliver the best potential scenarios for stakeholders and the ecosystem as a whole which is able to obtain these pillars to life.

3 Machine Vision Applications for Smart Cities

Smart cities use a combination of low-power sensors, cameras, and artificial intelligence algorithms to observe the city's efficiency. Governments benefit greatly from the use of machine vision and other associated technology. Sophisticated techniques enable city administrators to easily integrate and manage assets. Machine vision plays an important role in smart city management since it serves as the city's "eyes" [7, 8]. The following are some of the most important machine vision applications for smart cities (Fig. 4).

The concept of a smart city can be realized by the incorporation of multiple scalable technologies. Recent developments in artificial intelligence-based computer

Fig. 4 Main areas of machine vision applications for smart cities

vision combined with the IoT have enabled real-world computer vision systems to manage large amounts of complicated visual data, as well as fast processing, decentralization, and scalability. Computer vision is a subfield of artificial intelligence that encompasses technologies that allow computers to "learn" to recognize a picture or the features of an image. Objects, individuals, animals, or positions in an image or video feed can all be identified using this method. As a result, the goal of computer vision is for a machine to interpret the world and produce data that may be used to automate processes. New machine learning algorithms, most notably deep learning, have made substantial advances in the field of picture recognition, allowing AI vision to become more robust and helpful in mission-critical business applications.

Deep learning, unlike classical machine vision, does not require sophisticated imaging cameras and can get reliable results using practically any digital camera or even a webcam. As a result, network cameras (IP cameras or CCTV cameras) are frequently used in smart city applications of computer vision to give input for real-time video analytics using AI models. Edge AI pushes machine learning from the cloud to several linked edge devices that are attached to cameras, where the data is processed on device through edge intelligence. To handle various remote devices, this approach uses edge computing in conjunction with the Internet of Things (IoT). Edge AI overcomes cloud restrictions by enabling high-performing, resilient, real-time, and private computer vision applications, allowing for the deployment of computer vision at scale in the real world. Applications of machine vision in smart cities are discussed below.

3.1 Perimeter Intrusion Detection

In smart cities, people detection has a wide range of uses. Computer vision algorithms for real-time video analysis to analyse human activities are among the applications. Other computer vision applications are used to detect people and recognize activity in restricted locations such as airports and train stations [9]. Deep learning algorithms are used to detect intrusion events in real time and pinpoint the location of the target through the perimeter [10]. For safety and security purposes, such AI-based perimeter monitoring systems can efficiently cover huge monitoring regions.

3.2 Detection of Violence and Risky Circumstances

In smart city applications, computer vision is frequently used to automatically detect harmful scenarios with the purpose of ensuring the safety of citizens through smart video surveillance. Fighting, brawling, robbery, and other potentially risky situations are examples. The high variety of such situations is a difficult problem to solve. Algorithms are used to recognize the movements and interactions of people by detecting them, tracking them, and estimating three-dimensional human postures (human pose estimation). A logic flow is applied to the output of AI models to identify scenarios and events that should be triggered to automate human intervention or offer analytics [11]. People counting and detecting people who spend an extraordinary amount of time in specified regions are examples of such applications.

3.3 Vandalism Detection by Monitoring Object Activities

Different machine learning approaches can be employed in action recognition apps to identify vandalism by monitoring object actions. Machine learning and feature extraction approaches, for example, have been used to monitor and support human behaviour. Using one or multiple camera views, vision-based technology can distinguish human activities such as fighting and vandalism that may occur in a public domain. Even in complicated and busy surroundings, computer vision systems can detect and forecast suspicious and hostile behaviours' in real time.

3.4 Regulatory Compliance, Enforcement, and Inspections

In a smart city, computer vision can be used for a variety of compliance monitoring applications. Because manual inspection and site observations are difficult, time consuming, and expensive, AI vision approaches provide automated and scalable alternatives to manual inspection and site observations. Furthermore, computer vision approaches are not only faster but also more accurate. Because expert safety officers are not always present, inspection techniques are unreliable for routine use [12]. Computer vision algorithms can be used to monitor complicated, large-scale facilities in real time. Construction workers have the highest risk of injury or death on the job. It is 5 times higher than workers in any other business. To ensure that while working, workers should constantly wear suitable personal protective equipment (PPE) like helmets, safety glasses, vests, hand gloves, steel-toe boots, etc., compliance monitoring is critical. This will help in preventing accidents by detecting circumstances where workers are not wearing sufficient PPE or are consistently violating norms.

3.5   Video Surveillance in Suicide Prevention

By classifying visual elements, defining bodily joint movements, and spotting aberrant behaviour, camera systems are utilized to automate suicide prevention in public settings. At hotspots like metro stations, CCTV cameras with deep learning smart city apps can be utilized to assess crisis behaviours. As a result, automated computer vision programmes can assist in detecting ongoing self-harm attempts and initiating intervention [13]. The primary goal is to create automated detection systems for early intervention and pre-attempt behaviour recognition. As a result, computer vision applications can aid in the identification of risk factors, such as depression, inferred from facial expressions using face analytics. Another method is to look for aberrant behaviour, movement patterns, or wait periods in certain areas. In addition, forensic vision technologies are utilized to decipher elements following a failed effort.

3.6   Disaster Management

Computer vision surveillance in crowded settings is one of the most essential applications, owing to the growing number of people gathering in public spaces, where tragedies and stampedes are a possibility. To increase public safety, vision-based crowd disaster avoidance technologies are deployed. The majority of these applications are concerned with crowd scene analysis and behaviour analysis. Stampedes can happen as a result of unusual behaviour or unexpected circumstances [14]. With one or several cameras at scale, deep learning models are utilized to count people and estimate crowd density.

3.7   Inspection and Monitoring of Civil Infrastructure

Smart cities use sophisticated computing technology to monitor activity and avoid potential problems. Critical infrastructures are vital resources for society, and their failure would have significant consequences and costs. Roads, communications, water, energy, and other infrastructure are among them [15]. Traditional surveillance systems rely on the attention capacity of a human operator when confronted with a multitude of video streams. While closed-circuit television (CCTV) systems have become an essential element for security and law enforcement, traditional surveillance systems rely on the attention capacity of a human operator who is confronted with a multitude of video streams. Traditional video surveillance systems have a number of drawbacks, including high operational costs due to the requirement for humans to watch the footage, fatigue-related inaccuracies, and a large amount of video data requiring high bandwidths. Because centralized processing cannot grow, local processing (Edge AI) is essential. High latency and communication delays are major concerns in mission-critical real-time use cases, where fast decision-making is required to avoid risks. After deployment, rigid centralized systems lack flexibility and drive up maintenance costs. Smart edge AI computer vision systems enable a scalable and readily calibrated multi-camera system that automatically triggers alarms in the event of potential threats. In the same video pipeline, different machine learning and computer vision approaches might be used. On-device processing with local edge nodes (connected computers or edge devices) that are dynamically configured according to job conditions or metrics can considerably reduce network traffic. Any computer or embedded system capable of real-time high-performance video processing can be used as an edge computer. The NVidia Jetson TX2 is a popular edge device for edge computer vision. As a result, deep learning's automatic video analysis skills are extremely valuable. Perimeter monitoring, facial recognition, face tracking, and multi-person tracking with re-identification can all be utilized in conjunction with human detection using deep learning.

3.8 Weapon Identification and Reporting as a Means of Threat Monitoring

Controlling the rising crime rate in megacities with dense urban populations is a big task. As a result, deep learning vision apps can be designed to monitor public spaces autonomously and detect portable arms in real time. With all items visible in the camera streams, AI models examine video feeds to conduct object detection. When any form of weapon is visible, the weapon detection application sends out an alarm and can follow the object's progress. This method could be applied to the development of an automated firearm detection system. High-performance systems can detect and monitor the individual using the weapon and then utilize the data to conduct facial recognition. The automatic detection of suspicious objects placed in public spaces is a related smart city computer vision use case. AI models, such as the popular YOLOv3 [16] and the more recent YOLOR [17], can be used to detect placed things that potentially pose a concern and flag them for human inspection.

3.9 Social Distancing Monitoring and Violation Alert

Smart vision systems are capable of monitoring and enforcing social distancing amongst people in order to successfully prevent the development of the epidemic

[18]. Deep learning-based AI models, for example, can be used to build systems for social distancing monitoring with real-time object detection to locate humans in surveillance camera videos at scale. A deep neural network (DNN) model can automatically detect people in metropolitan public spaces using CCTV security cameras in crowd detection applications. The trajectories of people's movements are utilized to identify potentially high-risk places so that planners can restructure open and public spaces to make them more pandemic proof.

3.10 Automated Temperature Checking and Facial Mask Detection

Another example of the role of computer vision in the smart city is automated mask detection of people in public locations. Deep learning algorithms can be used to monitor the conditions and send alerts if people do not wear masks or do not comply with lockdown measures. This combination with compliance with social distancing protocols in public spaces such as metro stations can allow AI-enabled systems to facilitate mass surveillance [19]. The information may be used to improve cities' ability to predict pandemic patterns, promote a timely reaction, reduce virus spread, support specific industries, reduce supply chain interruption, and assure the continuation of key services and activities.

3.11 Infection Prevention Strategy Monitoring

Approaches based on computer vision and deep learning are useful for automating safety and compliance monitoring. Non-intrusive vision-based systems for tracking people's behaviour in public places and infrastructure, for example, are powered by such applications. When compared to proximity-based strategies, AI vision technologies have shown greater outcomes. Automated hand hygiene compliance monitoring using deep learning is one example of an application. This system uses spatial analytics to provide information about people's movements. For lowering hospital-acquired infections, there are some interesting application cases.

3.12 Traffic Monitoring and Control

Computer vision is used to assess and anticipate traffic conditions in traffic monitoring and control. Surveillance cameras provide traffic data such as car count, frequency, and direction. In high-traffic circumstances, vehicle counting uses deep neural networks to detect different vehicle kinds and use the information to optimize traffic management [20, 21]. As part of traffic monitoring, waiting times and traffic flows are also tracked.

3.13 Smart Parking

Real-time parking lot occupancy detection has recently garnered a lot of traction. Accuracy-wise, computer vision-based solutions performed well and could be applied to existing camera networks. Using several cameras, deep learning is utilized to discover vacant parking lots. Edge AI capabilities enable visual analysis close to sensing devices without the need to send video streams to a central controller for acquisition, encoding, analysis, and processing [22]. As a result, distributed parking lot occupancy monitoring outperforms traditional methods in terms of overall energy efficiency and precision.

Few other applications which indirectly contribute towards the achievement of a smart city are discussed below.

3.14 Productivity Analytics

It monitors the impact of workplace changes, how people use their time and resources, and how various tools are implemented. This type of information can help with time management, workplace cooperation, and staff productivity. With camera-based vision systems, computer vision lean management solutions strive to objectively quantify and assess processes [23].

3.15 Visual Inspection of Equipment

In smart manufacturing, computer vision for visual inspection is a significant strategy. Automated inspection of personal protective equipment (PPE), such as mask detection or helmet detection, is also becoming more popular using vision-based inspection systems. On building sites or in smart factories, computational vision can help monitor adherence to safety procedures [24].

3.16 Quality Control

In smart factories, smart camera apps offer a scalable way to automate visual inspection and quality control of production processes and assembly lines. In this case, deep learning uses real-time item detection to get better outcomes (detection accuracy, speed, objectivity, and reliability) than time-consuming manual inspection. Machine learning approaches are more resilient than classic machine vision systems, and they do not require expensive cameras or regulated settings [25]. As a result, machine vision approaches can be used in a variety of places and factories.

3.17 Skill Training

Optimizing assembly line operations in industrial manufacturing and human–robot interaction is another application sector for vision systems. Human action evaluation can aid in the development of standardized action models for various operation steps as well as the evaluation of trained workers' performance. Improving worker performance, boosting productive efficiency (LEAN optimization), and, most critically, detecting unsafe behaviours to reduce accident rates can all be aided by automatically measuring the quality of their actions. Lean aims to reduce costs, or non-value-added elements, from any operation. Until a process has been through lean several times, it will have some garbage. When implemented appropriately, lean can result in significant gains in efficiency, throughput times, profitability, cost of materials, and junk, lowering costs and increasing flexibility. Understand that lean is not just for manufacturing. It can help a team work better together, monitor inventories, and even engage with clients.

3.18 Health Care

Machine vision technologies that are the most promising and forward thinking have applications in health care and smart cities. Machine vision in health care can lead to a variety of solutions that can save patients' lives by speeding up and enhancing the reliability of medical diagnostics, as well as supporting medical personnel in crucial situations. To lessen the risk of transmission, face recognition technologies are employed to reduce the necessity for touching surfaces and objects, as well as physical contact between people. Smart solutions can assist in limiting the risk of human exposure, particularly in high-risk environments like airports and hospitals [26, 27].

In addition, telemedicine is increasingly being used to reduce risks by reducing hospital visits by vulnerable groups such as the elderly. Telemedicine can also aid in the recuperation process and provide remote monitoring. Human fall detection is a feature of healthcare monitoring systems that uses convolutional neural networks to detect a person falling. Surveillance footage combined with processing rules can be used to identify a fall and automatically warn caretakers [28, 29]. Figure 5 presents machine vision applications in the well-being and healthcare sector of smart city.

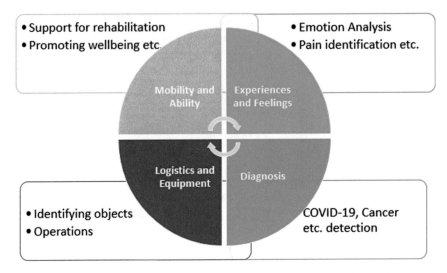

Fig. 5 Machine vision applications in the well-being and healthcare sector of smart city

4 Challenges of Machine Vision Applications

Machine vision is a fascinating tool for smart city automation; it elevates machine sensing to a new level. Vision systems, on the other hand, immediately revealed their limitations. To address those issues, tools were required, and they were deployed, but at the expense of processing speed. With careful engineering, some of the constraints of machine vision applications can be solved. Until new engineering and software solutions are discovered, these constraints will limit the full potential of machine vision in smart cities. Companies that deal with automation should come up with answers to following issues.

- There is a lack of public and commercial data for training and retraining, and the data that is accessible is of low quality and/or requires additional modifications.
- Machine vision applications in the real world need hardware support, sensors to supply visual input, and processing power to perform intelligence inference. There are latency limits, specifically for mission-critical machine vision applications that rely on close to real video analytics.
- Data privacy and security guidelines awareness. In machine vision, data encryption at transit and at rest, as well as secure access management, are a must.
- Various environmental conditions, such as temperature, have an impact on the performance of machine vision-based approaches. Few most influencing factors are listed below:

  a. Poor illumination
  b. Sensor noise
  c. Low lighting condition

    d. Blurs

    e. Geometric distortions

- A motion and vision system's component selection, set-up, and configuration are difficult processes. Integrators frequently select motion and vision equipment from a variety of manufacturers.
- It is possible that deciding where to install the camera will be difficult. Since industrial uses frequently have small range for a camera, size is an important consideration. When it comes to machine vision, it is common to have trouble assuring that there is enough contrast seeing the objects.
- Machine vision camera event capturing necessitates image storage and a powerful computer.

5 Conclusion

Machine vision applications driven by AI deliver real-time information to assist smart city decision-making. In the context of smart cities, machine vision and AI play a crucial role, serving as the city's "eyes". Smart cities can examine the motion of cars in the metropolitan area, recognize and forecast traffic jams, detect crowded residential streets, and so on, going beyond the ability to distinguish things and people. It is an important development for machine vision in smart cities, pushed mostly by rising pressure as e-commerce keeps growing as the dominating method of product purchase. Furthermore, smart manufacturing is becoming more versatile and suitable for adoption in a growing number of specialist or redeployable uses, which necessitate flexible and user-friendly vision technologies. Finally, the autonomous vehicle industry is a promising source of machine vision innovation, as the success of the whole endeavour is dependent on the effectiveness of cameras and the computational power, devices, firmware, and intelligent systems that enable them to not only observe but also understand the smart city world.

References

1. Steger C, Ulrich M, Wiedemann C (2018) Machine vision algorithms and applications. John Wiley & Sons
2. Nandini V, Vishal RD, Prakash CA, Aishwarya S (2016) A review on applications of machine vision systems in industries. Indian J Sci Technol 9(48):1–5
3. Gallego G, Delbruck T, Orchard GM, Bartolozzi C, Taba B, Censi A, Leutenegger S, Davison A, Conradt J, Daniilidis K, Scaramuzza D (2020) Event-based vision: a survey. In: IEEE transactions on pattern analysis and machine intelligence
4. Beyerer J, León FP, Frese C (2015) Machine vision: automated visual inspection: theory, practice and applications. Springer
5. Angelidou M (2015) Smart cities: a conjuncture of four forces. Cities 47:95–106
6. Ahvenniemi H, Huovila A, Pinto-Seppä I, Airaksinen M (2017) What are the differences between sustainable and smart cities? Cities 60:234–245

7. Dong CZ, Bas S, Catbas FN (2020) A portable monitoring approach using cameras and computer vision for bridge load rating in smart cities. J Civ Struct Heal Monit 10(5):1001–1021
8. Montemayor AS, Pantrigo JJ, Salgado L (2015) Special issue on real-time computer vision in smart cities. J Real-Time Image Proc 10(4):723–724
9. García CG, Meana-Llorián D, GBustelo BCP, Lovelle JMC, Garcia-Fernandez N (2017) Midgar: detection of people through computer vision in the Internet of Things scenarios to improve the security in smart cities, smart towns, and smart homes. Future Gen Comput Syst 76:301–313
10. Dinakaran RK, Easom P, Bouridane A, Zhang L, Jiang R, Mehboob F, Rauf A (2019) Deep learning based pedestrian detection at distance in smart cities. In: Proceedings of SAI intelligent systems conference. Springer, Cham, pp 588–593
11. Ryabchikov I, Teslya N, Druzhinin N (2020) Integrating computer vision technologies for smart surveillance purpose. In: 2020 26th conference of open innovations association (FRUCT). IEEE, pp 392–401
12. Ramirez-Lopez A, Cortes-González A, Ochoa-Ruiz G, Ochoa-Zezzatti A, Aguilar-Lobo LM, Moreno-Jacobo D, Mata-Miquel C (2021) A drone system for detecting, classifying and monitoring solid wastes using computer vision techniques in the context of a smart cities logistics systems. In: Technological and industrial applications associated with intelligent logistics. Springer, Cham, pp 543–563
13. Aydin I, Othman NA (2017) A new IoT combined face detection of people by using computer vision for security application. In: 2017 international artificial intelligence and data processing symposium (IDAP). IEEE, pp 1–6
14. Shirazi MS, Patooghy A, Shisheie R, Haque MM (2020) Application of unmanned aerial vehicles in smart cities using computer vision techniques. In: 2020 IEEE international smart cities conference (ISC2). IEEE, pp 1–7
15. Yaman O, Karakose M (2019) New approach for intelligent street lights using computer vision and wireless sensor networks. In: 2019 7th international istanbul smart grids and cities congress and fair (ICSG). IEEE, pp 81–85
16. Zhao L, Li S (2020) Object detection algorithm based on improved YOLOv3. Electronics 9(3):537
17. Elango S, Ramachandran N (2021) Novel approach to autonomous mosquito habitat detection using satellite imagery and convolutional neural networks for disease risk mapping
18. Punn NS, Sonbhadra SK, Agarwal S, Rai G (2020) Monitoring COVID-19 social distancing with person detection and tracking via fine-tuned YOLO v3 and Deepsort techniques. arXiv preprint arXiv:2005.01385
19. Juang CF, Chang CM (2007) Human body posture classification by a neural fuzzy network and home care system application. IEEE Trans Syst Man, Cyber-Part A: Syst Humans 37(6):984–994
20. Bortnikov M, Khan A, Khattak AM, Ahmad M (2019) Accident recognition via 3d cnns for automated traffic monitoring in smart cities. In: Science and information conference. Springer, Cham, pp 256–264
21. Ho GTS, Tsang YP, Wu CH, Wong WH, Choy KL (2019) A computer vision-based roadside occupation surveillance system for intelligent transport in smart cities. Sensors 19(8):1796
22. Baroffio L, Bondi L, Cesana M, Redondi AE, Tagliasacchi M (2015) A visual sensor network for parking lot occupancy detection in smart cities. In: 2015 IEEE 2nd world forum on internet of things (WF-IoT). IEEE, pp 745–750
23. Khan MM, Ilyas MU, Saleem S, Alowibdi JS, Alkatheiri MS (2019) Emerging computer vision based machine learning issues for smart cities. In: The international research and innovation forum. Springer, Cham, pp 315–322
24. Bhattacharya S, Somayaji SRK, Gadekallu TR, Alazab M, Maddikunta PKR (2020) A review on deep learning for future smart cities. Internet Technol Lett e187
25. Gade R, Moeslund TB, Nielsen SZ, Skov-Petersen H, Andersen HJ, Basselbjerg K, Dam HT, Jensen OB, Jørgensen A, Lahrmann H, Madsen TKO (2016) Thermal imaging systems for real-time applications in smart cities. Int J Comput Appl Technol 53(4):291–308

26. Hossain MS, Muhammad G, Alamri A (2019) Smart healthcare monitoring: a voice pathology detection paradigm for smart cities. Multimedia Syst 25(5):565–575
27. Nasralla MM, Rehman IU, Sobnath D, Paiva S (2019) Computer vision and deep learning-enabled UAVs: proposed use cases for visually impaired people in a smart city. In: International conference on computer analysis of images and patterns. Springer, Cham, pp 91–99
28. Solanas A, Patsakis C, Conti M, Vlachos IS, Ramos V, Falcone F, Postolache O, Pérez-Martínez PA, Di Pietro R, Perrea DN, Martinez-Balleste A (2014) Smart health: a context-aware health paradigm within smart cities. IEEE Commun Mag 52(8):74–81
29. Pacheco Rocha N, Dias A, Santinha G, Rodrigues M, Queirós A, Rodrigues C (2019) Smart cities and healthcare: a systematic review. Technologies 7(3):58

Emerging Non-invasive Brain–Computer Interface Technologies and Their Clinical Applications

Cory Stevenson, Yang Chang, Congying He, Chun-Ren Phang, Cheng-Hua Su, Ro-Wei Lin, and Li-Wei Ko

Abstract Brain–computer interfaces (BCIs) are a continuously evolving technology of great importance to society and human wellbeing. With a wide range of applications and the integration of many emerging technologies, BCIs have the capacity to change many fields, in particular, the field of clinical medicine and patient health. This chapter covers current developments in non-invasive BCIs and their use for a variety of clinical applications. It provides an overview of EEG hardware and non-invasive BCI systems and covers common electrophysiological recording techniques and signal processing algorithms often employed in BCIs. It then details examples of how these are implemented for particular clinical applications, including attention-deficit hyperactivity disorder identification, stroke rehabilitation, and sleep enhancement, highlighting the potential capabilities of BCI to address such current and future clinical challenges.

Keywords Brain–computer interfaces (BCI) · Electroencephalography (EEG) · Rehabilitation · Clinical technology

1 Introduction

Brain–computer interfaces (BCIs) are a collection of technologies which facilitate interaction between users and devices or computers by interpreting signals generated in the brain. Such technologies have a wide range of applications, from attention detection in marketing [1] to assisting clinical rehabilitation [2], and commensurately many different approaches and technologies have been utilized in the development of BCIs. Fundamentally, for a BCI to function properly, a specific signal or set of signals associated with a particular cognitive process must be detected, identified,

C. Stevenson · Y. Chang · C. He · C.-R. Phang · C.-H. Su · R.-W. Lin · L.-W. Ko (✉)
National Yang Ming Chiao Tung University, Hsinchu, Taiwan
e-mail: lwko@nycu.edu.tw

C. Stevenson
e-mail: cesteven@nycu.edu.tw

© The Author(s), under exclusive license to Springer Nature Singapore Pte Ltd. 2023   269
M. A. Chaurasia and C.-F. Juang (eds.), *Emerging IT/ICT and AI Technologies Affecting Society*, Lecture Notes in Networks and Systems 478,
https://doi.org/10.1007/978-981-19-2940-3_19

interpreted, and responded to. Each one of these steps presents particular technological challenges which can be addressed by various technological developments, though for each system, the intended application directs the implementation and selection of technologies. According to a survey report by Allied Market Research, the global BCI market has reached 1.488 billion U.S. dollars in 2020 and predicts that the BCI market will reach 5.463 billion U.S. dollars in 2030 (compound annual growth rate of 13.9%) [3]. In addition, according to a survey report by Grand View Research, the main driving factors of the BCI technology market include the prevalence of neurological diseases, the growth of elderly populations, the development of technology to promote communication and activity of paralyzed individuals, the use of virtual games, home control systems, and military communications [4].

Generally, BCIs can be characterized into various categories depending on use cases and which technologies are utilized. Conscious intent of the user is one major use case differentiator often employed, separating BCIs into active and passive systems. An active BCI executes actions consciously intended by the user, whereas a passive BCI responds to brain signals in a way that voluntary action on behalf of the user is not needed for the BCI to function [5]. Another common aspect for categorizing BCIs by implementation is whether the signal generated is induced or endogenous, where induced signals are produced in the brain by an external stimulus and consequently detected, and endogenous signals are produced by the user's brain based on cognitive processes that require no direct sensory induction. BCIs can be implemented using any combination of user intention and signal generation, though particular combinations are more applicable to specific utilizations.

Additionally, BCIs may be categorized by the technology employed for acquiring brain signals. Generally, these are split into two groups, invasive and non-invasive, with the general distinguishing factor being whether the signal acquisition devices penetrate the body and physically interact with the physiological processes of the brain. Invasive technologies used in BCIs include such technologies as electrocortiography (ECoG), which places arrays of electrodes on the surface of the brain, or intracortical approaches, which use probes to place sensors and/or stimulators deeper within the structures of the brain. Invasive technologies have benefits such as improved signal clarity and spatial resolution, but have all the limitations and challenges associated with neurological surgery, implant power and signal transduction, and immunological interactions to foreign bodies [6]. Non-invasive BCI systems consist of a range of technologies which can detect neurophysiological signals from outside the head. The major technologies which have been developed for non-invasive brain recording include functional near-infrared spectroscopy (fNIRS), which uses near-infrared light projected through the skull to assess changes in blood flow and oxygenation in certain brain regions as local neural activity changes, functional magnetic resonance imaging (fMRI), which uses magnetic fields to assess similar blood perfusion and oxygenation changes related to brain activity, and electroencephalography (EEG) which measures the electrical field activity of collections of neurons in the cortex from outside the skull. While non-invasive BCI systems have been developed based on each of these technologies, the physical properties endemic to fNIRS and fMRI have constrained their adoption into BCI systems [7, 8], as they are both expensive and have comparably low temporal resolutions relative to EEG, which are critical factors for the functional application of BCI systems. Thus, much of the advancements in non-invasive BCI incorporate EEG, which is increasingly inexpensive, portable, and relatively easy to use.

This chapter focuses on current developments in non-invasive BCIs and their use for a variety of clinical applications. The first parts of the chapter provide an overview of EEG hardware and non-invasive BCI systems, followed by a review of common electrophysiological recording techniques and signal processing algorithms utilized in BCIs. The latter parts of the chapter cover examples of how these are implemented for particular clinical applications and the potential capabilities of BCI to address such clinical challenges.

2 Hardware

The general design of a modern BCI consists of two components, an EEG system capable of detecting and transmitting brain signals and a computer interface capable of analyzing that signal and producing a response. This section provides an overview of various EEG systems that have been utilized and an introduction of common interactive components.

2.1 Acquisition Hardware

There are currently many designs and options for EEG systems. Non-intrusive EEG systems can be divided into two types based on the design and operation of their electrodes, wet or dry. Wet-electrodes are dependent on the presence of electrically conductive gel between the scalp and the metal electrodes, typically Ag/AgCl. To maintain signal transduction through the course of an experiment, these gels are designed to be viscous and dry out slowly. While providing relatively reliable signal, electroconductive gels require sufficient application to each electrode-scalp gap before recording and washing of both the electrodes and scalp after the experiment, a process that can be time consuming and uncomfortable, and typically requires administration by an experienced individual. Provided here is a compilation of eight different commercially available wet-electrode EEG devices, providing an overview of current options, illustrated in Fig. 1. A simple comparison of their technical specifications is presented in Table 1. Wet-electrode EEG is available in both wired and wireless setups and has varied electrode designs, including pin, cup, metal ring, and pre-applied gel electrodes. Applications of wet-electrode systems include research, games, education, medical treatment, and clinical evaluation, typically conditions which justify longer setup and cleanup times by having longer signal acquisition periods.

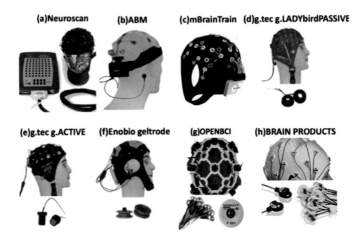

Fig. 1 Example set of commercially available wet-electrode EEG devices (*Sources* listed in Appendix 1)

In contrast, dry electrodes do not utilize conductive gel as the primary method of conduction. A newer set of technologies than wet-electrodes, these systems have a variety of electrode designs, ranging from completely dry systems to aqueous systems, that need to be pre-soaked in water or saline (sometimes called semi-dry), that all have the common feature of direct contact with the scalp without need of intervening conductive gel. Provided here is a compilation of eleven commercially available dry-electrode EEG systems, illustrated in Fig. 2, with corresponding specifications in Table 2. These dry-electrode designs include saline-soaked felt, polymer, sponge, comb, and metal electrodes. Such systems have been applied in research, games, education, emotion detection, and perception monitoring. Due to the convenience and speed of setup, along with minimal cleanup, dry-electrode systems are well suited for more casual use BCIs, which may have short use times and are more likely to be self-administered.

The development of dry electrodes has been an ongoing development for the expansion of BCI usage. As an example of such progress, earlier designs we developed used spring-loaded pins made of a conductive metal alloy [9]. This approach allowed the recording of EEG without the need of conductive fluids. However, the pins caused serious discomfort for users, thus making it impractical to use. The next iteration used hygroscopic sponges for skin–electrode contact [10]. The sponge design provided significantly more comfort and produced signal quality comparable to that of conventional gel-dependent electrodes; however, signal quality degraded over time as the sponges dried out. Our latest design made the EEG electrodes out of silicone, metal flakes, and graphene/graphene-oxide [11]. The metal flakes and graphene/graphene-oxide granules were blended into the silicone during production, and the whole mixture was solidified in a heater. The resulting electrodes are soft and flexible, making them more comfortable. To test signal quality, graphene-silicone electrodes and conventional gel-dependent wet-electrodes were used to record signal

Table 1 Overview of wet-electrode EEG system technical specifications

Wet-electrode EEG system	Electrode	Channels	Sampling rate (Hz)	Transfer	Application
(a) Neuroscan	Passive gel-based, Ag/AgCl ring	64/128/256	20,000	Wired (USB)	Medical, neuroimaging, clinical and research
(b) ABM	Electrode cream	10/24	256	Wireless	Medical, accelerated learning, consumer neuroscience, and sleep quality
(c) mBrainTrain	Gel-based, ring	24	250–500	Wireless	Game, psychology, sport research, drowsiness, and fatigue
(d) g.tec g.LADYbird PASSIVE	Passive gel-based, Ag/AgCl ring	16	32–38,000	Wired (USB)	Research, medical, and education
(e) g.tec g.ACTIVE	Active gel-based, Ag/AgCl circular plate	16	32–38,000	Wired (USB)	Research, medical, and education
(f) Enobio geltrode	Passive gel-based, Ag/AgCl coated core	8/20/32	500	Wireless	Medical, clinical, sleep monitoring, and epilepsy
(g) OPENBCI	Foam, solid gel, Ag/AgCl Gold cup electrodes	16	250	Wireless	Medical, R&D, and classroom study
(h) Brain products	conductive gel, paste	8/16/32/ 64/128/256	250/500/1000	Wired	Research, medical, and physiological monitoring

See Appendix 1 for sources

simultaneously. Signals from the graphene-silicone electrodes were highly correlated with the wet-electrode signals, showing that the new electrodes could reliably record EEG over sufficiently long periods.

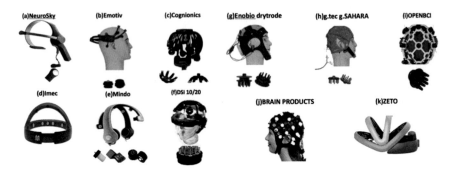

Fig. 2 Example set of commercially available dry-electrode EEG devices (Sources listed in Appendix 2)

2.2 Interactive Components

While the core component of any BCI system is the sensing technology and the interpretation of brain signals, an interactive component is critical to the function of a BCI, differentiating BCIs from the more general field of neurophysiological recording. This interaction can take many forms, but is typically represented by a virtual, physical, or environmental response which is presented to the user as feedback. This feedback can be present in many forms, such as by observing a controlled unit performing an action [12] or presenting a representation of the user's neural activity [13].

2.2.1 Mechanical Control

One of the most common applications of BCI is controlling mechanical objects in the environment. Various implementations of mechanical control have been produced, with different levels of finesse and abstraction. Typically, active BCIs are employed in mechanical control, where a user directs the activity of an object consciously, such as a robotic vehicle moving across a room, moving to a specific point or in a continuously updated trajectory [14]. Both induced and endogenous methods have been used for this purpose [5, 15].

A specific case of mechanical control particularly important to clinical systems is anthropomorphic control, where the user directs a human or body part-shaped robotic device. These systems vary in complexity, from a single degree-of-freedom joint extensor-flexor, through directing the movements of near-whole body facsimiles with many degrees of freedom. Such systems provide a wide array of possibilities for subjects who may have limited mobility, augmenting their physiological limits with BCI-controlled agency, such as directing the movements of an exoskeletal robot's limb.

Table 2 Overview of dry-electrode EEG system technical specifications

Dry-electrode EEG system	Electrode	Channels	Sampling rate (Hz)	Application
(a) NeuroSky	Embeddable biosensor	1	512	Game and education
(b) Emotiv	Saline-soaked felt pads	14	128	Measure cognitive and emotional
(c) Cognionics	Semi-dry polymer electrode, Active Ag/AgCl electrodes	16/24/32/64/128	250–2000	Research, driver, game, and brain training
(d) Imec	Cylinder-shaped conductive polymer dry electrodes	8	1024	Emotion detection
(e) Mindo	Moisture sponge electrode; Foam-based electrode and Spring-loaded metal	8	128/256/512	Game, emotion, fatigue, and attention
(f) DSI 10/20	Hybrid resistive-capacitive sensors	23	240/960	Medical training and warfighter training
(g) Enobio drytrode	Passive dry, 10 Ag/AgCl coated pins	8/20	500	Medical, clinical, sleep monitoring and epilepsy
(h) g.tec g.SAHARA	Active dry 8 gold-alloy coated pins	16	32–38,400	Stroke therapy and disorder consciousness
(i) OPENBCI	Dry EEG comb electrodes	8/20	250	Medical, R&D, and classroom study
(j) Brain products	Active dry electrode, gold plated electrode	8/16/32	250/500/1000	Research, medical, and physiological monitoring
(k) ZETO	Dry electrode	10/20	N/A	Research, medical, and clinical

See Appendix 2 for sources

2.2.2   Virtual and Augmented Reality

Virtual and augmented reality (VR/AR) technologies are also playing an increasing role in the development of BCIs. These allow for additional forms of interaction with a user that are especially relevant to BCI implementation. VR can provide environments for users which are suitable for the generation of BCI signals, allow precise spatial and temporal control of virtual cues in environments of varying complexity, and can monitor movement and gaze, which are important for giving context to BCI signals. Such a virtual environment interacts with the BCI directly, by providing spatially localized induction stimuli [16] or by simply just responding to endogenous signals that the subject may not even be aware of. The recent advent of portable, commercial VR systems has greatly expanded the capabilities of utilizing this technology with BCIs, leading to the development of integrated systems with physical BCI systems built directly into VR headsets [17].

3   Signal Acquisition/Techniques

In addition to the various hardware developments propelling the evolution of BCI, how these signals are collected has been integral to how BCIs have developed. All forms of neurophysiological recording require cleaning and processing of signals at multiple levels; BCIs inherit the numerous signal cleaning techniques that have been developed and used for general electrophysiological research. These start at the most basic, with bandpass filters and signal smoothing, and progress to more complicated implementations of unsupervised learning algorithms, such as the well-known family of blind source separation techniques and time–frequency analyses [5]. One complication of adapting general neurophysiological recording techniques to BCIs is that by their very nature BCIs are interactive and thus must function at the timescales at which users' responses occur and without full knowledge of the incoming signal. This limits signal processing techniques to those which can be conducted in near-real time, restricting some common EEG methods which utilize knowledge of future signal sequences or the averaging of many trials, and necessitating the evolution of machine learning algorithms which can run quickly with limited information [18].

How BCIs acquire their target signal is entwined with the nature of the target signal, design of the BCI system itself, and its intended use case. This results in the development of specific paradigms which produce specific conditions to optimally detect and respond to particular signals. Here, we will present the implementation and new developments of two commonly used BCI paradigms, steady-state visual evoked potential-based (SSVEP) BCI, which relies on induced signals, and motor imagery-based (MI) BCI, which relies on endogenous signals.

3.1 Steady-State Visual Evoked Potential

The brain responds to a suddenly changed visual stimulus by producing a regular, corresponding change in the visual cortex, identifiable by electrodes on the scalp, called a visual evoked potential. In experimental settings, these can be averaged together over multiple trials to assess things such as awareness [19]. Similarly, when visual stimuli are flickered rapidly and repeatedly at a fixed rate above 4 Hz, an SSVEP resonating at the same frequency can be detected [20]. The amplitude and frequency of this signal are dependent on the visual focus and attention of the viewer, and consequently, this signal has been used to identify on which targets a subject is concentrating [19]. SSVEPs are established rather quickly, thus allowing for a relatively high information transfer rate (ITR) as subjects change attentional focus [21]. This clear, high ITR signal, along with the fact that it requires minimal user training to be effectively used, has resulted in its widespread use as a method for BCI.

SSVEP-based BCI has generally been implemented using a screen or VR headset, which displays a visual scene to a subject. Within this scene, certain locations or virtual objects are flickered to induce an SSVEP. As long as the distances between the flashing visual targets are sufficiently far apart and/or have flicker rates at non-resonant frequencies to each other [16], then multiple locations can be utilized simultaneously. Thus, the visual target which is being attended to will be detectable in SSVEP. Often, canonical correlation analysis (CCA), a computational method which determines linear correlations between multidimensional variables, is used to isolate the target frequency. This implementation of SSVEP-induction from multiple attended targets has been embedded into games (see Fig. 3) [22, 23], virtual keyboards for text input [21], visual diagnostics [24], and robotic controllers [16].

Fig. 3 Match-three game with an embedded SSVEP-based BCI system

3.2 Motor Imagery

When producing a particular movement, certain parts of the brain are activated. Furthermore, when that movement is intended or imagined, there is similar activity in those regions that produce characteristic signals of motor imagery. Isolating and decoding these MI derived activities is the basis for utilizing them for BCIs. Brain connectomics has been proposed as a technique to analyze functional brain activities. Brain activities recorded from functional imaging modalities such as EEG or fMRI can be used to generate functional brain networks, which are the non-localized functional interactions between various brain regions. Functional brain networks could further describe synchrony between activities generated in distant areas of the brain. These functional brain connectivity techniques have been used to study neuropsychiatric disorders [25–27] and motor-related activations [28–31], in particular, connectivities associated with particular MI. In recent investigations, increases in midline to lateral connections, between electrode locations Cz to C3 and Cz to C4, and beta band sensorimotor outflow connections were observed during upper limb MI [28–30]. In addition, phase synchronies between electrodes C3-FCz and C4-FCz were found to be increased during upper limb MI [31].

Creating a BCI based on lower-limb MI poses a particular challenge, however, as the motor areas directly involved in lower-limb movement are difficult to resolve directly with EEG. Brain connectomic techniques present a particularly useful tool to address such challenges, potentially allowing the development of brain-based, active, lower-limb rehabilitation systems. Furthermore, by comparing the functional connections between healthy subjects and patients with neuromotor conditions, such as stroke, more effective neurofeedback rehabilitation paradigms could be developed. Toward this goal, we proposed a method of applying global cortical networks to distinguish MI of the lower limbs [32]. In this study, Pearson's correlations were calculated between each pair of 32-channel EEG signals, thereby generating a 32 × 32 connectivity matrix. Each element of the connectivity matrix represents the connectivity strength between two distant brain areas, with strengths ranging from −1 to 1. The extracted connections were fed into a support vector machine (SVM), a type of machine learning algorithm, to classify the lower-limb brain activities (see Fig. 4).

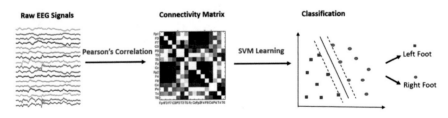

Fig. 4 Proposed functional connectivity classification system for the discrimination of lower-limb MI

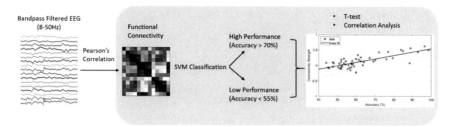

Fig. 5 Process for determining which brain connections are responsible for the performance of MI classification

Variability of MI performance in different subjects has been a considerable challenge to the development and general usability of MI-BCIs. Expanding on the connectomic work in [32], we investigated the functional brain networks between high-performing and low-performing subjects [33]. To precisely identify the brain connections responsible for MI performance, brain networks were extracted from spatially dense, 64-channel EEG signals. The brain networks were computed as before, thereby generating a connectivity matrix of 64×64, in which each element was the pairwise correlation between EEG signals. The performance of the MI was evaluated by the classification accuracy of MI tasks by SVM algorithm. The relationship between connectivity strengths and classification accuracy was investigated to pinpoint the brain interactions responsible for MI performance (see Fig. 5).

The global cortical network was shown to distinguish left or right foot MI with promising accuracy [32], demonstrating the feasibility to distinguish left and right foot motor intention using the global cortical network. Moreover, functional parietal connections were found correlated to the subjects' performance in MI [33]. Intra- and inter- lobular parietal connectivities within the alpha frequency range, brainwaves which occur between 8 and 12 Hz, were shown to be correlated to the classification accuracy of 48 subjects. Particularly, connections between left parietal areas and frontocentral region (P7-FC3, P7-Cz, and P7-C4) were significantly correlated to subjects' performance. These findings further elucidate neural network mechanisms underlying MI performance, contributing to the generalizability of brain-based lower-limb activities for controlling BCIs.

4 Applied Technologies

BCIs have a multitude of clinical applications. As the technologies that compose BCIs advance, their capacity for usage continues to expand. Current developments include their usage in diagnostics, therapeutics, and treatments. The following section presents three particular implementations of clinically applied BCIs representative of each of these uses: attention-deficit hyperactivity disorder (ADHD) identification, rehabilitation enhancement, and applied sleep monitoring.

4.1 Identification of Attention-Deficit Hyperactivity Disorder (ADHD)

Currently, clinical diagnoses of ADHD, a neurodevelopmental disorder which interferes with functioning and productivity, are based upon the behavioral criteria delineated in the *Diagnostic and Statistical Manual of Mental Disorders, Fifth Edition* (DSM-5). Diagnosis is based upon indicative symptoms of either or both delineated patterns, inattention and hyperactivity/impulsivity [34, 35]. While ADHD is most commonly diagnosed in childhood, symptoms often affect the patients persistently, through adolescence and into adulthood [36]. Finding robust, physiologically based objective approaches to identifying ADHD has been an ongoing challenge. A group of behavioral assays that have been found to be highly indicative are continuous performance tests (CPTs); however, these tests, by themselves, lack a physiological component. One of these, the Conners' Kiddie continuous performance test (K-CPT) is a standardized task-oriented computerized neuropsychological test and a useful adjunct in the process of early identification of ADHD and other attention-related conditions in children aged 4–7 years [37]; K-CPT uses varied time intervals between stimuli, creating a paradigm that presents more of a challenge than the fixed intervals used in many other CPTs. Some investigations into the relationships between quantitative EEG and ADHD have been attempted in the past [38], particularly looking at resting EEG, where data is collected while the subject is not performing a task. One such metric, which has been commonly used to aid identification of ADHD subjects, is theta-beta ratio. Previous research has demonstrated that ADHD subjects exhibit significantly lower relative beta powers, higher relative theta powers, and higher theta/beta ratios [39–42], and this ratio has been approved by the FDA for some diagnoses [43]. However, the accuracy of theta/beta ratio has gradually become more controversial [44] and might be indicative only in children with ADHD, but not in adults [45].

To expand the capabilities of EEG for ADHD characterization, we aimed to use electrophysiological markers to clarify heterogeneity of the ADHD population and classify ADHD subgroups according to their clinical and behavioral data. The first step was to quantify EEG oscillations during CPT in preschoolers using a sponge-electrode, wireless EEG system and define neural biomarkers of ADHD preschoolers, then identify correlations between CPT and EEG in these children. The methodological details can be found in Chen et al. [46]. Significant differences were observed between the ADHD and typical development (TD) groups in task-related EEG spectral power, and CPT with longer inter-stimulus intervals (slow-rate task) and shorter inter-stimulus intervals (fast-rate task). The spectral powers were distinct only in the slow-rate task condition (Fig. 6). In task conditions, the alpha powers were negatively correlated with specific CPT scores mainly on perseveration and omission.

These findings suggested that task-related neural dynamics were related to cognitive proficiency in preschool children with ADHD and that EEG profiles contained specific neural biomarkers that could assist clinical intervention [46]. Based on these

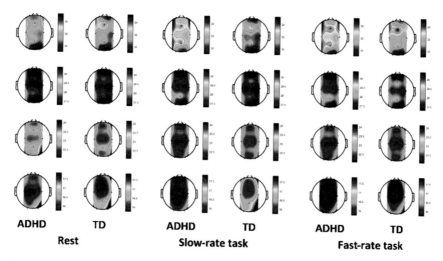

Fig. 6 Topography of four frequency band powers under varied conditions (resting state, slow-rate task, and fast-rate task)

findings, the investigation continued toward searching for additional discriminative biomarkers available in the EEG. We aimed to differentiate preschool children with ADHD from typically developed peers. ADHD diagnosed and TD preschool students completed K-CPT under the three previously described task states. The results showed that the slow-rate task-related central-parietal delta and central alpha and beta band powers between groups with ADHD and TD were significantly distinct and possessed the discriminative validity to identify preschool children with ADHD, with a classification rate reaching 85.71% [47].

These studies could help form a basis for a BCI for children with ADHD. In a preliminary development, we constructed an indicator using EEG oscillations to present indicative aspects of neurophysiological states: attention, fatigue, stress, and changes in brain activities between left and right hemispheres. When tested using a target detection task, it reported that attention levels were significantly higher during tasks than resting states, while the fatigue level increased, and became steady during the third resting state. This development provides a feasible reference for BCI control systems and has been applied to monitor the real-time physiological state changes and evaluate the quality to improve the BCI performance [48].

In 2020, the USFDA permitted the first game-based digital therapeutic, EndeavorRX, to be used to improve attention function in children with ADHD [49]. Such previously described research into neural mechanisms and the approval of innovative strategies encourages researchers who are committed to seeking non-drug interventions. By changing traditional, tedious treatments and diagnostics into engrossing games, there is great potential for expanding clinical access and improving involvement for children with ADHD. In the near future, children with ADHD could be more precisely assessed with indicative neural signals as ADHD is a highly complex and heterogeneous disorder. Combining this with neurologically guided, game-based treatments may improve outcomes and lead to better achievement and quality of life.

4.2 BCI Enhanced Rehabilitation

A major component to proper and timely recovery after bodily injury is the process of rehabilitation, and this is particularly true for damage to the central nervous system, such as occurs in stroke or traumatic brain injury. When damage to the brain occurs, depending on the location, particular functions may be impaired. Under proper conditions, neuroplasticity allows for new neural connections to be formed, regaining some or all functionality over time. Unfortunately, the ideal window for recovery is limited; therefore, the timely utilization of appropriate rehabilitation is critical for recovery.

Recently, there have been many technological developments in motor rehabilitation. Traditional functional metrics are parameters based on physiological and behavioral outputs, such as muscle tone, range of motion, and grip strength. These are critical for determining the effectiveness of and progression through rehabilitation treatment, however, in such conditions such as stroke, where the inciting damage was to brain tissue and where the recovery results from neuroplasticity, these metrics do not directly represent the neural processes occurring during recovery. Thus, one significant trend expanding beyond functional metrics, incorporates real-time neurophysiological monitoring to better inform the process of rehabilitation. Another major trend is the incorporation of advanced technologies into the rehabilitation process, such as VR interfaces and robotic assistive devices. Many such systems have been developed and researched [50–52]; if subjects have sufficient ability to initiate and direct movement, tasks can be presented in VR/AR; however, for subjects with more limited motor capacities, robotic exoskeletons have been used to assist in the performance of movement tasks, such as reaching [53] or leg flexion [54]. Most such systems that have real-time responsive capabilities utilize some form of human–machine interface based off of non-brain signals, such as kinematics, motion tracking, or muscle activations [52, 54].

Combining these developments forms a basis for creating enhanced rehabilitation utilizing BCIs. Systems combining neurological brain recording and advanced interactive devices have been gaining usage; many of these are still in the developmental stage of offline characterization, where neurophysiological recording and interactive devices are used simultaneously, but not necessarily directly connected [52]. An example of this was in our development and application of a mixed-reality music rehabilitation (MR^2) system composed of three integrated systems: an EEG monitoring system, for collecting neurophysiological data, a mixed-reality interface system, presenting both visual and auditory stimuli in augmented reality, and a gait analysis system, using inertial measurement units on the lower limbs (see Fig. 7, top) [2]. Using this system, a gait-triggered mixed-reality (GTMR) task was employed as an interactive game for ambulatory rehabilitation, employing a dual cognitive-motor task which uses a music rhythm game embedded in a real-time gait monitoring and

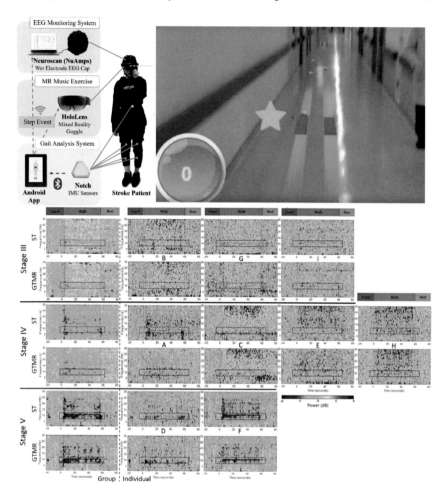

Fig. 7 Gait-triggered mixed-reality (GTMR) task for ambulatory stroke rehabilitation presenting virtual targets and the associated spectral powers of EEG signals among the nine stroke patients in different stroke stages

feedback system. The target cadence was set to approximate the natural walking cadence of a healthy elder, and the task was designed to prompt subjects to lift their feet. Moving targets were generating on alternating sides according to the beats, and participants had to perform a corresponding step, with sufficient amplitude and timing precision to intercept the target, and score a point. Kinematic and neurophysiological data was collected before, during, and after a 1-min walking task. Each participant repeated five standard walking trials without GTMR and five GTMR trials with self-determined lengths of rest in between. Significant increases in cadences during GTMR trials were observed, and spectral analysis of the EEG signals showed the power changes in different frequency components over the time course of the walking tasks (see Fig. 7, bottom).

Most instances where BCIs have been deployed in rehabilitation involve upper limb movements [52], as the neural representation and motor control of the arms are readily accessible by EEG. Developing BCIs for lower-limb rehabilitation has had additional obstacles, as lower-limbs support weight and require balance, walking necessitates mobile collections systems and introduces unavoidable movement artifacts, and the associated brain regions are deeper and less detectable [55]. Thus, systems incorporating online neural recording with interfaces, such as the GTMR task, are critical in the move toward creating fully integrated BCI systems for ambulatory rehabilitation. Combining these advancements with classification algorithms, such as those for distinguishing MI with connectivity networks in real time, are the basis for the development of rehabilitation BCIs.

4.3 Applied Sleep Monitoring

Traditionally, objective sleep quality is evaluated through polysomnography (PSG), which usually consists of multiple channels of EEG, electrooculogram (EOG), electromyogram (EMG), and electrocardiogram (ECG) signals. A sleep technician will then divide the PSG record into multiple 30-second epochs and classify each as one of five stages: wake (W), rapid eye movement (REM), and three stages of non-REM sleep (N1, N2, and N3), more colloquially recognized as falling asleep, light sleep, and slow-wave or deep sleep. The percentage and distribution of each class of sleep stages are then used to evaluate sleep quality. As PSG requires multiple channels and manual labeling, it is demanding in both resources and time. In addition, the numerous biosensors attached to the subject may also cause sleep disturbances by being overly intrusive.

Recent advances in technology have enabled two-channel/single-channel sleep stage classification through applications of conventional machine and deep learning. In conventional machine learning, sleep stage features are extracted manually from either EOG or EEG signals, often chosen based on current assumptions about human sleep stages. For example, Khalighi et al. [56] used a variant of discrete wavelet transformation (DWT) to extract features. In [57], Alickovic and Subasi used DWT to decompose single-channel EEG signals. The authors specifically decomposed the EEG into sub-bands roughly representing commonly recognized sleep EEG rhythms, such as delta and beta. After features are extracted, they are then selectively fed into classifiers, such as random forests [58, 59] or support vector machines [60] to determine the sleep stages. In deep learning approaches, instead of defining features manually, a neural network is used to automatically obtain features from biomedical signals. Most studies have used convolutional neural networks (CNNs), with variations, such as different window sizes, to better capture structural and frequential features [61, 62]. Because individual sleep stages often have temporal dependencies with neighboring stages, some studies have combined CNNs with variants of recurrent neural networks (RNNs) to learn inter-epoch relations [61, 63]. Both traditional machine and deep learning techniques greatly reduce the resources required for sleep stage classification and by extension sleep quality evaluation.

While sleep quality evaluation is critical for addressing parasomnias, it has generally been constrained to post hoc clinical treatment, typically via the use of hypnotic (sleep inducing) drugs. Automatic sleep stage classification is a critical step toward implementing real-time sleep evaluation, a typical prerequisite for the development of BCIs. Uses of BCIs during sleep have additional technical limitations beyond computational sleep evaluation. Typically, EEG components of PSG utilize wired, wet-electrode systems, which can be intrusive enough to disrupt sleep themselves and are constrained to use in dedicated sleep labs with administration by clinicians. Maintaining a reliable and consistent signal over the sleep period is a challenge, precluding the use of sponge electrodes which would dry out and degrade the signal over such a long duration. As previously discussed, the development of wireless, dry-electrode systems alleviate these constraints. With the advancement of dry electrodes, such as soft silicone-graphene, sleep quality evaluation can be reliably performed in the home or community, which would allow the automatic sleep analysis algorithms to show their maximum potential.

An additional component of developing BCIs for sleep improvement is the ability to respond to the user in a meaningful way. BCIs are typically envisioned as requiring a conscious user, precluding their use during sleep, but passive BCIs can be used to directly adjust the environment in ways that improve sleep quality. Over the years, various methods have been explored as a complement or substitute for hypnotic drugs, including light therapy and aromatherapy. In aromatherapy, essential oil was either applied on skin or inhaled before or during sleep; however, most findings on this topic have been challenged due to limitations of methodology, namely a lack of proper placebos in blinded study designs, as the presence of such an aerosolized chemical is unmistakable to a conscious user. Previous research has been done using lavender essential oil, released only after the first sleep cycle, while the user was asleep [64]; this produced improvements in duration of N3 sleep. Such types of sleep enhancement could be integrated into sleep improvement BCIs, directing the timing or intensity of subtle environmental manipulations with EEG signals indicative of sleep quality and stage.

5 Conclusion

This chapter presents the current state of non-invasive BCIs, their underlying technologies, and their use in exemplary clinical applications. As was shown, the field of non-invasive BCIs continues to grow, involving new technologies and techniques, which in turn expands the potential utility of such devices. With this growth, the capacity for BCI systems to impact the lives of people increases, changing everything from how people play cutting edge games to how we interact with the Internet of Things [65]. This impact may be most notable in clinical applications; as BCIs are relatively cheap and adaptable, they could be used to leverage these advancements to expand diagnoses and treatments and extend the capabilities of clinical professionals and medical systems.

Appendix 1: References for Wet-electrode EEG Systems

a) http://compumedicsnuroscan.com/
b) https://www.advancedbrainmonitoring.com/neurotechnology/
c) https://mbraintrain.com/smarting/
d) https://www.neuroelectrics.com/products/enobio/enobio-8/
e) http://www.gtec.at/Products
f) https://openbci.com/
g) https://www.brainproducts.com/products_by_apps.php?aid=5

Appendix 2: References for Dry-electrode EEG Systems

a) http://neurosky.com/
b) https://www.emotiv.com/
c) https://www.cognionics.net/us-price
d) https://vandrico.com/wearables/search/node/Imec
e) http://www.bri.com.tw/product_br8plus.html
f) http://www.quasarusa.com/products_dsi.htm;   https://bio-medical.com/fre edom-24d-wireless-eeg-headset-w-brainavatar-acquisition-software.html
g) https://www.neuroelectrics.com/products/enobio/enobio-8/
h) http://www.gtec.at/Products
i) https://openbci.com/
j) https://www.brainproducts.com/products_by_apps.php?aid=5
k) http://zeto-inc.com/

References

1. Lee YC, Lin WC, Cherng FY, Ko LW (2016) A visual attention monitor based on steady-state visual evoked potential. IEEE Trans Neural Syst Rehabil Eng 24(3):399–408. https://doi.org/10.1109/TNSRE.2015.2501378
2. Ko L-W et al (2021) Integrated gait triggered mixed reality and neurophysiological monitoring as a framework for next-generation ambulatory stroke rehabilitation. IEEE Trans Neural Syst Rehabil Eng 29:2435–2444. https://doi.org/10.1109/TNSRE.2021.3125946

3. Brain Computer Interface Market Size and Industry Trends|2030. Allied market research. https://www.alliedmarketresearch.com/brain-computer-interfaces-market. Accessed 16 Dec 2021

4. Brain Computer Interface Market Size Report, 2020–2027. https://www.grandviewresearch. com/industry-analysis/brain-computer-interfaces-market. Accessed 16 Dec 2021

5. Gu X et al (2021) EEG-based brain-computer interfaces (BCIs): a survey of recent studies on signal sensing technologies and computational intelligence approaches and their applications. IEEE/ACM Trans Comput Biol Bioinform 18(5):1645–1666. https://doi.org/10.1109/TCBB. 2021.3052811

6. Schalk G (2010) Can electrocorticography (ECoG) support robust and powerful brain-computer interfaces? Front Neuroengineering. https://doi.org/10.3389/fneng.2010.00009

7. Naseer N, Hong K-S (2015) fNIRS-based brain-computer interfaces: a review. Front Hum Neurosci 9(JAN):1–15. https://doi.org/10.3389/fnhum.2015.00003

8. Sitaram R, Weiskopf N, Caria A, Veit R, Erb M, Birbaumer N (2008) fMRI brain-computer interfaces. IEEE Signal Process Mag 25(1):95–106. https://doi.org/10.1109/MSP.2008.440 8446

9. Liao L-D et al (2014) A novel 16-channel wireless system for electroencephalography measurements with dry spring-loaded sensors. IEEE Trans Instrum Meas 63(6):1545–1555. https://doi. org/10.1109/TIM.2013.2293222

10. Ko L-W et al (2019) Development of a smart helmet for strategical BCI applications. Sensors 19(8):1867. https://doi.org/10.3390/s19081867

11. Ko L-W, Su C-H, Liao P-L, Liang J-T, Tseng Y-H, Chen S-H (2021) Flexible graphene/GO electrode for gel-free EEG. J Neural Eng 18(4):046060. https://doi.org/10.1088/1741-2552/ abf609

12. Wolpaw JR, Birbaumer N, McFarland DJ, Pfurtscheller G, Vaughan TM (2002) Brain–computer interfaces for communication and control. Clin Neurophysiol 113(6):767–791. https://doi.org/10.1016/S1388-2457(02)00057-3

13. Nan W et al (2012) Individual alpha neurofeedback training effect on short term memory. Int J Psychophysiol 86(1):83–87. https://doi.org/10.1016/j.ijpsycho.2012.07.182

14. Mousavi M, Krol LR, de Sa VR (2020) Hybrid brain-computer interface with motor imagery and error-related brain activity. J Neural Eng 17(5):056041. https://doi.org/10.1088/1741-2552/ abaa9d

15. Parikh D, George K (2020) Quadcopter control in three-dimensional space using SSVEP and motor imagery-based brain-computer interface. In: 2020 11th IEEE annual information technology, electronics and mobile communication conference (IEMCON), Vancouver, BC, Canada, Nov 2020, pp 0782–0785. https://doi.org/10.1109/IEMCON51383.2020.9284924

16. Wang R et al (2020) Design and implement the continuous flickering SSVEP-BCI in augmented reality. J Phys Conf Ser 1631(1):012172. https://doi.org/10.1088/1742-6596/1631/1/012172

17. Wen D, Liang B, Zhou Y, Chen H, Jung T-P (2021) The current research of combining multi-modal brain-computer interfaces with virtual reality. IEEE J Biomed Health Inform 25(9):3278–3287. https://doi.org/10.1109/JBHI.2020.3047836

18. Lin C-T et al (2008) Development of wireless brain computer interface with embedded multitask scheduling and its application on real-time driver's drowsiness detection and warning. IEEE Trans Biomed Eng 55(5):1582–1591. https://doi.org/10.1109/TBME.2008.918566

19. Morgan ST, Hansen JC, Hillyard SA (1996) Selective attention to stimulus location modulates the steady-state visual evoked potential. Proc Natl Acad Sci 93(10):4770–4774. https://doi.org/ 10.1073/pnas.93.10.4770

20. Nakanishi M, Wang Y, Chen X, Wang Y-T, Gao X, Jung T-P (2018) Enhancing detection of SSVEPs for a high-speed brain speller using task-related component analysis. IEEE Trans Biomed Eng 65(1):104–112. https://doi.org/10.1109/TBME.2017.2694818

21. Nakanishi M, Wang Y, Wang Y-T, Mitsukura Y, Jung T-P (2014) A high-speed brain speller using steady-state visual evoked potentials. Int J Neural Syst 24(06):1450019. https://doi.org/ 10.1142/S0129065714500191

22. Nayak T, Ko L-W, Jung T-P, Huang Y (2019) Target classification in a novel SSVEP-RSVP based BCI gaming system. In: 2019 IEEE international conference on systems, man and cybernetics (SMC), Bari, Italy, Oct 2019, pp 4194–4198. https://doi.org/10.1109/SMC.2019.891 4174

23. Zhang H-Y, Stevenson CE, Jung T-P, Ko L-W (2020) Stress-induced effects in resting EEG spectra predict the performance of SSVEP-based BCI. IEEE Trans Neural Syst Rehabil Eng 28(8):1771–1780. https://doi.org/10.1109/TNSRE.2020.3005771

24. Nakanishi M et al (2017) Detecting glaucoma with a portable brain-computer interface for objective assessment of visual function loss. JAMA Ophthalmol 135(6):550. https://doi.org/10.1001/jamaophthalmol.2017.0738

25. Mohan A et al (2016) The significance of the default mode network (DMN) in neurological and neuropsychiatric disorders: a review. Yale J Biol Med 89(1):49–57

26. Rubia K et al (2019) Functional connectivity changes associated with fMRI neurofeedback of right inferior frontal cortex in adolescents with ADHD. Neuroimage 188:43–58. https://doi.org/10.1016/j.neuroimage.2018.11.055

27. Phang C-R, Noman F, Hussain H, Ting C-M, Ombao H (2020) A multi-domain connectome convolutional neural network for identifying schizophrenia from EEG connectivity patterns. IEEE J Biomed Health Inform 24(5):1333–1343. https://doi.org/10.1109/JBHI.2019.2941222

28. Hu S, Wang H, Zhang J, Kong W, Cao Y (2014) Causality from Cz to C3/C4 or between C3 and C4 revealed by granger causality and new causality during motor imagery. In: 2014 International joint conference on neural networks (IJCNN), Beijing, China, Jul 2014, pp 3178–3185. https://doi.org/10.1109/IJCNN.2014.6889769

29. Kuś R, Ginter JS, Blinowska KJ (2006) Propagation of EEG activity during finger movement and its imagination. Acta Neurobiol Exp (Warsz) 66(3):195–206

30. Pfurtscheller G, Graimann B, Huggins JE, Levine SP, Schuh LA (2003) Spatiotemporal patterns of beta desynchronization and gamma synchronization in corticographic data during self-paced movement. Clin Neurophysiol 114(7):1226–1236. https://doi.org/10.1016/S1388-245 7(03)00067-1

31. Wang Y, Hong B, Gao X, Gao S (2006) Phase synchrony measurement in motor cortex for classifying single-trial EEG during motor imagery. In: 2006 International conference of the IEEE Engineering in Medicine and Biology Society, New York, NY, Aug 2006, pp 75–78. https://doi.org/10.1109/IEMBS.2006.259673

32. Phang C-R, Ko L-W (2020) Global cortical network distinguishes motor imagination of the left and right foot. IEEE Access 8:103734–103745. https://doi.org/10.1109/ACCESS.2020.2999133

33. Phang C-R, Ko L-W (2020) Intralobular and interlobular parietal functional network correlated to MI-BCI performance. IEEE Trans Neural Syst Rehabil Eng 28(12):2671–2680. https://doi.org/10.1109/TNSRE.2020.3038657

34. American Psychiatric Association (2013) Diagnostic and statistical manual of mental disorders. 5th edn. American Psychiatric Association. https://doi.org/10.1176/appi.books.978089042 5596. https://web.archive.org/web/20220113074628/. https://dsm.psychiatryonline.org/doi/book/https://doi.org/10.1176/appi.books.9780890425596

35. Sroubek A, Kelly M, Li X (2013) Inattentiveness in attention-deficit/hyperactivity disorder. Neurosci Bull 29(1):103–110. https://doi.org/10.1007/s12264-012-1295-6

36. Moffitt TE et al (2015) Is adult ADHD a childhood-onset neurodevelopmental disorder? Evidence from a four-decade longitudinal cohort study. Am J Psychiatry 172(10):967–977. https://doi.org/10.1176/appi.ajp.2015.14101266

37. Conners C (2015) Conners kiddie continuous performance test 2nd edition (K–CPT 2). Multi-Health Syst Inc.MHS Tor

38. Lenartowicz A, Loo SK (2014) Use of EEG to diagnose ADHD. Curr Psychiatry Rep 16(11):498. https://doi.org/10.1007/s11920-014-0498-0

39. Lansbergen MM, Arns M, van Dongen-Boomsma M, Spronk D, Buitelaar JK (2011) The increase in theta/beta ratio on resting-state EEG in boys with attention-deficit/hyperactivity disorder is mediated by slow alpha peak frequency. Prog Neuropsychopharmacol Biol Psychiatry 35(1):47–52. https://doi.org/10.1016/j.pnpbp.2010.08.004

40. Ogrim G, Kropotov J, Hestad K (2012) The quantitative EEG theta/beta ratio in attention deficit/hyperactivity disorder and normal controls: Sensitivity, specificity, and behavioral correlates. Psychiatry Res 198(3):482–488. https://doi.org/10.1016/j.psychres.2011.12.041
41. Loo SK, Cho A, Hale TS, McGough J, McCracken J, Smalley SL (2013) Characterization of the theta to beta ratio in ADHD: identifying potential sources of heterogeneity. J Atten Disord 17(5):384–392. https://doi.org/10.1177/1087054712468050
42. Shi T et al (2012) EEG characteristics and visual cognitive function of children with attention deficit hyperactivity disorder (ADHD). Brain Dev 34(10):806–811. https://doi.org/10.1016/j.braindev.2012.02.013
43. U.S. Food & Drug Administration., Device Classification Under Section 513(f)(2)(De Novo). https://www.accessdata.fda.gov/scripts/cdrh/cfdocs/cfpmn/denovo.cfm?ID=DEN110019. Accessed 13 Jan 2022
44. Arns M, Conners CK, Kraemer HC (2013) A decade of EEG theta/beta ratio research in ADHD: a meta-analysis. J Atten Disord 17(5):374–383. https://doi.org/10.1177/1087054712460087
45. Markovska-Simoska S, Pop-Jordanova N (2017) Quantitative EEG in children and adults with attention deficit hyperactivity disorder: comparison of absolute and relative power spectra and theta/beta ratio. Clin EEG Neurosci 48(1):20–32. https://doi.org/10.1177/1550059416643824
46. Chen I-C, Chang C-H, Chang Y, Lin D-S, Lin C-H, Ko L-W (2021) Neural dynamics for facilitating ADHD diagnosis in preschoolers: central and parietal delta synchronization in the kiddie continuous performance test. IEEE Trans Neural Syst Rehabil Eng 29:1524–1533. https://doi.org/10.1109/TNSRE.2021.3097551
47. Chen I-C, Lee P-W, Wang L-J, Chang C-H, Lin C-H, Ko L-W (2021) Incremental validity of multi-method and multi-informant evaluations in the clinical diagnosis of preschool ADHD. J Atten Disord 108705472110457. https://doi.org/10.1177/10870547211045739
48. Chang Y, He C, Tsai B-Y, Ko L-W, Multi-parameter physiological state monitoring in target detection under real-world settings. Front Hum Neurosci 793
49. Kollins SH et al (2020) A novel digital intervention for actively reducing severity of paediatric ADHD (STARS-ADHD): a randomised controlled trial. Lancet Digit Health 2(4):e168–e178. https://doi.org/10.1016/S2589-7500(20)30017-0
50. Teo WP et al (2016) Does a combination of virtual reality, neuromodulation and neuroimaging provide a comprehensive platform for neurorehabilitation?—a narrative review of the literature. Front Hum Neurosci 10(June):1–15. https://doi.org/10.3389/fnhum.2016.00284
51. Calabrò RS et al (2017) The role of virtual reality in improving motor performance as revealed by EEG: a randomized clinical trial. J NeuroEngineering Rehabil 14(1):53. https://doi.org/10.1186/s12984-017-0268-4
52. Li M, Xu G, Xie J, Chen C (2018) A review: motor rehabilitation after stroke with control based on human intent. Proc Inst Mech Eng 232(4):344–360. https://doi.org/10.1177/0954411918755828
53. Heo P, Gu GM, Lee S, Rhee K, Kim J (2012) Current hand exoskeleton technologies for rehabilitation and assistive engineering. Int J Precis Eng Manuf 13(5):807–824. https://doi.org/10.1007/s12541-012-0107-2
54. Shi D, Zhang W, Zhang W, Ding X (2019) A review on lower limb rehabilitation exoskeleton robots. Chin J Mech Eng 32(1):74. https://doi.org/10.1186/s10033-019-0389-8
55. Shafiul Hasan SM, Siddiquee MR, Atri R, Ramon R, Marquez JS, Bai O (2020) Prediction of gait intention from pre-movement EEG signals: a feasibility study. J NeuroEngineering Rehabil 17(1):50. https://doi.org/10.1186/s12984-020-00675-5
56. Khalighi S, Sousa T, Oliveira D, Pires G, Nunes U (2011) Efficient feature selection for sleep staging based on maximal overlap discrete wavelet transform and SVM. In: 2011 Annual international conference of the IEEE Engineering in Medicine and Biology Society, Boston, MA, Aug 2011, pp 3306–3309. https://doi.org/10.1109/IEMBS.2011.6090897
57. Alickovic E, Subasi A (2018) Ensemble SVM method for automatic sleep stage classification. IEEE Trans Instrum Meas 67(6):1258–1265. https://doi.org/10.1109/TIM.2018.2799059
58. Koley B, Dey D (2012) An ensemble system for automatic sleep stage classification using single channel EEG signal. Comput Biol Med 42(12):1186–1195. https://doi.org/10.1016/j.compbiomed.2012.09.012

59. Klok AB, Edin J, Cesari M, Olesen AN, Jennum P, Sorensen HBD (2018) A new fully automated random-forest algorithm for sleep staging. In: 2018 40th annual international conference of the IEEE Engineering in Medicine and Biology Society (EMBC), Honolulu, HI, Jul 2018, pp 4920–4923. https://doi.org/10.1109/EMBC.2018.8513413
60. Wu H, Talmon R, Lo Y-L (2015) Assess sleep stage by modern signal processing techniques. IEEE Trans Biomed Eng 62(4):1159–1168. https://doi.org/10.1109/TBME.2014.2375292
61. Phan H, Andreotti F, Cooray N, Chen OY, De Vos M (2019) Joint classification and prediction CNN framework for automatic sleep stage classification. IEEE Trans Biomed Eng 66(5):1285–1296. https://doi.org/10.1109/TBME.2018.2872652
62. Supratak A, Dong H, Wu C, Guo Y (2017) DeepSleepNet: a model for automatic sleep stage scoring based on raw single-channel EEG. IEEE Trans Neural Syst Rehabil Eng 25(11):1998–2008. https://doi.org/10.1109/TNSRE.2017.2721116
63. Yuan Y et al (2019) A hybrid self-attention deep learning framework for multivariate sleep stage classification. BMC Bioinformatics 20(S16):586. https://doi.org/10.1186/s12859-019-3075-z
64. Ko L-W, Su C-H, Yang M-H, Liu S-Y, Su T-P (2021) A pilot study on essential oil aroma stimulation for enhancing slow-wave EEG in sleeping brain. Sci Rep 11(1):1078. https://doi.org/10.1038/s41598-020-80171-x
65. Putze F, Weiß D, Vortmann L-M, Schultz T (2019) Augmented reality interface for smart home control using SSVEP-BCI and eye gaze. In: 2019 IEEE international conference on systems, man and cybernetics (SMC), Bari, Italy, Oct 2019, pp 2812–2817. https://doi.org/10.1109/SMC.2019.8914390

Artificial Intelligence-Monitored Procedure for Personal Ethical Standard Development Framework in the E-Learning Environment

Rabia Abhay⦿, **Abirami Abi**⦿, **Poornima Kapadan Othayoth**⦿,
Joseph Varghese Kureethara⦿, **and Jiran Kurian Puliyanmakkal**⦿

Abstract The changes in the lifestyle of human beings due to the pandemic COVID-19 have affected all walks of human life. As a pillar of human development, the arena of education has a vital role to play in this changing world. The humongous and disruptive technologies that had made inroads into the educational scene as E-learning paved the way for ethical concerns in an unimaginable manner. Artificial intelligence is prudently incorporated for developing an ethical lifestyle for students all over the world. The Personal Ethical Standard Framework would work as a vaccine for the pandemic of the cancerous growth of the unethical habits of learners.

Keywords Personal ethics · Holistic life · Professional ethics · Ethical development

1 Introduction

Humans are evolutionary beings. In the course of history, every aspect of human beings has been transformed both mentally and physically. In comparison with other living beings, human beings have undergone a change that is seen in their living styles, dwelling places, eating habits, interpersonal relationships, individuality, societal behaviour, learning systems, and even in knowledge advancement

R. Abhay · A. Abi · P. K. Othayoth · J. V. Kureethara (✉) · J. K. Puliyanmakkal
Christ University, Bangalore, India
e-mail: frjoseph@christuniversity.in

R. Abhay
e-mail: abhay.rajendran@res.christuniversity.in

A. Abi
e-mail: abirami.av@res.christuniversity.in

P. K. Othayoth
e-mail: poornima.purushotham@res.christuniversity.in

J. K. Puliyanmakkal
e-mail: jiran.kurian@res.christuniversity.in

methods from time to time. While other living beings do not have recorded history, human beings have both recorded and unrecorded history. Among all things, how human beings have changed their individual behaviour and social responses towards decision-making situations based on their cognitive faculties matters of great interest to philosophers, anthropologists, and sociologists. Differentiating binaries of various actions and terming them as good, bad, neutral, and so on have been part of general concern throughout the history of human beings. Societies, religions, organisations, and institutions were formed and are being created based on the actions their members are desired to have. This mutually desirable code of conduct, commonly called ethics, is what governs the actions of the individuals that constitute any given society. This chapter explains how ethical behaviour can be developed by the support of gamification with mundane activities.

This chapter has five more sections. In the Sect. 2, the definition of ethics and how it is associated with personal, professional, and society are discussed. The Sect. 3 throws light on the relevance of the panopticon and how an individual behaves in a closed or monitored environment. The next section touches on the features of learning and discusses mechanisms of unethical behaviour in E-learning. Finally, the Sect. 5 introduces a framework using AI's features to resolve and self-train individuals to abide by ethical values. The conclusion sums up the issue raised in this paper with relevant examples and literature.

2   Ethics

Ethics is a discipline of human thought that gives good or bad hues for human actions. Ethical codes contain a collection of information that collectively helps a group of people who are agreeing upon it. Societal rules and regulations, norms and practices, values, and expectations are founded on these ethical codes. At the same time, ethical codes written and unwritten have a strong impact on humans too. Society in various forms is the guardian and provider of ethical codes. An ethical society is a society rooted in ethical codes. Ethical codes help individuals to develop into holistic persons. An individual who does not respect the ethical codes of a society or system would naturally be a codebreaker and become unfit. In all such cases, many societies have established corrective measures. However, there is no mechanism in societal terms for internal moral corruption. This type of corruption leads to societal disruptions through a cascading effect. As hinted, ethical codes are nurtured in an individual. Although the fundamental nature of human beings may be contrary to some of society's ethical codes, the codes nurtured in an individual will help them live a harmonious life in tandem with the social norms present in the society that they live. Nurturing the ethical codes involves oneself, family, educational institutions and the state in general.

2.1 Ethics in Day to Day Life

Are human beings inherently ethical in their daily life? Are they moral and ethical only if a surveillance system constantly supervises them? Let us try to dive into the nuances of this statement by taking a look into various aspects of day-to-day life of an individual.

2.1.1 Personal Ethics

Ethics and morality originate from an individual's personal life experiences. An individual's lived experiences can be better through continuous reinforcements. They get reflected into other avenues of that person's life. A very common example would be the fact that in a household with a male child and a female child, the differential treatment is given to both of them in the lines of women should look after the household and men should take care of the financial workings and should always be bold. This goes a long way in moulding an individual's ethical and moral foundations. A man who has been taught in his/her house that dominating over a woman is a normal thing to do then when he forms a family of his own would consider his domination over his partner as a normal thing to do and will not see an unethical flavour for the same.

The rigorous religious indoctrination through which an individual goes like a child in most households in the form of ritualistic practices and compulsory praying patterns can be seen as another mode through which ethics and morality are tried to be instilled into young minds. It is true that to some extent, it does good but what most people fail to see is that the rigidity in ethics and morals and views that these practices provide significantly undermine the flexibility of a person to see the good in other faiths and beliefs and try to establish the ethics and morals bestowed by his/her faith is much superior to the rest. This rigidity in ethics and morals followed by such an individual in his/her private life is seen to be the root cause of various communal revolts and violence. Instilling fluidity and accepting the unacceptable should also be provided to young minds while instilling a sense of ethicality and morality. The initial steps for the same should begin from one's personal life.

2.1.2 Professional Ethics

The professional life is of great importance as an individual will spend a considerable amount of time in this area. The ethics and morality captured by an individual in formative years both in the personal and external environment [10] perhaps have spillover to the professional sphere of the lives during the productive years, for instance, a child who is not discouraged when he/she brings home a pencil from their classmates without their knowledge or the child has not taught how to be accommodative of the differences within his/her friend circle from a very early age. It is

bound to affect later into their professional life. A manager taking up the credit for
the work to which his entire team has contributed significantly and a researcher using
botched data or non-scientific methods to collect the data are a few instances in which
the ethical foundations of individuals can be assessed in their professional lives. This
leads to the age-old question: Is an individual's character or ethical construct solely
based on the environment or is the individual born with an inclination towards certain
ethical and personality traits?

2.2   Ethics: Nature Versus Nurture

The question whether humans are born inherently with an inclination towards the
personality that they will manifest in their later ages or is the external environment
the deciding factor on the nature of the personality that will manifest has been an
ongoing debate within the circles of philosophers as well as psychologists. Acclaimed
philosopher and thinker John Locke has put forward the concept of *Tabula Rasa*,
which refers to the concept of a child being born as a "Blank Slate" devoid of any
inclinations or predetermined paths for development. Regardless they attain ethical
and moral codes that ultimately sculpt his/her personality as they interact and mingle
with other external elements.

In this context, a new angle to this age-old debate based the claims on the theories
of Sigmund Freud. According to him, each human psyches are divided into Id, Ego,
and Superego. The constant interaction between these constituents is what leads to
an individual's personality. The Id part of the psyche comprises all the primal, arcane
needs and drives. It puts forth a sense of immediate gratification of carnal desires
such as sex, fear, hunger, and anger are highly influenced by the same. Superego is
the moral compass that each one possesses. It orients our internal desires with the
realities and the restraints of our external world. It is ego the moderator which works
in keeping the balance between Id and superego. The Id part of the human psyche
that demands instant gratification is to be an inbuilt inclination that every human
being is born with. Crying when hungry and feeling discomfort when in pain are a
few of the inbuilt tendencies humans show right from birth. It is true that these cannot
be directly quoted to the ethical spectrum of a human being, but they can be seen
as the potential seeds in which the person builds up his/her personality and ethical
skeleton.

The superego part is the human psyche's highest ethical and reality-oriented part.
It acts as a reality check between the internal and external environment of a human
being. It assesses various aspects and factors existing around a person and fine-
tunes the "socially ethical approaches" that one may undertake during exposure to
such occurrences. Unlike Id, the superego is not ingrained in a human during birth. It
evolves slowly over time, influenced by the environment and immediate surroundings
to which the individual is exposed in his/her formative years. The society in which
one is born and brought up moulds the superego constituent one's psyche. Thus,

society does indeed play a major role in sculpting the superego component of an individual.

The ego part of the psyche is seas be the orchestrator, orchestrating the demands and needs of Id and superego and maintaining a balance between them as any imbalance could lead to cognitive dissonance. The evolution of the ego part of the psyche is not entirely external or internal in origins like superego and Id, respectively. An individual's perception of both internal and external stimuli to which he/she is exposed goes a long way in moulding the characteristics of an individual's ego. It is this foundation of ego that becomes the basis of its orchestration of Id and superego demands and needs.

To give a partial conclusion to this time-tested debate, one could say that the ethical and moral framework of a human being is influenced both by the internal and external environment, but the decisions and frameworks thus devised by an individual would be independent of these external and internal influences.

2.3 Ethics in Society

Inherently accustomed way to rightfully do a procedure is seen as the ethics imbued within that process. Society has a great hand in moulding these ethical structures that govern the individuals forming the society. It is said society and ethical structures are self-fulfilling systems that complement and evolve as a single unit. Denoting theft and murder as socially unethical practices comes from the social construct where most of the elements in society see the same as unethical. The Konyak tribe of Nagaland in old times were known as *headhunters* [9] as they were known to cut the heads of their enemies and hang them outside their houses as tokens. This practice within their tribes was seen as normal and thus was not seen to be unethical. Lately, in modern days with the changing zeitgeist, this practice of theirs has stopped owing to their changed notion that what they did was not ethically and morally right. This example directly links how society and ethics are intricately intertwined.

Coming to a consensus on what is ethically right and what is ethically wrong for the society has a long history. For instance, during the French revolution (the 1780s) where thinkers such as Rousseau put forward the concept of "social contract" it states that humans should wilfully give up few of their freedom and wilfully restrict their activities for the sake of the common good of the entire society.

This generalistic view of ethics is prevalent within each society. The bottom-up approach would be that individualistic ethics prevalent within the people in a society would ultimately fine-tune the ethical approach of the entire society. Even if one were to see in an evolutionary perspective, the importance of ethical norms that govern the societal structures remains salient for the collective co-existence of all individual elements within that society. If the concept of headhunting of the Konyak tribe was prevalent even now, then it could have posed a great threat to the very fabric of the entire society. Murder and killings would have become rampant, and the peaceful co-existence of the elements within the society would be jeopardised (Fig. 1).

Fig. 1 Based on Freud's concepts: differential influence on Id, ego, and superego through the different stages of one's life

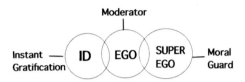

Therefore, it could be safely said that the invisible limitations and restraints of such ethical and moral unsaid codes prevalent within the society actually keep the fabric of the society from unravelling and collapsing into anarchy.

3 The Relevance of Panopticon

The word Panopticon is a portmanteau that originated from the words pan and opticon. It has a deep meaning, as pan means all and opticon refers to eyesight. It is evident that the idea is drawn from the Greek mythological character Argus Panoptes, who has many eyes considered to be an apt watchman. It was utilitarian philosopher Jeremy Bentham who founded the idea of the panopticon. It is a structure he created for a higher level of prison. This was crafted in a way where there is a watchtower separated from the centre of the building. This tower has access to all the inmates' cells from the centre. The structure replaces the traditional relationship of inmates and multiple guards to one guard watching n numbers of cells and inmates from the watchtower. It is developed exclusively to monitor the behaviour of all individuals with a single guard.

The idea of the panopticon was further expanded and established by critic and philosopher Michale Foucault in his work titled "Discipline and Punish: The Birth of the Prison" [15]. Foucault took forth Bentham's idea as an archetype in the modern era to monitor and note the graph of discipline. According to him, this concept highlights how the role of power works and the change in the behaviour is seen. Moreover, he exemplifies how the binaries of discipline and behaviour work together in a monitored space.

There is a transformation in the behaviour pattern when someone is being watched, from being lethargic to shifting as productive, being disobedient to obedient, from breaking the rules to following the instructions, there is a change in the behaviour subconsciously adapting while being watched. For instance, in the case of inmates, guards from the watchtower monitor the inmates in prison where the inmates believe they are let free in contrast they are in their best behaviour when they see the guard in person.

In Foucault's accounts, the manner in which the human body works while being monitored differs from unmonitored space. The panopticon concept for him is how the human mechanism undergoes a behavioural change and reshapes the approach, showing the significant role of authority and power.

3.1 Applying Panopticon-Derived Ideas into Everyday Life

In contemporary times, this concept is applicable in institutional places such as schools, offices, hospitals, police stations, and asylums and even in everyday life being watched without knowing who is watching similarly, behaving in their obedient manner as someone is watching them. In contemporary times, we encounter the presence of the panopticon in many forms such as CCTV, mobile cameras, higher officers, teachers, parents, and pet dogs. This is visible in the figure mentioned. Popular taglines and slogans such as "smile you are under the camera" and "this property is protected by a video camera" subconsciously make one project the best behaviour. In the case of unmonitored spaces, there is a breakage of rules which lead to ethical violations. The panopticon concept will examine how there is a lack of ethical behaviour and the obedient pattern of the individual being violated; this will be analysed in detail through e-learning. The role of ethics and how the deviation took place will be explained.

Depending on the affordability, everyone is upgrading his/her electronic devices. Governments and educational institutions all over the world have come with free and affordable e-contents to make learning in the lockdown time easier. These e-contents proved to be a game-changer. This helped education reach many doors. This is going to last forever. In the hands of technology and the blooming variety of users on the Internet, the scope of e-learning plays a pivotal role in contemporary times. In fact, the budget allocation for this e-learning has been increased in all institutions around the globe. Similarly, students and teachers are in the process of accepting and welcoming the online education medium. However, there is an ingrained questioning blooming when it comes to students' ethical behaviour in the process of e-learning classroom space. The daily schedule of a student, despite the age and class, has a standard time, for instance, from 8.30 to 3.30. In schools, students go late and blame traffic and the mode of transport they come by, and they come up with lame reasons for their unfinished work. Similarly, when a student gets up late to attend the e-learning classrooms, he/she blames the power supply or Internet. This shows how the ethical violation is changing even when everything is presented in the comfort space (Fig. 2).

The ethical violation happens subconsciously and consciously in a student while involved in e-learning programmes and classes. With reference to the concept of the panopticon, it is evident that there is a change in the mannerism, and slight obedient behaviour is observed. In contrast to this, while applying the lens of the panopticon in e-learning classes, for example, when there is no compulsion for the student to switch on the video, there is a comfortable space for the students where no one is there to monitor them. The living "panopticon" figures like parents or guardians when not at home or not being physically present, the student's behaviour alters. Hence, they sit on their bed or couch in the worst case they sleep. This shows how unconsciously they violate the behaviour.

Fig. 2 The adaptation of the Panopticon diagram resembles when teachers and parents monitor a change in students' behaviour

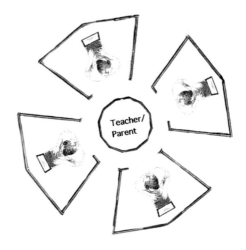

4   Unethical Behaviour and E-Learning

In the ancient period, humans were learning through oral tradition. In contemporary times, there are mobile-enabled E-learning that can be used to teach people even from the opposite end of the world. In between the caves and clipped, the significance of education is connected to this medium. Despite the place and environment, one connects through the educational medium using E-learning platforms which connect everywhere and anywhere. This helps one to learn and resolve information, and maintain and sustain living practices. Further, this type of learning can bring awareness and acknowledgement to global issues. Proposing a question, can social, cultural and academic values be successfully transmitted in computer-mediated education? Is it attainable for the E-learning platforms to maintain the ethical behaviour of students and teachers? This chapter focuses on the different kinds of unethical behaviour practised by students, and the chapter tries to develop a framework that can help resolve the unethical behaviour.

Teaching and learning at a distance raise a lot of ethical issues, particularly noticed when the learning is outside classrooms. When it comes to e-learning platforms, the situation becomes more complex [2]. Ethical values such as honesty, trustworthiness, and responsibility help guide us along a pathway to deal more effectively with ethical dilemmas by eliminating those behaviours that do not conform to our sense of right and wrong, our best rational interests without sacrificing others. As a student, following ethical behaviour is significant. It is the stage where humans mould their way of behaviour. Being ethical will help students to become better human beings.

During the pandemic, teachers and learners moved to online platforms where they lost the fear of surveillance [6]. E-learning platforms give an opportunity to question self-ethics as it leads the students to commit unethical behaviour and practices. Chances for unethical behaviours through E-learning are more than in traditional learning, which led students to behave more unethical in the E-learning platform.

The unethical behaviour of students can crucially influence their qualification, future employment and manners in their professional career [2].

One of the main unethical behaviours found in the E-learning platform is cheating in examinations with a false identification card, proxy behalf of the candidate, pre-writing the answers as the mistake there will not be any immediate repercussions for that behaviour. These situations show the lack of fear among the students for immediate consequences. As long as the students feel E-learning platforms do not provide efficient surveillance, they will continue their unethical behaviour and misuse the resources in order to obtain good grades. They cultivate a misunderstanding and believe that their lives will be cut short as a result of this. If they continue to practice this, they will fail in life, and this leads to failure caves. This will start affecting their personal lives. As they strive to cheat in their lives and in workplaces, they are actually cheating on themselves.

With the advent of virtual learning, this teaching angle was taken away from the students leading to a lack of respect for the teachers. The teacher is essentially a spiritual entity greeted with salutations meant for God and is the essence of Bliss [12]. Respect towards the teacher is one of the most important moral values for a student.

Alternative disruptions in E-learning are improper citation, paraphrasing in research papers case studies seeking help from friends, family, seeking help from paid services for writing assignments, browsing the Internet during online examinations for finding appropriate answers or proxy-writing of examinations, sharing passwords, disturbing classes by anonymous logins, multitasking while attending classes and intruding the privacy of others, etc. A study conducted by the American Council on Education [5] identified the following categories of E-learning practices: inappropriate guidance on examinations, misuse of references on papers and projects, writing help and support and other improper mentoring, false representation in data collection and reporting, inappropriate use of academic resources, disgracing others' work, breaches of computer ethics, and lack of compliance to copyright and copy protection, and lack of adherence to academic regulations. These things during the studies lead to the chaos of the goals of the study. In the long run, students will fail to acquire the intended skills and knowledge and will have an increased inclination towards adopting malpractices and immoral practices to attain their goals. This will lead students to apply malicious working methods that will cause damages to the employee's reputation as well as companies. Therefore, it is important for the teachers to teach and give awareness on how to be ethical in the E-learning platform and how to set up their own effective learning strategies.

When E-learning platforms are left unchecked in a similar way, in days to come when virtual learning becomes the new norm, the quality and efficiency of education would be undetermined. In the ever-increasing E-learning environment, unethical practices such as copying the works of others without their permission, pirating software, duplicating the personal assignments and submitting to multiple teachers, sending junk, spam and unethical materials to others, viewing immoral files, and visiting unethical websites have been on the rise [1]. Many teachers are of the opinion that any kind of fraudulent activity by participants of online courses should be dealt with proper punitive measures, including public display of such punishments [17]. Even in an online mode, understanding the learning patterns of the learners can help a lot in helping them maintain the ethical codes of learning [14, 16]. Online learning activities seemed to help students to use ethical theories to develop arguments which they did not hold themselves [7].

Almseidein and Mahasneh [1] recommend including a mandatory subject on ethical values as a seminar course within the student academic plan to curb the pandemic of unethical practices. Much before the widespread use of online classes, Bušíková and Melicheríková [5] noticed that academic fraud cases in daily programmes have fallen down over the years, whereas it has significantly increased in online programmes. Teachers experience trolling and bullying from students by taking photos of the teacher from personal accounts and posting them sarcastically with no responsibility [1].

The E-learning process is quite different from campus-based learning and provides greater opportunities for academic misrepresentation [4]. Holmström [8] suggests that irrespective of the type of education, virtues may be developed through education and practice rather than necessarily being innate. Muhammad et al. [11] believe that parents and community leaders should be involved in the ethical building mechanism for university students. The next topic will exemplify how unethical behaviours can be self-corrected and altered with a different approach. The framework provided below will help an individual to create a habit and keep track of their ethics.

5 Artificial Intelligence as a Self-trainer for Ethical Development

A pandemic is a situation that entails the entire society to go through significant changes in its core structure. The present COVID-19 crisis is an actual example of the same. One among the main changes thus occurring is the shifting and increased inclination towards adoption of online technologies. Contemporary examples of these are companies moving from traditional offices to work from home, educational institutions moving towards online classes, government offices switching towards online platforms, etc. These changes would have taken years of work if not for a pandemic situation. However, the question here is how these changes affect society's ethical foundations.

For example at the time when cars were invented, initially there would have been no ethical issues that happened during the time of its development. There would have been no traffic or overspeeding. However, once these technologies are accessible to ordinary people, the above-mentioned ethical issue would have become a significant concern to the society. Furthermore, there are rules and regulations which are considered to be the supporting pillars that uphold the ethical concerns which get involved while the people engage in various activity. If there is no external force overseeing them, then most probably they will stop following these ethical standards that need to be adhered to. In that case, it cannot be said that these rules and regulations are sustainable solutions for addressing ethical issues. It is believed that these pioneer supporting pillars did a great job in solving the ethical problems societies face or did they solve these issues?

The better way to define the conundrum will be to assume that these supporting pillars solved these problems only when the people realised that they were being watched. Still, these rules and regulations are just two words under the cover of a mask. This means that relying entirely on rules and regulations as the final solution for ethical problems faced by society is similar to the common belief that no crime will happen under the presence of the sun.

A good habit is one of the crucial pillars that help one grow and support the development of personal ethical standards. One of the most efficient ways to include these ethical structures and moral values into one's daily life is to find space for practising these structures on a regular basis until they turn into habits [3]. Thus, they will be followed even if there is no surveillance system. We have in this section put forward a framework denoted as the "Personal Ethical Standard Development (PESD) Framework" which aims at achieving the same.

5.1 Personal Ethical Standard Development Framework

To understand the PESD Framework, certain questions need to be addressed. Why is learning for hours boring and playing a game for hours, much more enjoyable? How can a person spend extended hours on entertainment platforms such as Netflix and Amazon Prime but unable to do the same on his/her studies? Furthermore, why is scrolling Instagram or Facebook pages more exciting than turning the pages of the books need to be studied. This shows that when it comes to entertainment platforms, hours of spending time are easy in comparison with working on an assignment or project for an hour, which makes one exhausted.

A thorough study could show that those boring works are forcefully done, and the exciting works are done due to the satisfaction that one derives from instant gratification of getting likes in the social media post, achievements in gaming, emotional satisfaction when seeing a film or having tasty fast foods rather than nutritious foods and so on. These are similar to obeying the rules when watched and disobeying them when not monitored. These instant gratifications have the potential to make such activities into habits in no time. These newfound habits can directly contradict one's personal ethical values, making them think that it is hard for them to do their responsibilities such as studies and it is much easier to do activities that provide instant gratification. As discussed above, an external force for enforcing ethical concerns is not a sustainable solution, which means the ethical part needs to be developed into a habit in such a way that it survives the normal cognitive extinction and persists throughout the life of that individual.

Digital Tools, Gamification Concepts and Artificial Intelligence-Monitored Systems are the primary pillars on which the PESD Framework is based. These three pillars will help a person develop new habits that gel along with his/her personal ethical standards. Once the habits are developed, it does not need to be checked or reinforced promptly.

5.1.1 Digital Tools

Digitals tools are day-to-day tools like Calendars, Reminders, Tasks, etc., which can be integrated with sandbox tools like Notion with the help of automation tools like Automate.io, Zapier, IFTTT, and so on. These sandbox tools will be configured in such a way that it incorporates the concept of gamification. These digital tools help gather and provide required information to and fro from the users at the right time. The speciality of the sandbox environment is that the users can develop their systems without even knowing sophisticated programming languages. These features allow every user to use such systems with basic computer knowledge. For example, software like Notion and Google Calendar can be linked with the aid of Zapier. The Notion is a typical example of a sandbox application where anyone with zero exposure to coding can customise the settings to create an environment to implement their day-to-day work and routine. Linking it with Google calendar or other similar applications would likely give a holistic overview of that person's day-to-day life. The main highlight of such software is that it can be customised and used in a gamified way. One could keep daily goals that need to be attained and assess the extent to which these goals have been achieved without the individual having any advanced knowledge of coding.

5.1.2 Gamification

There are two types of motivations, Extrinsic Motivations and Intrinsic Motivation. Extrinsic motivations are driven by rewards, for example: getting likes on Facebook or Instagram, which encourages posting more content. In comparison, intrinsic motivations are driven by satisfaction or enjoyment of doing an activity without rewards, for example: engaging in hobbies or professions just because it feels good. Extrinsic motivators are typically employed to facilitate the enactment of something that is not interesting. However, reward value diminishes over time, which means much bigger rewards are required in order to keep motivated. In the case of Intrinsic Motivation, the person will do an activity if and only if he/she is interested in it. This is the reason why both types of motivation are very much required in a balanced way to accomplish day-to-day activities. Game developers make sure that they have included all the required elements that will trigger both these motivations. This is why games are fun and much more enjoyable than other day-to-day activities. This is exactly the same kind of thinking behind Gamification. Gamification is the process of implementing gaming concepts in real-life scenarios. Unknowingly, most people are used to these gamification concepts; for example, different enterprises are using the gamification concept to engage both their customers and employees. They engage

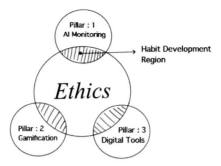

Fig. 3 PESD framework, which consists of three pillars 1—artificial intelligence monitoring system, 2—gamification concept, 3—digital tools. These pillars assist in transforming ethical standards into habits, making the user follow their ethics even in the absence of surveillance

customers through incentives, reward programmes, leader boards, badges, points, etc. These elements motivate people to complete activities, one of the examples is getting back a reward when doing a bank transaction through different UPI applications like GPay, Paytm, and Phonepe. Sailer et al. [13] show how gamification motivates an individual positively and has an impact on the individual (Fig. 3).

These are a few principles that can be used to gamify our day-to-day life:

1. Well-defined **Objectives and Rules** that show why and how to do a task.

   Boring tasks can be converted to challenging and exciting tasks by clearly defined rules and objectives. Rules are the primary element that increases the difficulty of a game, for example, when it comes to basketball, if there are no rules mentioned, the player can just take the ball and drop it in the basket using a ladder. It is the rules which make the game more challenging.

2. Implement **Quests/Questioners** to ensure that the individual is doing the right thing.

   This can be considered as the constant reminders or small milestones that need to be completed or answered while doing the actual tasks. For example, while doing the work, the following questions or quests can be considered for the self-audit.

   a. Not using the social media apps while working for 30 days. (Quest)
   b. Only open the required tabs while browsing. (Quest)
   c. To be punctual on time for 30 days. (Quest)
   d. Has one procrastinated anything today? (Question)
   e. Is one involved in a time-consuming conversation? (Question)
   f. Is one forwarding unwanted materials? (Question)

3. Get **Rewards** for each successful completion of tasks and quests.

   Getting rewards is the crucial component of building new habits, and if the reward is quite exciting, then the chances of doing the tasks again are much higher. For instance, if having sweets is more enjoyable than consuming red chillies, our brain tends to crave more sweets again. If this concept can be implemented in the day-to-day responsibilities, getting exciting rewards when doing the duties, then the brain tends to do similar activities again. Once it becomes a habit, then there is no need to get the reward again to do that task.

4. Evaluate the **Feedback** in each cycle to correct the trajectory.

This is the crucial principle through which the players find an efficient way to win the game. For example, if a child touches the fire unknowingly while doing a task, his/her hands get hurt as the feedback of touching the fire and next time they would avoid the fire or use some methods which will help them not to get hurt from the fire while doing their tasks. If they did not receive the feedback, then getting injured from the fire is much higher.

5. Sharing the **Achievements** in the related community pages.

Publishing the achievements will give extrinsic motivation to the individual and encourage him/her to accomplish more tasks.

The gamification concept can be implemented in one's life by following the principles mentioned above using the available digital tools. Though these tools are interconnected, the overall progress can be measured easily, and the reward system can be activated, which encourages the person to do it again until it becomes a habit. Some examples of how these digital tools could help implement these principles are given below.

- Set Calendar pop-ups for regular intervals during the day.
- Set Calendar pop-ups for shorter regular intervals during the night.
- Daily/Weekly/Monthly Self-Audit on Screen Activities.
- Use digital services to restrict social media app usage during work/study time, etc.

5.1.3 Artificial Intelligence-Monitored System

When a person is supposed to do any new activity other than his/her interest list, they will perform the tasks based on their extrinsic motivation. The biggest problem here of extrinsic motivation is that it is not driven by personal interest or not done because it feels good. The activity that is accomplished is only in expecting a reward as it may not last long. For example, if a person gets two dollars for doing some activities, the person would not be interested in the same rewards after a certain point. This means the rewards need to be increased to keep him/her motivated to do the same activity. What if the same individual increases the rewards multiple times for performing the same action. This will nullify the fundamental aim of gamification daily activities. Perhaps after a few days, he/she will get bored of the entire gamification process.

A bored process can be made more exciting when an external entity manages the factors like rewards and feedback systems; for instance, in the game of football, players get trained with the help of their coach. The coach helps the players to maintain discipline and train the player during the training period. When it comes to a match or tournament, since the players have developed the habit of playing according to the climate, they could perform well in the tournaments. This shows the habits have been formed already by the coach, and he knows how and when to motivate the players because he continuously monitors, motivates, appreciates, and creates a disciplined team. Similarly, suppose there is an external entity to monitor and manage the complete gamification process and the individual's day-to-day activities. In that case, the entity could easily make sure the individual is motivated the entire time and stays on the right track. In the PESD Framework, this external entity will be an Artificial Intelligence-Monitored System. The AI Monitoring System will assign the tasks for a certain time span till it forms into a habit. The individual will receive exciting rewards initially to gear up the process; however, the AI makes sure the reward does not become a practice.

In the PESD Framework, available digital tools need to be interconnected among each other. It plays a vital role when it comes to configuring these tools to incorporate gamification concepts and provide a platform for an individual to utilise and abide by the ethical value. The AI-Monitored System collects the data from the digital tools and the available sensors such as camera, microphone, and ambient light sensors. The collected data will examine to ensure that the individual sticks to the track. For instance, it will be assigned in the schedule for the individual to wake up in the morning and to organise the bed. If that activity is not followed, AI (in this case the concerned apps) will detect the failure to keep up with the schedule through various data such as how long it took to switch off the alarm or how many times was the alarm snoozed and accordingly motivate the individual to perform the activity on time until it becomes a habit. Under this loop, the individual has to constantly see their advancements to attain their daily goals and correct the trajectory if necessary. This can be to some extent rectified, and this entire system can be made free from first-hand human interference by creating a skeleton surveillance system that works at the backdrop and never shows its presence and is activated only when the individual does not seem to meet their daily levels of goals or if the individual strays off the path he/she has set for that day. This can be attained by keeping an eye on the statistics provided by the customised digital tools used by the person.

Once the system arrives at a conclusion, if it is not up to the expectations, it can provide checks such as reminders that they are not living up to their expectations or wasting away their future, for such behaviours. With its constant reminder, the person will probably be instigated to rectify the wayward behaviour and get back to the practice. This shall continue until it becomes a habit for that individual. Once the system decides that the behaviour has been sufficiently reinforced by keeping track of the frequency and stats of desired behaviour happening in a given period of time, it will eliminate the process. To achieve the transformation of one's ethical standard to a habit, the framework will make a person undergo a habit development loop.

5.2 Habit Development Loop

The procedure described in the diagram can be utilised to mould new habits on the basis of the ethical foundations that need to be taken care of while enlisting into a new job. For example, when a new employee is recruited into a firm where the work atmosphere and the ethics involved are alien to him/her, this loop can help them get well versed in the prevalent ethics and moral code that company or organisation uses. Thus, this loop described above can be used to instil ethics and required moral code within a person with the aid of the available digital tools.

The inner loop uses various simple tasks in the direction of the habit that needs to be instilled by gamifying the same into various quests. A person needs to complete them before proceeding into next levels. This finally leads to reaping awards in the form of small appreciations or encouragement when he/she attains the relative level of advancements in changing or adapting an ethical or moral value into his/her behaviour. This shall be continued until it finally becomes a reinforced habit. The gamification principles along with the digital tools (various apps, etc.) which were discussed above will be incorporated into this loop, in order to reinforce the habit which is being formed. Firstly, the objective of the habit as well as the rules on how to develop the habit will be defined. Secondly, quests and questions that align with the habit will be created. Thirdly, rewards will be shared with the individual whenever he/she completes the tasks and quests. Finally, the entire process will be reviewed from the feedback of the system and then decide whether to shift to a new habit or stick with the current habit. Developing a new habit will take multiple cycles in the inner loop.

The outer loop acts as a surveillance system that tests the person's progress on a routine basis. This is by asking questions or conducting short tests to assess the progress of the individual in the right direction. This loop shall be governed and evaluated by artificial intelligence and free of human involvement. Here, the system can be considered as a personalised mentor. There will be two types of reviews happening in this loop. First, the system review will review the entire gamification system to make sure that it is perfectly aligned to contribute to the habit development of each individual. Second, ethical review: this review will make sure that the developing habits are aligned with the personal ethical standard of the person. If any changes are required from the ethical or system perspective, the AI will automatically incorporate those changes into the system.

When the individual successfully completes the loop in a definite manner and finally incorporates the ethical value thus learned into his daily life, it becomes a reinforced habit that has an increased chance of being followed even when there is no surveillance overseeing the enactment of the same (Fig. 4).

6 Conclusion

One of the significant problems and challenges modern universities and educational institutions face in today's world especially in this pandemic situation is ensuring the moral and ethical development of their students more than the sole development of their educational standards. This becomes a serious issue, especially in E-learning platforms. When these platforms were introduced, they provided opportunities for

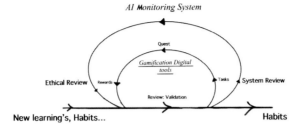

Fig. 4 Habit development loop: the core part of PESD framework is to create a habit that will sustain lifelong to be ethical. The first loop helps one to create a habit in daily life using gamification concepts. The second loop helps in monitoring the development of habits using AI

students to learn and skill up themselves. However, on the contrary, it comes with the burden of academic-cheating and related ethical concerns. When learning comes to an online platform, students tend to feel it is less personal as one cannot see or touch. This is where one tends to break their ethical codes. It is in lieu with this fact that the concept and features of the panopticon can be used in everyday life by parents and teachers to ensure the strengthening of the moral and ethical foundations of their children and students, respectively.

Early patterns of unethical behaviour in E-learning platforms which the supervisor (Parent/teacher) has to be in lookout for such as inappropriate guidance on examinations, misuse of references on papers and projects, writing help and support and other improper mentoring, false representation in data collection and reporting. This chapter summarises how unethical behaviour is found with suitable examples in E-learning, and with the help of an Artificial Intelligence-monitored PESD Framework and HD Loop, the situation could be improved. The daily activity done by any ordinary individual shows that in self-ethics can be nurtured and thus rooted with incorporating good habits.

References

1. Almseidein T, Mahasneh O (2020) Awareness of ethical issues when using an e-learning system. Int J Adv Comput Sci Appl 11(1):128–131
2. Anderson B, Simpson M (2007) Ethical issues in online education. Open Learn J Open Distance eLearn 22(2):129–138
3. Ayasrah MN, Yahyaa SM, Al-Mahasneh OM (2020) Educational values included in story collections for fourth and fifth grade students in Jordan. Univ J Educ Res 8(11B):5859–5868
4. Brown T (2008) Ethics in eLearning. Rev Educação Cogeime 17(32/33):211–216
5. Bušíková A, Melicheríková Z (2013) Ethics in e-learning. In: International association for development of the information society international conference e-learning, pp 435–438
6. Fox S (2021) Optimal class size for online education in Pasifika contexts. Micronesian Educ 30:69–72
7. Higgs A (2012) E-learning, ethics and 'non-traditional' students: space to think aloud. Ethics Soc Welfare 6(4):386–402

8. Holmström C (2014) Suitability for professional practice: assessing and developing moral character in social work education. Soc Work Educ 33(4):451–468

9. Hutton JH (1928) The significance of head-hunting in Assam. J R Anthropol Inst Great Br Irel 58:399–408

10. Isa PM, Samah SAA, Jusoff K (2008) Inculcating values and ethics in higher education e-learning drive: UiTM i-learn user policy. In: World academy of science, engineering and technology conference, pp 452–455

11. Muhammad A, Ghalib MFMD, Ahmad F, Naveed QN, Shah A (2016) A study to investigate state of ethical development in e-learning. J Adv Comput Sci Appl 7(4):284–290

12. Prakasha GS, Jayamma HR (2012) Professional ethics of teachers in educational institutions. Artha J Soc Sci 11(4):25

13. Sailer M, Hense JU, Mayr SK, Mandl H (2017) How gamification motivates: an experimental study of the effects of specific game design elements on psychological need satisfaction. Comput Hum Behav 69:371–380

14. Salhab R, Hashaikeh S, Najjar E, Wahbeh D, Affouneh S (2021) A proposed ethics code for online learning during crisis. Int J Emerg Technol Learn 16(20):238–254

15. Sargiacomo M (2009) Michel Foucault, Discipline and punish: the birth of the prison. J Manag Gov 13:269. https://doi.org/10.1007/s10997-008-9080-7

16. Tawarah HM, Mahasneh OM (2020) Learning patterns for the students of Shoubak University College. Univers J Educ Res 8(10):4700–4706

17. Toprak E, Ozkanal B, Aydin S, Kaya S (2010) Ethics in e-learning. Turk Online J Educ Technol 9(2):78–86

The Horizontal Handover Mechanism Using IEEE 802.16 E Standard

Fahmina Taranum

Abstract Mobility management protocols deal with location and handover management, which are designed to cater to the need of an increasing demand of mobile devices, mobile networks, and roaming across different standards. Overabundance mobility management protocols have been proposed to handle complex issues related to connectivity and compatibility viz. IEEE 802.21 (medium-independent handover) to create a heterogeneous network along with IEEE 802.11 and IEEE 802.16 d for fixed wireless network. The aim of the article is to focus on a scenario with the implementation of high mobility in multi-cell environment when a subscriber station hands over from one base station to another, depending on the available signal quality. The purpose is to maintain connectivity in mobility, for which different threshold values have been experimented on receiver signal strength. The idea is to design a heterogeneous network or an enhanced version of a 3G network. The performance metrics used for analysis include throughput, which was calculated using receiver signal strength for different data rates, average jitter, and end-to-end delay. The purpose of the proposal is to discuss the communication difficulties and issues related to 3G plus generation devices, and its solutions for mobile adhoc networks are proposed in the article.

Keywords BS—Base station · MS—Mobile station · RSS—Received signal strength · 802.16e · WiMAX

1 Introduction

1.1 Technological Requirement

As the technology is advancing, heterogeneous wireless networks are becoming more dominant; therefore, the integration and interchangeability of various wireless access

F. Taranum (✉)
Muffakham Jah College of Engineering and Technology, Osmania University, Telangana, India
e-mail: ftaranum@mjcollege.ac.in

© The Author(s), under exclusive license to Springer Nature Singapore Pte Ltd. 2023
M. A. Chaurasia and C.-F. Juang (eds.), *Emerging IT/ICT and AI Technologies Affecting Society*, Lecture Notes in Networks and Systems 478,
https://doi.org/10.1007/978-981-19-2940-3_21

networks need to be addressed. Adhoc networks are collaborated with WLAN and WiMAX architecture to create an economically viable solution for wide deployment of high speed, scalable, and ubiquitous wireless Internet services. The proposition is to design an inter-working architecture of wireless mesh with horizontal handover. In today's distributed environment and changing technologies, there are certain issues, which need to be taken care of viz-a-viz operability of 4G+ devices and its incompatibility, requirement for the handover of devices with different architectures, challenges of adhoc networks and its solutions. With the simulation results, the performance of the handover can be analyzed using metrics like throughput, average end-to-end delay, and average jitter for uni-cast transmission. Comparative analysis is made for homogeneous architecture using different protocols to understand the requirement of mobility and handover.

1.2 Emerging Demands

With the enhancement in the current communication technologies for infrastructure of the adhoc network, the importance is given to handle issues encountered during mobility. Various mobility protocols have been designed to handle multiple problems in the overall mobility environment. The WiMAX forum focus is on IEEE 802.16 standard to promote and certify interoperability and compatibility of broadband wireless products for large network. In general, WiMAX is designed for transmission of multimedia services at high-data rates. These standards are targeted for line-of-sight channel conditions and work in a spectrum of 10–66 GHz with a bit rate of 32–134 Mbps.

1.3 IEEE Architectures

The primary task of the IEEE 802.16 WiMAX MAC layer standard is to provide an interface between the higher transport layer and lower physical layer. The 802.16 MAC is designed for point–multi-point communication based on collision-sense multiple access transmission with collision avoidance (CSMA/CA). The IEEE 802.16 is the most flexible and reconfigurable selection across a large area of frequency and bandwidth, which is our reason for using it, inferred from the characteristics listed out in Figure 1. The IEEE 802.16 PHY specification uses OFDM and specific modulation and encoding combinations. The standard IEEE 802.16e (mobile WiMax) is released for nomadic and mobile use for handover.

The handover mechanism is used to show how the handing over occurs among the cells. The same is shown in Fig. 2. The designed scenario works on high mobility nodes in heterogeneous environments having multiple substations and performing multiple handovers using IEEE 802.16e protocol. These handover decisions are based on radio signal, QOS, security support, economic cost, and user personal preferences.

Protocol	Multiplexing methods þr Radio technique	Channel access	Radio range	Network	Band	Bit rate	Band width efficiency	Deployment
802.11	OFDM,DSSS	DCF (CSMA/CA), PCF	30-100 meters	LAN<100mi	2.4GHz	1 or 2Mb/s	2.7M b/s /Hz	Fast delivery, simultaneous communication
802.16	QPSK,16QAM,64QAM,S OFDMA	TDMA(TDD way),FDD	Upto50km	MAN 1-3 mi	10-66GHz	32-134 Mb/s	5Mb/s Hz	Fast and cost effective
802.16e	OFDMA, 64 QAM+BPSK, 16QAM+QPSK	TDD,FDD(FULL DUPLEX /HALF DUPLEX)	10 km	MAN, 1-3 mi	<6 GHz	Up to 15Mb/s	5Mb/s /Hz	802.16e defines a series of sleep and idle mode to enable power conservation and preserve battery life

Fig. 1 Characteristics of IEEE architectures

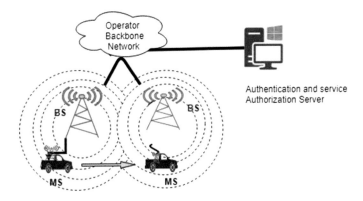

Fig. 2 A handover of BS for MS

A valuable motive for handover is channel quality, which is measured using downlink power control process, adaptive modulation, and adaptive code rate. These measurements are received signal strength indicator (RSSI) and signal-to-interference plus noise ratio (SINR), and they are transmitted to the BS by the SS.

They occur if the RSS threshold of the current base station is less, following which the subscriber station scans and gets connected to the closest neighboring base station with the highest RSS value. The recent enhancements in signal processing techniques have led to a faster and efficient mode of communication. In today's smart environment, people with smart devices (nodes) can freely organize and configure scenarios to receive and send packets to destination over multiple nodes via intermediate connectives in different networks. An attempt is made to design a heterogeneous and compatible network for enhanced communication along with handoff mechanism. Handover is the ability of the subscriber station to retain the connectivity in mobility, while transiting from one radio channel to another. It enables the

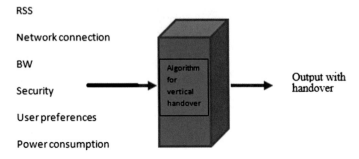

Fig. 3 Components used in handover

network to operate at low power levels, thereby providing high capacity. Handoff management enables the network to maintain the ongoing connection when a mobile terminal switches its access point or base station. Handover is the key technology in the research of the next generation of wireless mobile communication systems.

A random waypoint is to show the path of mobility. In the *random waypoint mobility* model, a destination node is randomly selected to move to the destination at a constant speed which is again, randomly selected. At the destination, the node pauses for a given length of time after which the process is repeated throughout the simulation for the specified path. The mobility pattern of a mobile object (such as a device, satellite, or weather pattern) can be specified by setting waypoints. The speed between those waypoints is determined by their locations and time. The prerequisites of handover are defined in Fig. 2 (Fig. 3).

Handover mechanism aims at reducing the changeover and dis-connectivity when the devices are in mobility and requires changing the base station because of low receiver signal strength (RSS). The handover decisions are based on radio signal, QOS, security support, economic cost, and user personal preferences. Both IEEE 802.11 and 802.16 have integrated media-independent handover (MIH) functionality in the MAC layer. Handover mechanism aims at reducing the changeover and dis-connectivity, when the devices are in mobility and requires changing the base station because of low receiver signal strength (RSS). Other types of IEEE standards can also be used in order to perform changing of the station because of RSS or signal-to-noise ratio.

1.4 Mobility Management Protocols

Mobility management protocols deal with location and handover management, which are designed to cater to the need of an increasing demand of mobile devices, mobile networks, and roaming across different standards. We do so to reduce dis-connectivity issues when devices are in mobility. Overabundance mobility management protocols have been proposed to handle complex issues related to connectivity and compatibility, viz. IEEE 802.21 (medium-independent handover) to create a heterogeneous network along with IEEE 802.16a and IEEE 802.16d for fixed wireless network. This paper focuses on a scenario with the implementation of high mobility in multi-cell environment when a subscriber station hands over from one base station to another, depending on the available signal quality.

The purpose is to maintain connectivity in mobility, for which different threshold values have been experimented on a received signal strength. The idea is to design a heterogeneous network or an enhanced version of a 3G network.

The chapter is organized with the subsections as follows: Sect. 1 portrays on introduction, Sect. 2 discusses the literature survey, Sect. 3 is about the proposed system, Sect. 4 discusses the results followed with conclusion.

2 Related Work

Author Nikhat in [1] has proposed a horizontal handover for IEEE architectures. The proposal is to check the efficiency using IEEE 802.11 and IEEE 802.16 architectures. Author has shown the area of development as handing over from IEEE 802.16 to 802.11 or vice versa. The paper [2] by Fahmina Taranum is to discuss the various modes of transmission in node mobility. The modes of power consumption are experimented on different types of batteries extended with the calculation of residual energy for mobile nodes.

The approach defined by author Cai l in [3] reduces network energy consumption and enhances the lifetime of wireless sensor networks. The author has improved the cluster-head node selection in LEACH. The cluster-head node may be far from the base station, and the remaining energy may be poorly distributed in the system, leading to less time to live value of the node.

In paper [4], Aleemuddin et al. have used a relay node in between base and mobile stations to check the power consumption in different modes. The proposal was extended on different routing protocols to check the most energy-efficient adhoc routing protocol.

Similarly, [5] Mandeep Kaur is about mobility management for all IP mobile networks. These papers were used as a reference to existing strategies and implemented a few of them toward the enhancement of available approaches.

The idea to work on the hop concept for wide spreading of signal in the network using CDMA is adapted from Jain [6], in which single hop relay network has been modeled, and its battery consumption performance has been analyzed. The experimental analysis shows that there is improvement in coverage of existing cells and reception quality along with the concept of relay, owing to reduced average power consumption. The proposed system has taken an initiative to implement multi-hop concept and its comparison with other systems. Venkateswaran and Sarangan [7] work on the concept of showing the dependency of relay node on underlying network layers.

In Nguyen et al. [8] specified the use of multi-hop relay networks to improvise the range of the existent cells and analyze the power saving obtained but have not been able to shed enough light on highly mobile environments. In Jha and Dalal [9], a discussion on a comparative study of homogeneous and heterogeneous mobile device adhoc networks has drawn attention to integrate the network for our proposal.

The concept of saving energy becomes an important factor in the algorithm designed by De Rango et al. in [10]. Chandwani et al. [11] talk about how data forwarding function eliminates data loss during VHO execution and resolves abrupt disconnection to source network. The Author Hu et al. in [12] have implemented Mobility using IEEE 802.21 in a heterogeneous IEEE 802.16/802.11-based IMT advanced (4G) network. In [12], the authors Jing Hu, Xiaoxiang Wang, Hongtao Zhang provide a location-aware relay station selection scheme in the opportunistic relay communication technique to help reduce the number of contending nodes. The relay station selection mechanism involves timer logic, which depends on channel state information (CSI) of both source-to-relay and destination-to-relay channels.

New selection method to enhance the lifetime of network is proposed in [13] Song W., Chung J., Lee D., Lim C., Choi S., and Yeoum in which the energy consumed between nodes in the clusters and the power consumed for transferring in between the cluster head and the base station are considered.

In [14], Takeo Ohseki, Naoki Fuke, Hiroyasu Ishikawa, and Yoshio Takeuchi have used multi-hop networks to counter the dead spots in urban areas. The usage of fixed relay station helps in concluding the reduction of signaling overhead, which has been utilized in this paper.

The idea of using relay nodes for storing the transmission in between nodes is extracted from Tang and Mc Kinley [15]. In a nutshell, relay nodes are used for minimizing power consumption in adhoc and sensor networks. The basics of cooperative communications in wireless networks are targeted with a class of multi-hop relay networks that are meant to utilize the spatial diversity of the network along with the advantages of improving spatial diversity utilization, increased coverage area, and reduced average power consumption in [16] by Aria Nosratinia, Todd E. Hunter, and also mentions how certain relay station algorithms such as decode-forward and amplify-forward behave in a cooperative environment.

In paper [17], Kolio Ivanov, Gustav have proposed a handover management in integrated WLAN and mobile WiMAX networks using MIH to provide seamless handover. Peppino et al. in [10] along with the effect of multi-hop relay networks can be a lucrative opportunity to actually improve coverage of the existing cells as well as reception quality demonstrated in Ivanov and Gustav [17].

3   Proposed System

The configuration consists of a homogenous network of IEEE 802.11 with a handover of access points, a homogenous network of 802.16 with a handover from one base station to other demonstrating horizontal handovers, and finally, a heterogeneous network for handover from access point to base station for vertical handover is taken as an enhancement. The handover decision algorithms, combined with admission control can guarantee QoS support to the existing traffic flows in WLAN, by transferring new calls to the other network whenever necessary providing an un-interrupted communication with interoperability. The aim of the proposal is to manage and maintain the connectivity of a subscriber station in mobility. The architecture is designed using mobile nodes, correspondent node, Ethernet, and a base station which are connected via subnet with different radio frequencies, using point to multi-point connections. The CBR traffic is activated from the base station to the mobile stations and traffic flows from base station to subscriber stations. The hardware configuration is listed in Table 1 and Fig. 4 shows the architecture.

a. Horizontal handovers are accomplished when a mobile node moves using waypoint path and gets connected to the base station with the strongest signal strength. Proposed IEEE 802.11 Handover characteristics and prerequisites:

In 802.11, the data exchange is initiated by associating an access point with a wireless host. The access point is located by the host using active and passive scanning. In active mode, the host transmits a probe request frame, and access point responds to it with 'probe response frames,' whereas in passive mode, host locates access points through 'broadcast beacon messages' sent via access points. The aim is to demonstrate handover mechanism between both APs including beacons, active scanning, authentication, and association processes. After discovering available APs, host decides which AP to join, thereby initiating an authentication and association of frame exchange. The agreement on encryption type is carried out in association phase. The final phase is to grant host an access to the network for beginning the data exchange process. Once the host is handed over to other AP using authentication and association process, it retains this connectivity till APs depart. Since base stations are not available in 802.11, access points are used for communication with respect

Parameter	Value
Simulator	Qualnet 7.4
Terrian size	1500 * 1500 m in ($X * Y$)
Routing algorithm	Bellman–Ford for subnets and AODV
Seed	1
Mobility	Random waypoint
Packet reception model	PHY Model
Application traffic	
Simulation time	CBR
Handover-RSS trigger	240 s
Neighbor BS scanning	−78 dBm
RSS trigger	−76 dBm

Table 1 Parameters used in the design

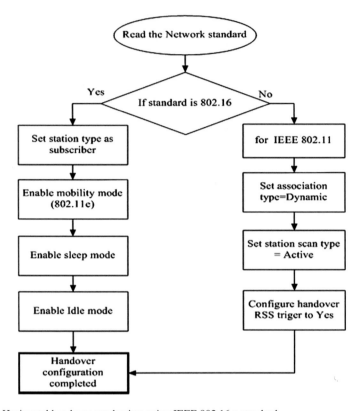

Fig. 4 Horizontal handover mechanism using IEEE 802.16 e standard

to subscriber stations. The 'scan type' needs to be activated to set the scan-channel-time, so as the station waits for the probe response on the channel for a maximum of 1024 micro-seconds (time unit) and the handover receiver signal strength can be activated. In 802.11 b, the threshold is estimated to be a sum of signal strength and RSS margin. If the RSS of the serving base station is less than the threshold, then STA needs to scan and reassemble for a new access point. It selects the route with the closest BS in the random waypoint and then performs a handover to the next closest BS. This continues till it comes back to the first BS.

b. The algorithm used to manage handoff in mobile WiMax is defined below

Algorithm for handing over Base Station in WiMax
Step 1: Cell Reselection.

i. Neighbor BS information advertisement: The information broadcasted by BS for neighboring BS is used by MS to proceed with BS scanning
ii. MS performs scanning and association with one or more neighboring BSs in order to determine its suitability as target BS.

Step 2: Handoff decision and initiation.

i.   The MS sends a MOB_MSHO-REQ to a BS indicating one or more BSs as handoff target
ii.   The BS responds by indicating the suitable BS to be used for the handoff by a MOB_MSHO-RSP message.
iii. The MS sends a MOB_MSHO-IND indicating the BS selected for the handoff

from the ones specified in the MOB_MSHO-RSP.
Or this process can also be initialized by the BS.

i.   BS initiates sending a MOB_BSHO-REQ to the MS indicating one or more BSs for the handoff target.
ii.   The MS responds by indicating its choice through a MOB_BSHO-IND.
iii. The BS performs paging (MOB-PAG-ADV) to reach an inactive MS (i.e., idle or sleep mode)
iv.   BS defines a handover-RSS-trigger and handover-RSS-margin to initiate handover

Step 3: Synchronization to the target BS, after the target BS determination.

i.   MS analyzes the DL frame preamble to obtain time and frequency synchronization
ii.   it obtains information about the ranging channel by analyzing the DL-MAP, DCD, and
iii. UL-MAP

Step 4: Ranging with target BS-The MS synchronizes its UL transmission with the BS. Maintains the QoS of link between SS and BS.

Step 5: Termination of context with previous BS-After the establishment of the connection with the target BS, the MS terminates the connection with the serving BS, sending a MOB_HO-IND message to the BS.

Step 6: Authentication and service authorization (ASA) server: which helps in providing access control to MS's.

Step 7: Backbone functionalities: BS uses backbone to communicate with each other for services such as network/BS-assisted handovers.

4   Result Analysis and Discussions

The original IEEE 802.16 specification is to provide high-data rate, point to multipoint communication, and line-of-sight (LOS) conditions between fixed locations. A high-data rate can be achieved using large constellations such 64-QAM and less robust error codes like 3/4 convolutional. The transmitter has to know the channel signal-to-interference plus noise ratio (SINR) to determine the optimum modulation and transmit power. The different data rates depend on the channel bandwidth, modulation, and code rate that are presented. The received data rate is calculated using data rate formulae; i.e., the system performance degrades when a handover occurs. The FFT size determines the number of available subcarriers and OFDM symbol duration. In general, for a given bandwidth, a larger FFT size results in a greater number of available subcarriers and a longer OFDM symbol duration. The achievable data rate 'R' is calculated using the following parameters for a simulation time of 500 s, along with other formulas as listed below.

$$R = (N * b * c)/T \tag{1}$$

$$Fs = \text{floor}\left(\text{sampling factor} * \frac{\text{bandwidth}}{8000}\right) * 8000 \tag{2}$$

$$\text{Tb} = \frac{\text{FFT}}{\text{Fs}} \tag{3}$$

$$\text{Tg} = G * \text{Tb} \tag{4}$$

$$T = \text{Tb} + \text{Tg} \tag{5}$$

Number of subcarriers = 1440, C = coding rate = 1/2, 2/3 or 3/4 sampling factor = 8/7 for OFDMA, G = Cyclic prefix = 4,8,16 and 32, b = bits per modulation for QAM = 6, Channel Bandwidth = 20 MHz, FFT = 2048.Using the above equations, the data rate achieved are 46,733, 45,370, 42857, and 38,570 Mbps for a cyclic prefix value of 32, 16, 8, and 4, respectively. Tb and Tg are in microseconds, whereas in Wi-Fi the data rates are 1, 2, 5.5, and 11 Kbps.

The client and server are connected by CBR traffic generator to generate traffic at a constant rate by transmitting packets of a fixed size at a fixed rate using different precedence. The format of CBR represents source destination items-to-send item-size interval start-time end-time precedence value.

In general, a total of six handovers are performed in the complete simulation of 240 s. All the BSs are connected to a correspondent node − node 25 via wired point-to-point link as shown in Fig. 5.

In our findings, we observed that the addition of IEEE 802.16 e increases the mobility support for the nodes. There is the possibility of an intra-base station handover in 802.16 architecture. The statistics obtained at the subnet level are shown in Table 2, assuming data rate is inversely propositional to radio range. The parameters set are Channel Index = 0, Frequency = 2400.000000 MHz, Transmitting Node = 2, Interface Index = 0, Receiving Node = 1, and Interface Index = 0.

Figure 6 gives details of signal transmitted at the PHY and MAC layer for 802.16. All the signals received and forwarded to MAC are locked by PHY layer. The number of uplink and downlink mobile application messages received by MAC layer is same.

The simulation results show that mobile WiMAX outperforms in many parameters with respect to fixed WiMax. Therefore, the integration of these two technologies can benefit WiMAX operators through a large coverage area to avoid frequent disruptions

Fig. 5 Architecture with one relay node

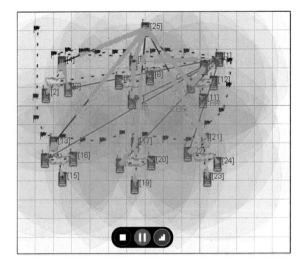

Table 2 Data rate verses radio range

Radio range (m)	Data rate (Mbps)
421.41	14.28
298.33	21.43
223.72	28.57
149.52	42.85
118.77	42.85
94.34	57

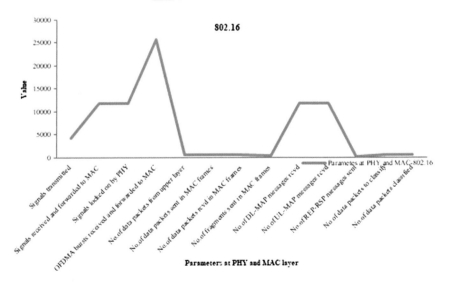

Fig. 6 Signal transmission at physical and MAC layer

due to handover with high mobility.

$$\text{Average End to End delay} = \frac{\text{transmission delay}}{\text{no of packets received}} \tag{6}$$

$$\text{Average Jitter} = \frac{\text{Total packet jitter for all received packets}}{\text{no of packets received - 1}} \tag{7}$$

The data rate is derived using Eqs. (1) to (5) with varying values of RSS ranging from −80, −85, and −90 dBm to calculate the throughput as depicted in Fig. 7. The analysis interprets that the throughput is linearly propositional to data rate, which in turn increases with the increase in RSS value. Figure 8 depicts the results generated when all three scenarios have been experimented at network layer.

The number of packets sent as source using AODV, IPV4 protocol is more when compared to the one received at network layer by node 1. The number of control bytes

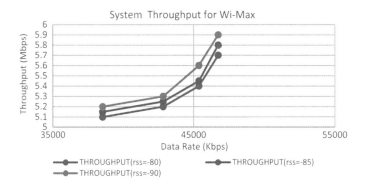

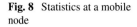

Fig. 7 System throughput with varying RSS value

Fig. 8 Statistics at a mobile node

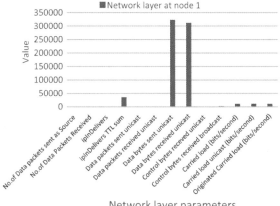

Fig. 9 Statistics at an application layer

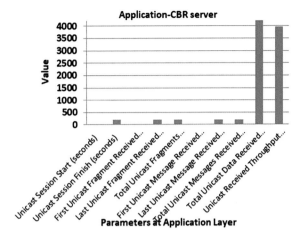

received by broadcast is almost double when compared to unicast. The data bytes sent and received by unicast are almost equal. The application CBR traffic of the server gives the details of unicast session, fragment, data, messages, and throughput. It can be concluded from the analysis that the throughput measured in terms of bits/second is quite high at the application layer as depicted in Fig. 9.

Simulation results show that WiMAX outperforms in many parameters with respect to Wi-Fi, and therefore, the integration of these two technologies can benefit WiMAX operators through a low-cost service deployment provided by Wi-Fi. In the designed scenario, if the mobile node moves at a high speed using 802.11 e or 802.16 e, then it is preferred to continue in a larger coverage area to avoid frequent disruptions due to handover, while moving across smaller WLAN coverage. In our findings, it is observed that the addition of IEEE 802.11e or IEEE 802.16e increases the mobility support for the nodes. There is a possibility of intra-base station handovers in 802.16 architecture.

The vertical handover occurs when there is a difference in the signal strength and reception level, i.e., when the connectivity is created from WLAN to WiMAX is vertical handover. The horizontal handover between 802.11 and 802.16 wireless access networks is investigated.

Figure 10 signifies that the handover and throughout are inversely propositional to each other, that is the system performance degrades when a handover occurs. As the data rate increases the number of handover increases. The graph is plotted for the mobile device which handoff from one base station to another.

5 Conclusion

The purpose of the experimentation is to measure the efficiency of IEEE architectures for the mobile adhoc networks. With the utilization of high mobility protocol of

Fig. 10 Throughput and
handovers in WiMax

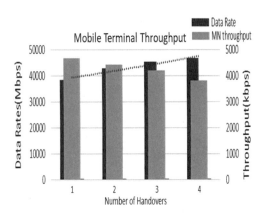

IEEE 802.16e, we observed that the data rate is far better when compared to the one obtained in [14] for different RSS values. Further, the relay nodes can also be connected by using a subnet or Internet rather than using the point–multi-point link for upgrading the level of capturing the transmission. The handover decision algorithms combined with admission control can guarantee QoS support to the existing traffic flows by transferring new calls to the other networks whenever necessary, providing an un-interrupted communication with interoperability.

Future Enhancement

The future enhancement can be done by using different energy models or a different version of 802.16 with an enhanced routing algorithm. The integration of Wi-Fi and WiMax to create a 3G plus heterogeneous network is also an enhancement. Furthermore, the aim is to explore MIH support for multi-hop heterogeneous networks, resource allocation, and appropriate routing algorithm that can be applied for generating optimal results with different IP protocols.

References

1. Taranum AF, Khan BK, Nikhat CR (2020) Handover management using IEEE 802.11 and IEEE 802.16 Standards in MANETs. SAMRIDDHI 12
2. Taranum F, Khan KUR (2019) Power management strategies in MANETs—a review. Int J Recent Technol Eng
3. Cai1 X (2019) Optimal LEACH protocol with improved bat algorithm in wireless sensor networks. KSII Trans Internet Inform Syst 13(5)
4. Taranum F, Khan KUR, Aleemuddin M (2018) Power consumption analysis in multi-hop networks in mobile environments. In: International conference on Futuristic trends in network and communication technologies. Springer Conference
5. Kaur M (2013) comparative study of homogeneous and heterogeneous mobile Adhoc networks. Int J Adv Res Eng Appl Sci

6. Jain A (2012) Reduction of power consumption at mobile station in multi-hop networks in mobile environments. In: 1st International conference on emerging technology trends in electronics communication and networking
7. Venkateswaran A, SaranganV (2012) MANET: network mobility as a control primitive. CRC press/Taylor & Francis Group
8. Nguyen TD, Berder O et al (2011) Energy efficient cooperative techniques for infrastructure-to-vehicle communications. IEEE Trans Intell Transp Syst 12(3):659–668, (2011)
9. Jha R, Dalal UD (2011) Location based WiMAX network optimization: power consumption with traffic load, ACWR. In: Proceedings of the 1st international conference on wireless technologies for humanitarian relief, ACM New York, pp 529–534
10. De Rango F, Guerriero F, Fazio P (2010) Link-stability and energy aware routing protocol in distributed wireless networks. IEEE Trans Parallel Distrib Syst
11. Chandwani G, Datta SN et al (2010) Relay assisted cellular system for energy minimization. In: Annual IEEE Indian conference (INDICON)
12. Hu J, Wang X, Zhang H (2010) Location aware relay selection scheme in opportunistic relay communications. In: IEEE 71st Vehicular technology conference (VTC Spring 2010)
13. Song W, Chung J, Lee D, Lim C, Choi S, Yeoum T (2009) Improvements to seamless vertical handover between mobile WiMAX and 3GPP UTRAN through the evolved packet core. IEEE Commun Mag 47(4):66–73
14. Ohseki T, Fuke N, Ishikawa H, Takeuchi Y (2006) Multi-hop mobile communications system adopting fixed relay station and its time slot allocation schemes. In: IEEE 17th International symposium on personal, indoor and mobile radio communication (2006)
15. Tang C, Mc Kinley PK (2006) Energy optimization under informed mobility. IEEE Trans Parallel Distrib Syst 17(9):947–962
16. Nosratinia A, Hunter TE, Hedayat A (2004) Cooperative communication in wireless networks. IEEE Commun Mag 42(10):74–80
17. Ivanov K, Spring G (1995) Mobile speed sensitive handover in a mixed cell environment. In: IEEE 45th vehicular technology conference, vol 2, pp 892–896